Lecture Notes in Economics and Mathematical Systems 559

Marcus Schulmerich

Real Options Valuation

The Importance of Interest Rate Modelling in Theory and Practice

Author

Dr. Marcus Schulmerich, CFA, FRM
Vice President
Allianz Global Investors Group
Nymphenburger Strasse 112–116
80636 Munich/Germany
marcus.schulmerich@alum.mit.edu

Library of Congress Control Number: 2005928811

ISSN 0075-8442
ISBN-10 3-540-26191-5 Springer Berlin Heidelberg New York
ISBN-13 978-3-540-26191-9 Springer Berlin Heidelberg New York

Springer is a part of Springer Science+Business Media

springeronline.com

Printed in Germany

Typesetting: Camera ready by author
Cover design: *Erich Kirchner*, Heidelberg

Printed on acid-free paper 42/3130Di 5 4 3 2 1 0

To my parents.

Foreword

Managerial decision-making during the lifetime of a project can have important implications on project handling and its contribution to shareholder value. Traditional capital budgeting methods (in particular methods based on net present value) fail to capture the role of managerial degrees of freedom and therefore tend to lead to a systematic undervaluation of the project. In contrast, the real options approach to investment analysis characterizes decision-making flexibility in terms of (real) option rights which can be evaluated analogously to financial options using contingent-claims pricing techniques widely used in capital markets.

The research carried out by Marcus Schulmerich analyzes real options for non-constant and stochastic interest rates versus constant interest rates. Analyzing stochastic interest rates in the context of real options valuation is of particular relevance given their long time to maturity which makes them more vulnerable to interest rate risk than short-term financial options. To date, there has not been a comprehensive review of this issue in the academic literature. The fact that interest rates have fluctuated widely over the recent years further highlights the need for studying this issue.

This study incorporates variable and stochastic interest rates into numerical approaches to real options valuation and analyzes the implications for the efficiency of these numerical methods. The author starts out by providing a critical review of the approaches taken in the literature to value complex real option rights and adopts a pragmatic approach in expanding them. He is specifically interested in assessing to what extent the assumption of a constant discount rate frequently observed in corporate practice leads to wrong investment decisions. Although capital market experts would at no point assume interest rates to be constant, this issue has only been marginally addressed in the real options literature. However, as the author points out, unexpected shifts in the interest rate curve will often exert a lasting influence on the value-oriented control of projects.

The main part of the study presents extensive numerical simulations and the historical backtests for various complex real option rights by combining standard numerical modelling techniques for real options and interest rate risk. Following introductory assessments on the efficiency of standard numerical valuation methods and the possibility of extending these methods to non-constant interest rates, the author looks at various equilibrium and no-arbitrage models for interest-rate modelling in real options valuation. A concluding section examines the additional benefit achieved by including non-constant interest rates in real options valuation through historical backtesting. Alongside stochastic interest rates, models with constant interest rates or implicit forward interest rates are analyzed. The simulation results provide important numerical findings for the first time indicating the extent to which the common assumption of constant interest rates in valuation practice can actually be justified. It turned out that it is important to adjust for the shape of the term structure in real options valuation, even if interest rates are modelled stochastically. In methodological terms, stochastic interest rate models should be preferred, although their additional benefit over using implicit forward rates is not verifiable. In fact, the use of implicit forward rates is overall preferable, especially given the ease of applying these models in corporate practice. Along the same lines, the assumption of constant interest rates should be rejected on principle.

This research study by Marcus Schulmerich generates new knowledge for capital market research and corporate valuation practice and develops guidelines for their practical implementation. The book closes an important gap in the literature and represents a valuable contribution to answering the question how real options insights can effectively be employed to improve the quality of valuation exercises and thereby real investment decisions. We hope that the study will be widely disseminated and that it will receive the attention it deserves by academics as well as practitioners.

May 2005

Cambridge, MA, USA — Prof. Stewart C. Myers, Ph.D.
Oestrich-Winkel, Germany — Prof. Ulrich Hommel, Ph.D.

Preface

This preface is dedicated to all the people and institutions without whom the completion of my doctoral thesis would not have been possible. I want to thank all the people involved in this thesis; I want to thank all the academic institutions that gave me my education and allowed me to use their facilities; and I want to thank Commerzbank for supporting my research by granting me a 10-week leave of absence in fall 2002 to work on my thesis full-time, a time I spent at Harvard Business School.

This thesis could not have become a reality without the ongoing support of several people. First of all, my thanks go to Prof. Ulrich Hommel, Ph.D., my thesis advisor at the EUROPEAN BUSINESS SCHOOL - **ebs** (*Endowed Chair for Corporate Finance and Capital Markets*) in Oestrich-Winkel, Germany. His critique during the last three years provided the right mixture of motivation, challenge, and support to bring the thesis to a successful end. He always allowed me to follow my own ideas as to how to pursue the research, yet put me back on the right track whenever necessary. Moreover, he demanded results, which pushed my research forward sufficiently to enable me to finish the thesis within the timeframe we both had in mind. Finally, I want to thank him for the opportunity to publish parts of this thesis in his latest book on real options in corporate practice (see Hommel, Scholich & Baecker [60]). Additionally, my thanks go to Prof. Dr. Lutz Johanning, head of the *Endowed Chair for Asset Management* at **ebs** for his valuable feedback during my research period and for being one of the evaluators of this thesis.

I also want to thank Mischa Ritter and Philipp N. Baecker, both doctoral candidates of Prof. Hommel. My long and fruitful discussions with Mischa Ritter helped to shape the focus of this thesis and gave several new twists to my research and the writing of the thesis. Philipp N. Baecker's experience in quantitative modelling of real options proved a great resource of new ideas and research directions as well, especially during several doctoral workshops at the **ebs**. Moreover, his help was extremely valuable in expressing my ideas in TEX, a passion of his, Prof. Hommel's, and mine.

Last but not least I would like to thank Gudrun Fehler, the secretary to the *Endowed Chair for Corporate Finance and Capital Markets* at **ebs** for proofreading the final version of this thesis with enthusiasm and diligence. Of course, all the remaining errors you will find on the following pages are entirely my fault.

Besides the people at **ebs** mentioned so far several others contributed tremendously to this thesis. First, I want to thank Prof. Stewart C. Myers, Professor of Economics, Finance, and Accounting at the Sloan School of Management at the Massachusetts Institute of Technology, MIT. He introduced me to the exciting world of Corporate Finance and Capital Markets during my studies at MIT's Sloan School where I studied for the MBA from 1999 to 2001. His course *15.401-Finance I* gave me the "first contact" with Finance Theory. He is also one of the evaluators of this thesis.

The introductory Finance course by Prof. Myers was followed by many others. Especially, I want to mention the courses *Options and Futures* as well as *Investment Banking* (which could also be called Fixed Income Securities Valuation, since this was its only content) taught by Prof. John C. Cox. His way of presenting the material as well as his openness to students and their questions, in and outside of the classroom, were among the lasting memories I took from my studies at MIT Sloan. When my ideas of pursuing a doctorate in Finance began to shape, his input in considering possible topics helped me a lot to find the direction of my research.

While at Harvard Business School during Fall semester 2002, I also got a great deal of input for my research from Prof. Li Jin, Assistant Professor of Finance at Harvard. He invited me to be his Research Assistant for one semester, allowing me to devote my time entirely to my research and thesis writing. He also proofread the initial work and gave me valuable suggestions that shaped the look of my thesis. Many thanks are dedicated to him, his efforts and support during the time I was working on the thesis at Harvard Business School.

Finally, I want to thank Markus F. Meyer, then Risk Manager at Commerzbank Asset Management. His support with financial data from the Commerzbank data providers *Datastream* and *Bloomberg* allowed me to conduct my simulations, case studies, and backtesting with historical data. My thanks also go to Abraham Eghujovbo for proofreading the early versions of this thesis for spelling and grammar. Finally, I want to thank Mohini Pahnke and Christian Theis for proofreading the final version of my thesis and giving me valuable input.

This preface would not be complete without mentioning the people that brought me to the world of stochastic interest rate modelling, the topic of my Master's thesis in Mathematics at the Johannes Gutenberg-University in

Mainz, Germany. First, I want to thank Prof. Dr. Claudia Klüppelberg, my advisor, who introduced me to the application of Mathematics for modelling interest rate movements. Second, I want to thank Michael Hallacker and Dr. Hans-Peter Rathjens, who both were at ADIG Investment, a subsidiary of Commerzbank, in 1996/1997 and offered me the chance to write my Master's thesis during that time in co-operation with ADIG. This Master's thesis laid the foundation of the knowledge on which I built for my doctoral research.

Besides the people mentioned above, I also want to thank four institutions that allowed me to receive an education that enabled me to write this thesis. First, I want to thank the Johannes Gutenberg-University, which gave me my valuable education in Mathematics, a backbone ever since in my academic and professional life. Second, I want to thank the MIT Sloan School of Management, which accepted me as an MBA student, another prerequisite for this thesis. Sloan paved my way to the academic world of Corporate Finance and Capital Markets. Third, many thanks to Prof. Hommel's *Endowed Chair for Corporate Finance and Capital Markets* and the **ebs** for accepting me as a doctoral candidate. Finally, I want to thank the Harvard Business School for allowing me to pursue my library studies as Prof. Li Jin's Research Assistant in fall 2002. I hope the inspiration I found while sitting and working in Harvard's Baker library is reflected in the following pages.

The challenge of my research not only lay in the topic itself but it was also an organizational challenge as an external doctoral candidate working full-time for asset management companies besides doing my research. Since September 2001 I worked at Commerzbank Asset Management in Frankfurt am Main, Germany, as a Risk Manager. In April 2003 I joined PIMCO (Pacific Investment Management Company), a subsidiary of Allianz Global Investors, as a Product Manager in the Business Development Team, based in London, U.K., and Munich, Germany.

Finally, I want to thank my parents for their ongoing support and motivation not only during my research period for this thesis but also throughout my previous studies.

Marcus Schulmerich

Contents

1

Introduction

1.1 Motivation for the Thesis

Real options are one of the most fascinating research topics in Finance today. There are many reasons for this. First, the weaknesses of the Discounted Cash Flow (DCF) method, the most common valuation tool in Corporate Finance, have become more and more obvious to practitioners and are therefore putting pressure on academic research to improve traditional valuation tools. Second, literature on real options, especially the pricing techniques (which are mathematically and numerically much more savvy than classical methods like the DCF method, Internal Rate of Return method, or Payback Period method) have become more accessible to the broad public since the mid 1990s. This has sharpened the awareness for evaluating investment projects with the real options approach and has highlighted its advantages over traditional valuation tools. Third, the very volatile world economic situation and the even greater flexibility being built into investment projects calls for valuation tools that can incorporate this volatility and flexibility and model it accurately. As will be seen, traditional methods are not capable of doing this. This insight was the starting point of the research area *real options* in 1977 when Stewart C. Myers from the MIT Sloan School of Management published his pioneering article on the subject in the Journal of Financial Economics[1].

Since 1977, which is seen as the birth year of the real options field, academics developed various highly mathematical models for real options, especially for problems that offered a closed-form mathematical solution. However, these problems were often too simplified to ever become a real investment issue in the world of Corporate Finance. Therefore, the practitioners' interest in these models was limited. This changed in the mid 1990s with the publication of the book *Real Options* by Lenos Trigeorgis in 1996 (see Trigeorgis [132]). Trigeorgis provided for the first time a thorough summary of valuation methods for real options that could actually be used in practice. The main tool he applies to real options pricing is the traditional Cox-Ross-Rubinstein binomial

[1] See Myers [99].

tree which is based on the idea of creating replicating portfolios. He also explains a new model for valuating complex multi-option investments, applying a logarithmic transformation of the underlying. Modifications of these two real options valuation tools are the core of this thesis.

When considering all the available literature, it is obvious how real options are impacting not only the academic world but increasingly applications in Corporate Finance. In the United States for instance the real options approach became increasingly famous in practice in the 1990s. This is seen in the work of Tom Copeland, who shows how to transfer academic knowledge to Corporate Finance practice. He served as Professor of Finance and is head of the Corporate Finance consulting division at the consulting firm Monitor Group, applying the real options theory in practice to value investment projects.

The real options approach is superior to traditional valuation tools like the DCF method since classical tools cannot incorporate project-inherent flexibility into the valuation process. In the words of Pritsch and Weber[2]: *Traditional valuation methods have often been criticized for not being able to include the strategic value of an investment project such that they give recommendations opposite to the managers' intuition. Negative NPV analyses are compensated by an intuitive reasoning revolving around a "strategic project value". Insofar as those ignored value components result from the interaction of flexibility and uncertainty, the real options approach offers a tool to express this intuition and to formally recognize it [...].*

Surveys among companies revealed the importance of tools to value flexibility as well. One of the best surveys available was carried out by Vollrath[3] in 2000 among a sample of firms that have their headquarter in Germany. This survey clearly shows how flexibility plays a critical role in the decision making process on new investment projects, see Figure 1.1. More than 50% of the investment decisions both at the company level and at the level of the operational unit contain flexibility. Therefore, the huge potential for real options valuation tools is without question.

There are many insightful books about real options on different levels of complexity and difficulty and with different goals. Two very popular introductory books are, as already mentioned, *Real Options* (1996) by Trigeorgis and the book *Real Options* (2001) by Copeland and Antikarov (see Copeland & Antikarov [33]). Trigeorgis' book is known for building a bridge between academic theory and its practical application even in the context of complex real options. Copeland and Antikarov's work is known as an introduction for Corporate Finance managers on a less quantitative level. However, both books

[2] See Pritsch & Weber [108], page 23, English translation by the author.
[3] See Vollrath [134].

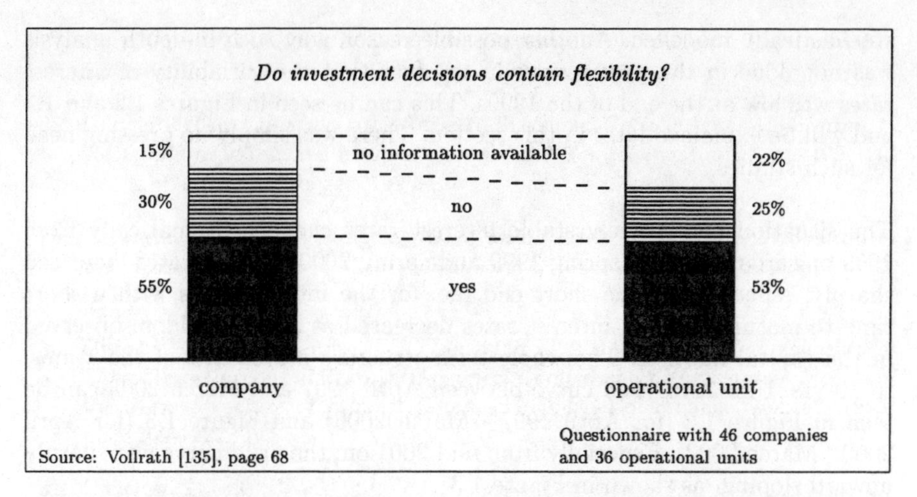

Fig. 1.1. Real options in practice: Do investment decisions contain flexibility?

only briefly mention an important aspect of the valuation process: the impact of non-constant interest rates on real options pricing, i.e., how non-constant interest rates influence the outcome of real options valuation and therefore the investment decision of Corporate Finance managers.

There are many analytical models and numerical methods to value financial options and real options. Some of these include a stochastic risk-free interest rate, for example: Hull & White [64], 1990; Ingersoll & Ross [69], 1992; Sandmann [115], 1993; Ho, Stapleton & Subrahmanyam [58], 1997; Miltersen & Schwartz [97], 1998; Miltersen [96], 2000; Alvarez & Koskela [1], 2002. In many cases these methods were developed to price financial options but do not consider or analyze real world projects that are characterized by complex real options situations.

Although the literature on real options in general is abundant[4], the real options literature that deals with non-constant interest rates in practical real options valuation is, at best, rare. Some of the few publications specifically on real options valuation including a stochastic interest rate are Ingersoll & Ross [69], Miltersen & Schwartz [97], Miltersen [96], and Alvarez & Koskela [1]. All of these models are highly mathematical and only applicable to specific real options. However, for practical application there is a lack of in-depth analyses of different complex real options which apply various term structure models and include simulations and historical backtesting. One obvious reason is the difficulty of undertaking such an analysis analytically. It is likely that no closed-form solutions exist for such complex real options if the risk-free rate is

[4] See, e.g., Brennan & Trigeorgis [21], Dixit & Pindyck [40], Trigeorgis [131], and the overview in Section 2.6.3.

stochastically modelled. Another possible reason why such in-depth analysis was not done in the past resides in the fact that the variability of interest rates was low at the end of the 1990s. This can be seen in Figures 1.2 and 1.3 and will be explained later in this section. There was simply no pressing need for such studies.

The situation of relatively stable interest rates changed dramatically from 1999 onwards. Between spring 1999 and spring 2000, interest rates increased sharply, especially at the short end, i.e. for the interest rates with a short time to maturity. Then, interest rates decreased at a speed seldom observed in the capital markets. The whole term structure movements of the 1-mos. to 10-yrs. U.S. Zero yield curve between April 1997 and March 2003 can be seen in Figure 1.2 (for April 1997 - March 2000) and Figure 1.3 (for April 2000 - March 2003). Especially from mid 2001 on, the yield curve was sharply upward sloping, as shown in Figure 1.3.

According to the unbiased expectations theory the yield curve at any given point in time reflects the market's expectations of the future spot rate. If, for example, the current yield curve is upward sloping (a so called *normal yield curve*), the market believes that interest rates will be higher in the future. If the current yield curve is downward sloping (a so called *inverted yield curve*), the market believes that the interest rates will fall. Both scenarios imply a

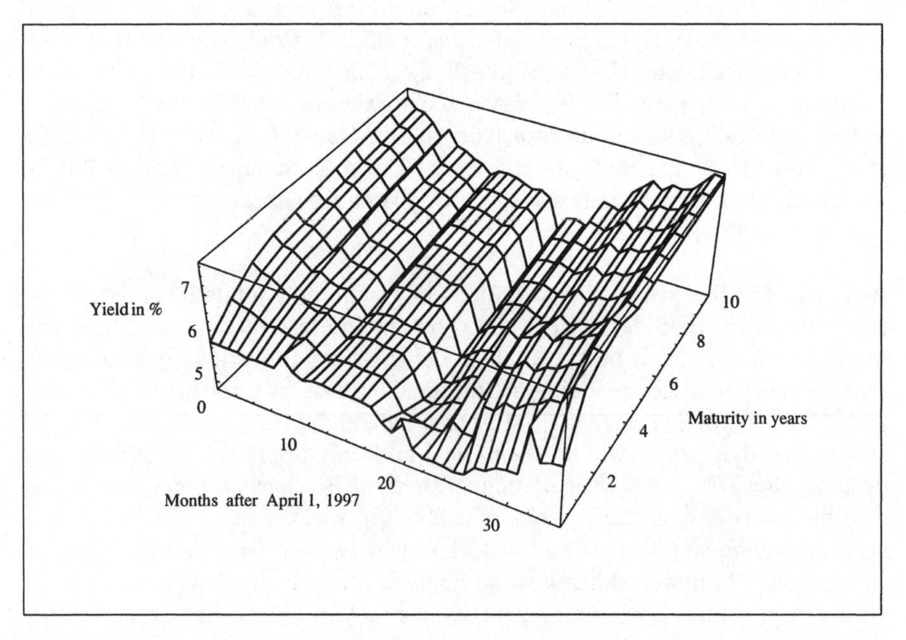

Fig. 1.2. Term structure of U.S. Zero yields, April 1997 - March 2000, monthly data.

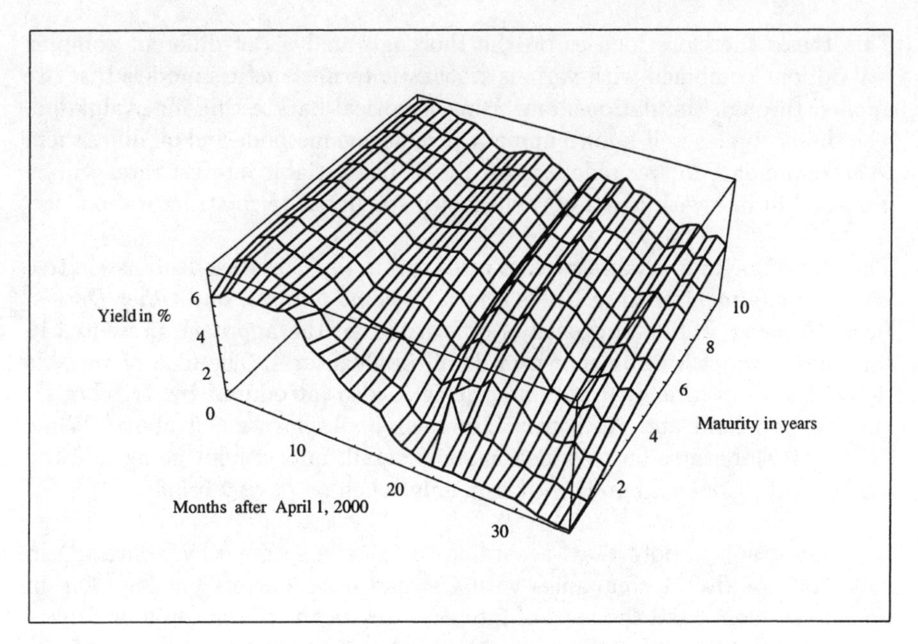

Fig. 1.3. Term structure of U.S. Zero yields, April 2000 - March 2003, monthly data.

future change of interest rates. Only in the case of a flat yield curve are the interest rates assumed to remain unchanged in the future, according to the unbiased expectations theory[5]. However, it should be pointed out that the unbiased expectations theory has its shortcomings[6] like all the other theories used to explain the term structure of interest rates[7]. These will not be discussed here.

When turning to the (real) options approach, which has become popular in the last couple of years due to its advantages over traditional capital budgeting methods, one has to be aware of an important implication that can best be expressed by citing Hull[8]: *The usual assumption when American options are being valued is that interest rates are constant. When the term structure is steeply upward or downward sloping, this may not be a satisfactory assumption. It is more appropriate to assume that the interest rate for a period of length Δt in the future equals the current forward interest rate for that period.*

[5] See Fabozzi [42], pages 291-333, which gives a very good introduction on the term structure of interest rates and various explanatory term structure theories.
[6] See Fabozzi [42], page 303.
[7] See Fabozzi [42], pages 291-333.
[8] See Hull [62], page 356.

This thesis therefore focuses on the thorough analysis of different complex real options combined with various stochastic term structure models that are applied through simulations, and using historical backtesting for evaluation. The thesis applies well-known numerical valuation methods and modifies them to accommodate for variable interest rates. The variable interest rates will be modelled in line with existing, well-known stochastic term structure models.

The idea of having variable interest rates within the Cox-Ross-Rubinstein tree is also briefly mentioned in Hull's book[9] *Options, Futures, and Other Derivatives.* However, Hull's approach is different from the approach presented in this thesis, which will be shown in detail in Chapter 4. The idea of variable risk-free rates within real options pricing is also introduced by Trigeorgis[10] and by Copeland and Antikarov[11] in their books mentioned above. While Trigeorgis elaborates on this idea in some detail, but without going into too much depth, Copeland and Antikarov only touch on it very briefly.

It is important to note that, according to the survey done by Vollrath[12] in 2000, 24% of the 21 companies with German headquarters participating in the survey mentioned the level of interest rates as one of the top four drivers of the real options value. However, the level of interest rates was not of any consideration for the 18 operational units asked. This is an astonishing finding. Moreover, the shape of the yield curve, something that according to Hull is the main reason for replacing the constant risk-free interest rate with a variable risk-free interest rate in real options valuation tools, was of no concern to companies at all, let alone to operational units. But it should be - and this thesis intends to raise the awareness of this real options driver.

To provide a better view on special maturity buckets, Figures 1.4 to 1.7 display the development of U.S. Zero yield for four maturity buckets between April 1997 and March 2003 with beginning-of-month data. These U.S. Zero yields are risk-free interest rates since the Zeros are backed by the full faith of the U.S. government and, therefore, always paid back.

Figure 1.4 displays the 3-mos. U.S. Zero yields. This yield almost drops from over 7% in December 2000 to below 2% in December 2001, a very dramatic decrease of over 1% per quarter. The same situation can be seen for other maturity buckets (see Figure 1.5 for the 1-year U.S. Zero yield). Also longer-term maturities like 5 years and 10 years in Figure 1.6 and Figure 1.7, respectively, display this movement, but it is not as dramatic as in the case of the 3-mos. U.S. Zero yield.

---

[9] See Hull [62], pages 356-357.
[10] See Trigeorgis [132], pages 197-199.
[11] See Copeland & Antikarov [33], pages 158-159.
[12] See Vollrath [134], page 71.

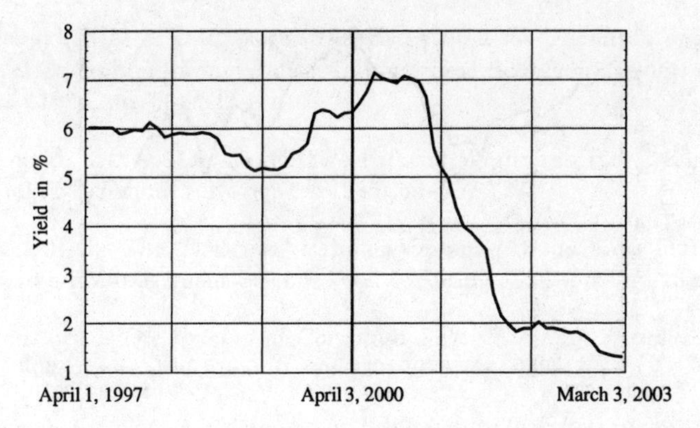

Fig. 1.4. 3-mos. U.S. Zero yield, April 1997 - March 2003, monthly data.

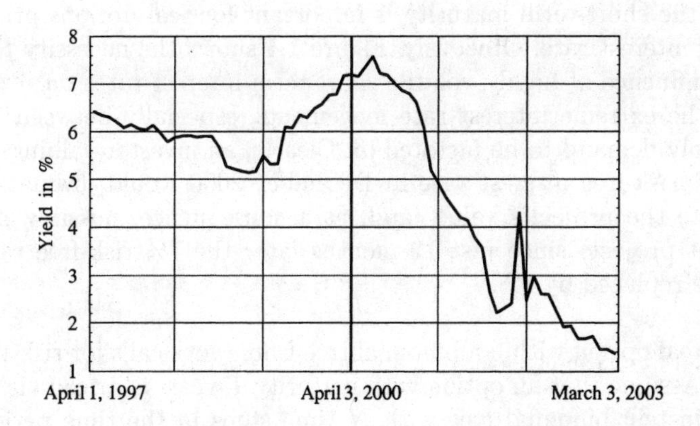

Fig. 1.5. 1-yr. U.S. Zero yield, April 1997 - March 2003, monthly data.

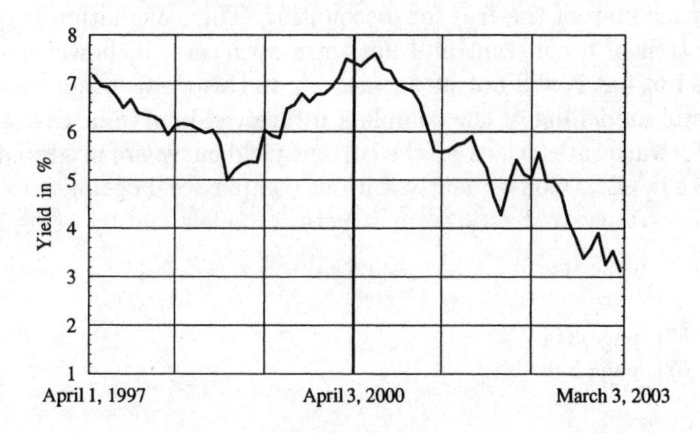

Fig. 1.6. 5-yrs. U.S. Zero yield, April 1997 - March 2003, monthly data.

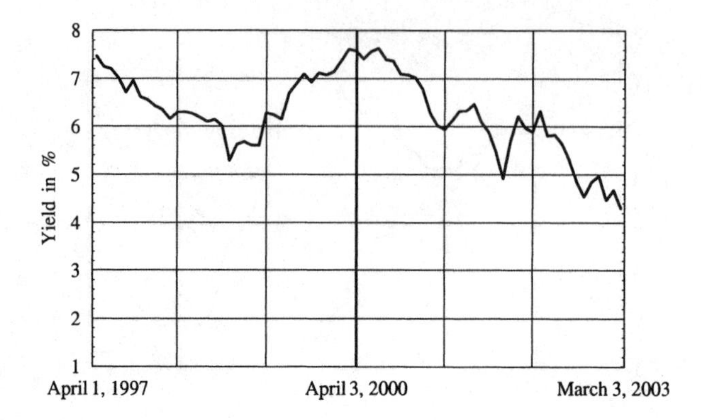

Fig. 1.7. 10-yrs. U.S. Zero yield, April 1997 - March 2003, monthly data.

However, the short-term maturity is important for real options pricing with a variable interest rate. Obviously, Figure 1.4 shows the necessity to investigate the influence of highly volatile short-term interest rates on real options pricing. The extreme interest rate movements, especially between 1999 and 2002, simply demand to be factored in: Clearly, an investor valuing a project with a 7% risk-free interest rate in December 2000 would dramatically underestimate the project's value (and, as a consequence, possibly reject the investment project) since just 12 months later the 7% risk-free rate would have to be replaced by 2%.

Pricing a real option within a binomial tree framework calls for risk-free interest rates. Assume that an option with maturity T years is priced via the Cox-Ross-Rubinstein binomial tree with N time steps in the time period $[0, T]$. Then the time step size is $\Delta t := \frac{T}{N}$. According to Hull, the risk-free rate (named $r_f(T)$) is usually set at the yield on a Zero bond maturing at the same time T as the option[13]. Then, $r_f(\Delta t) := r_f(T) \cdot \Delta t$ is the risk-free rate applied in each step of the tree for discounting. This calculation is correct on the assumption of a constant and flat term structure; if, however, the term structure is not flat it will not be accurate[14]. In this thesis other methods like the stochastic modelling of the complete future yield curve or the calculation of implied forward rates based on the current yield curve are used to derive the risk-free rate $r_f(\Delta t)$. Consequently, various complex real options are analyzed by applying various stochastic term structure models and by using historical backtesting.

[13] See Hull [62], page 344.
[14] See Hull [62], page 356.

The idea of a variable risk-free interest rate can be best described by an example provided by Ingersoll and Ross[15] in the first ever article to describe the existence of real options due simply to a stochastic interest rate despite the future cash flows being deterministic[16]. In the following, all values are given in US \$ (in short \$). It is assumed that a decision has to be made on an investment project that can be realized within a single day and which generates a deterministic cash flow of 112 \$ in exactly one year after the project was undertaken. The project is either undertaken today or in exactly one year from today. If it is undertaken today the immediate investment cost is $I = 100\,\$$ and a deterministic cash inflow of $C = 112\,\$$ will take place exactly one year from today. If the project is undertaken in one year from today it will give a deterministic cash inflow of 112 \$ exactly two years from today. The investment cost when investing one year from today is assumed to be 100 \$ as well.

In order to calculate the net present value (NPV), information is needed on the 1-year spot rate today and one year from today. The current yield curve is assumed to exhibit $r_f^{(0)}(1) = 10\%$ as the current 1-year spot rate. The superscript 0 indicates the time point t when the spot rate is considered (i.e., $t = 0$ for today) and argument 1 indicates that the 1-year spot rate is considered. It is further assumed that the 1-year spot rate in one year from today is known: $r_f^{(1)}(1) = 7\%$. This gives the current NPV_0 as:

Start of project today (indicated by the superscript 0):

$$\begin{aligned} NPV_0^{(0)} &= \frac{C}{1 + r_f^{(0)}(1)} - 100 = \frac{112}{1 + 0.1} - 100 \\ &= 1.82. \end{aligned}$$

Start of project one year from today (indicated by the superscript 1):

$$\begin{aligned} NPV_0^{(1)} &= \left(\frac{C}{1 + r_f^{(1)}(1)} - 100 \right) \cdot \frac{1}{1 + r_f^{(0)}(1)} = \left(\frac{112}{1 + 0.07} - 100 \right) \cdot \frac{1}{1 + 0.1} \\ &= 4.25. \end{aligned}$$

[15] See Ingersoll & Ross [69], pages 1-2.

[16] See Ingersoll & Ross [69], pages 2-3: [...] *even though the project itself has no option characteristics* [...] *the uncertainty in interest rates* [...] *gives it an option-like feature. We are not the first to recognize that delaying a project can be desirable, but we are the first to observe that this need have nothing to do with changes in the cash flows of the project itself or with the effects of certain changes in interest rates.*

Obviously, it is better to undertake the investment one year from today if the future spot rate is known today. Erroneously, it is often assumed that the NPV of a project in the case of a deterministic cash flow structure decreases with time because of the time value of money. This is only correct for a flat and constant yield curve. However, if the 1-year spot rate one year from today is known to be $r_f^{(1)}(1) = 13\%$ instead of 7%, the NPV in one year would be $NPV_1^{(1)} = -0.80\,\$$ and the project would start immediately and not one year from today. The huge impact of the interest rate is obvious even in this simple example.

If it is known today that the 1-year spot rate one year from today will be $r_f^{(1)}(1) = 7\%$ it is interesting to know how much someone is willing to pay to acquire the option that allows him to start the project either today or one year from today. The value R of this real option, an *option to defer*, would be

$$R = NPV_0^{(1)} - NPV_0^{(0)} = 4.25 - 1.82 = 2.43.$$

Although this is a simple example, it clearly shows the influence of the risk-free rate on the value of an investment project and the decision on whether or not to start/carry on with an investment project. This is an example of an option to defer, i.e. a simple option. This thesis considers complex real options mainly in the context of a stochastic interest rate. First, non-constant risk-free interest rates will be incorporated into the Cox-Ross-Rubinstein binomial tree and into a log-transformed binomial tree developed by Trigeorgis. These methods will then be used to value various complex real options common in Corporate Finance practice. The effects on the real options price depending on the chosen term structure model, the (complex) real option, and the input parameters of both will be thoroughly analyzed.

This raises the question as to what types of real options are worth thorough analysis. An excellent overview of the various real options types is provided by Vollrath[17] and displayed in Figure 1.8. According to Figure 1.8 the option to defer is most common for both the company and its operational units. Almost as common as this real option is the option to abandon the investment project.

Based on this survey three types of real options are analyzed, one being a simple and two being complex real options:

1. **Case 1:** Option to abandon the project at any time during the construction period for a salvage value, i.e. an American option.
2. **Case 2:** Complex real option comprising two simple real options: option to abandon the project at any time during the construction period for a

[17] See Vollrath [134], page 73.

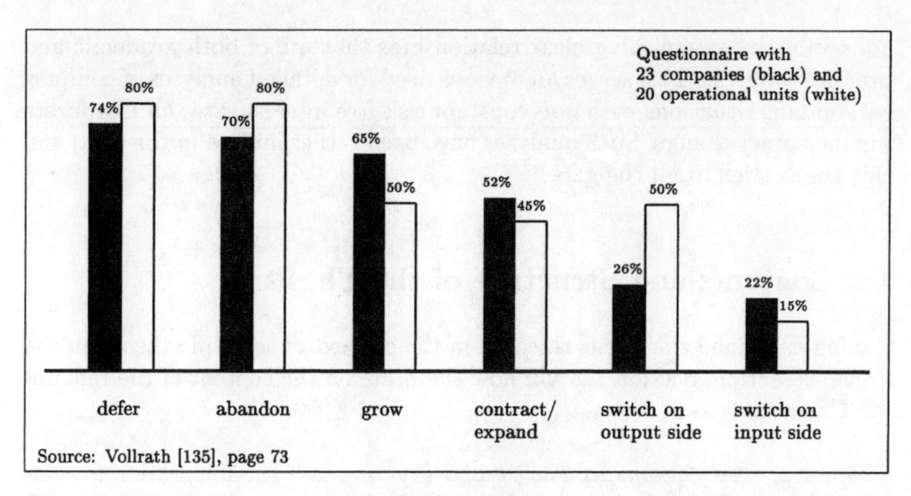

Fig. 1.8. Real options in practice: Which types of real options play an important role in the decision making process of companies and operational units?

salvage value (case 1) and option to expand the project once at the end of the construction period.

3. **Case 3:** Complex real option comprising three simple real options: complex real option in case 2 combined with an option to defer the project start by exactly one year. The project can start today or in exactly one year from today if the investment in one year has a positive NPV.

Who is the intended audience of this thesis? The thesis is written for two purposes. The first one is to enrich the field of academic research by providing reference literature that gives insight into a subject that has not yet been analyzed in detail, especially from the viewpoint of numerical analysis, simulation, and historical backtesting. To this end, the dependencies of the parameters in the stochastic term structure models and the various real options valuation tools will be described in detail. Moreover, the relationship of the real options value/project value in the various real options cases and the underlying value will be investigated thoroughly. The second purpose is to establish rules for practical application based on the relationships found. Expert knowledge of these relationships and rules is important for Corporate Finance practice in a world where interest rates are changing rapidly (as seen especially between 1999 and 2002).

This thesis is written as simply as possible to make it accessible to Corporate Finance practitioners. On the other hand, all significant theoretical topics are dealt with in depth in order to provide the mathematical foundations needed to understand each step in the argument. By proceeding this way, the thesis adds value by establishing clear relationships that are of both academic and practical interest. This serves an obvious need for detailed analyses of complex real options situations with non-constant risk-free interest rates for Corporate Finance practitioners. Such analyses have been rather limited in the past; and this thesis tries to fill the gap.

1.2 Contents and Structure of the Thesis

Having explained the thesis research in the context of scientific theory in the previous section, Section 1.2 will now elaborate on the content of the remaining Chapters 2 to 6.

Chapter 2, *Real Options in Theory and Practice*, lays the foundation for discussing *real options*. A thorough description of the chapter's contents (as well as for all other chapters) is given in the first section (2.1). The idea of real options is the topic of Section 2.2. Sections 2.3 and 2.4 are devoted to the mathematical foundations necessary to deal with real options from a valuation point of view. Stochastic processes that model the underlying of real options are diffusion processes. Therefore, Section 2.3 introduces the idea of diffusion processes in the framework of classical probability theory. The language used to describe these processes is *stochastic calculus*, also called *Ito calculus*, and is introduced in Section 2.4 to the extent needed in the thesis. In Section 2.5, *Discretization of Continuous-Time Stochastic Processes*, numerical simulation methods to generate discrete paths of continuous-time stochastic processes are presented. These methods are all implemented in the computer simulation program.

The quantitative description of real options models in the literature is the topic of Section 2.6, a section divided into four subsections: Section 2.6.1 deals with decision-tree analysis, a tool used a lot in practice to value project flexibility[18]. The second section (2.6.2) introduces the idea of contingent-claims analysis, the basic idea that is employed in the binomial tree approach of Cox, Ross, and Rubinstein. Section 2.6.3 categorizes the various real options valuation methods. Section 2.6.4 introduces the flexibility of an investment project due to interest rate uncertainty. Finally, Section 2.7 gives a summary of Chapter 2.

Chapter 3, *Stochastic Models for the Term Structure of Interest Rates*, elaborates on the various stochastic term structure models used in the thesis. The purpose is to introduce the theory of the term structure of interest rates, the historical development of these term structure models and their classification. Section 3.2 introduces the general idea of term structure models, their interpretation, and the short-rate model as the core of each term structure model

[18] See Vollrath [134], page 70, Figure 20.

(3.2.1). This section also gives the definitions of all the terms needed in the course of this thesis. In 3.2.2 an explanation of the cubic spline method is given. This is needed to interpolate the yield curve between given maturity buckets. Section 3.2.3 deals with stochastic interest rate models from a general point of view, describing their historical development over time and showing how the models differ from each other. This section will also provide a categorization for various interest rate models. Section 3.3 explains all models needed for this thesis in great detail, starting with an explanation of the mean reversion feature in stochastic term structure models (3.3.1). Section 3.3.2 is devoted to one-factor term structure models while Section 3.3.3 looks at two-factor term structure models. The final section, 3.4, then gives a summary of this chapter and provides a table with an overview of all the stochastic term structure models presented in Chapter 3.

Chapter 4, *Real Options Valuation Tools in Corporate Finance*, describes the valuation methods for real options to be used in the course of this thesis. First, well-known real options valuation tools that employ a constant risk-free interest rate are presented. Then, two of these methods, the Cox-Ross-Rubinstein binomial tree and the Trigeorgis log-transformed binomial tree methods, are modified to incorporate a stochastically modelled risk-free interest rate. Chapter 4 is organized in three main sections, with each of these sections being further divided into subsections:

1. Section 4.2: *Numerical methods for real options pricing with constant interest rates*
   - 4.2.1: *Lattice methods*

     The Cox-Ross-Rubinstein binomial tree and the Trigeorgis log-transformed binomial tree.
   - 4.2.2: *Finite difference methods*

     Log-transformed implicit finite differences and log-transformed explicit finite differences.
2. Section 4.3: *Schwartz-Moon model*
3. Section 4.4: *Real options pricing with stochastic interest rates*
   - 4.4.1: *Ingersoll-Ross model*
   - 4.4.2: *A modification of the Cox-Ross-Rubinstein binomial tree*
   - 4.4.3: *A modification of the Trigeorgis log-transformed binomial tree*

The final section (4.5) contains the summary of the chapter.

Chapter 5, *Analysis of Various Real Options in Simulations and Backtesting*, is the core of the thesis. It presents the simulation results for various term structure models, real options cases as well as term structure types (normal, flat, and inverted) and it presents a thorough historical backtesting to evaluate the effectiveness of the new, modified real options valuation tools. A computer simulation program was developed in C++ in order to simulate all term structure models described in Chapter 3, all real options pricing methods described in Chapter 4, and the historical backtesting procedures.

First, Section 5.2 introduces the calibration process of some stochastic interest rate models. Calibration methods for the Vasicek model and the Cox-Ingersoll-Ross model are presented in 5.2.1, while calibration methods for the Ho-Lee model are discussed in 5.2.2. The remaining of Chapter 5 is devoted entirely to simulation analysis and historical backtesting of various real options, organized in five consecutive *test situations*:

1. Section 5.4: *Test situation 1: the Schwartz-Moon model with a deferred project start*

   Variable interest rates in the case of the Schwartz-Moon model with a deferred project start.

2. Section 5.5: *Test situation 2: preliminary tests for real options valuation*

   Preliminary tests to investigate the parameters of the real options valuation tools and to compare the valuation methods.

3. Section 5.6: *Test situation 3: real options valuation with a stochastic interest rate using equilibrium models*

   Influence of the salvage and expand factors in cases 1, 2 and 3 on the real options (Section 5.6.1); analysis of all cases for equilibrium models (Vasicek model in Section 5.6.2 and Cox-Ingersoll-Ross model in Section 5.6.3).

4. Section 5.7: *Test situation 4: real options valuation with a stochastic interest rate using no-arbitrage models*

   Analysis of cases 1, 2 and 3 for the Ho-Lee model (Section 5.7.1), comparison of the Hull-White one-factor model with the Hull-White two-factor model (Section 5.7.2), and comparison of the Ho-Lee model with the Hull-White one-factor model (Section 5.7.3).

5. Section 5.8: *Test situation 5: real options valuation in historical backtesting*

   Historical backtesting for cases 1, 2 and 3 through a comparison of the real options pricing when using a stochastically modelled risk-free interest rate (Ho-Lee model), a constant rate, interest rates that equal the currently implied forward rates (Hull's approach), and the historical risk-free rates of the backtesting period.

The exact research strategy for each of these five test situations and their logical flow of ideas is elaborated in Section 5.3. At the end of each test situation, a recapitulation of the main research results is given. The final section (5.9) offers a summary of the chapter.

Finally, Chapter 6 provides a summary of the whole thesis: Section 6.1 restates the purpose of the research in this thesis, Section 6.2 then provides an overall summary of the research results; thereafter, Section 6.3 focuses on the economic implications of the results found; and finally, Section 6.4 completes the thesis with an outlook and a brief overview of possible future areas of research with respect to the topic of this thesis.

The way the chapters fit together is graphically presented in the flow chart in Figure 1.9. As already mentioned, all values are given in US $ (in short $) whereby the unit is omitted in most cases. For mathematical notation in general see page 343.

1.3 Main Results of the Thesis

This section summarizes the main results of the thesis without explaining them in detail. The summaries are emerged by test situations (see 1.2), but since test situation 1 (the Schwartz-Moon model with a deferred project start) only stresses the importance of non-constant risk-free interest rates in the context of real options valuation, it will not be further mentioned here.

Test situation 2 (preliminary tests for real options valuation) yields the following main results[19]:

- The valuation results of both the log-transformed explicit and the implicit finite difference methods are almost identical but the results for real options with longer times to maturity are inaccurate.
- The real options values derived from the Cox-Ingersoll-Ross method and the Trigeorgis log-transformed binomial tree method are almost identical for both a constant and a stochastic risk-free interest rate. However, the computational time for the Trigeorgis method combined with non-constant interest rates is extremely long while the Cox-Ross-Rubinstein binomial tree method offers quicker valuation results.

Test situation 3 (real options valuation with a stochastic interest rate using equilibrium models) yields the following main results[20]:

[19] For a detailed summary see Section 5.5.2.

[20] For a detailed summary see Section 5.6.4.

- The chosen salvage factor and the chosen expand factor have a very significant influence on the project's NPV.
- The level of the interest rate has an important impact on the NPV of a project. The higher the interest rate level, the higher the NPV, if the project contains many real options.
- The NPV of a project is virtually independent of the chosen volatility in the equilibrium model.

Test situation 4 (real options valuation with a stochastic interest rate using no-arbitrage models) yields the following main results[21]:

- Both Hull-White models contain a mean reversion force unlike the Ho-Lee model which is not mean reverting. The influence of the mean reversion force on the NPV is insignificant.
- For both Hull-White models the NPV is almost independent of the volatility parameters.

Test situation 5 (real options valuation in historical backtesting) yields the following main results[22]:

- The application of the implied forward rates in real options valuation results in (almost) the same NPV as the application of the constant risk-free rate.
- In more then two-thirds of the analyzed backtesting scenarios the application of either the implied forward rates or the Ho-Lee model is better suited to calculate the NPV of an investment project than using a constant risk-free interest rate.
- Using the implied forward rates approach for real options valuation never yields the worst NPV results in all historical backtesting scenarios, whereas using a constant risk-free rate leads to the worst NPV results in two-thirds of the scenarios.
- From the stand point of implementational convenience, Hull's implied forward rates approach is better suited in Corporate Finance practice than a stochastic term structure model.
- The overall result in the analyzed historical scenarios is that using the forward rates in real options valuation that are implied in the current yield curve has to be preferred over using a constant risk-free interest rate.

[21] For a detailed summary see Section 5.7.4.
[22] For a detailed summary see Section 5.8.4.

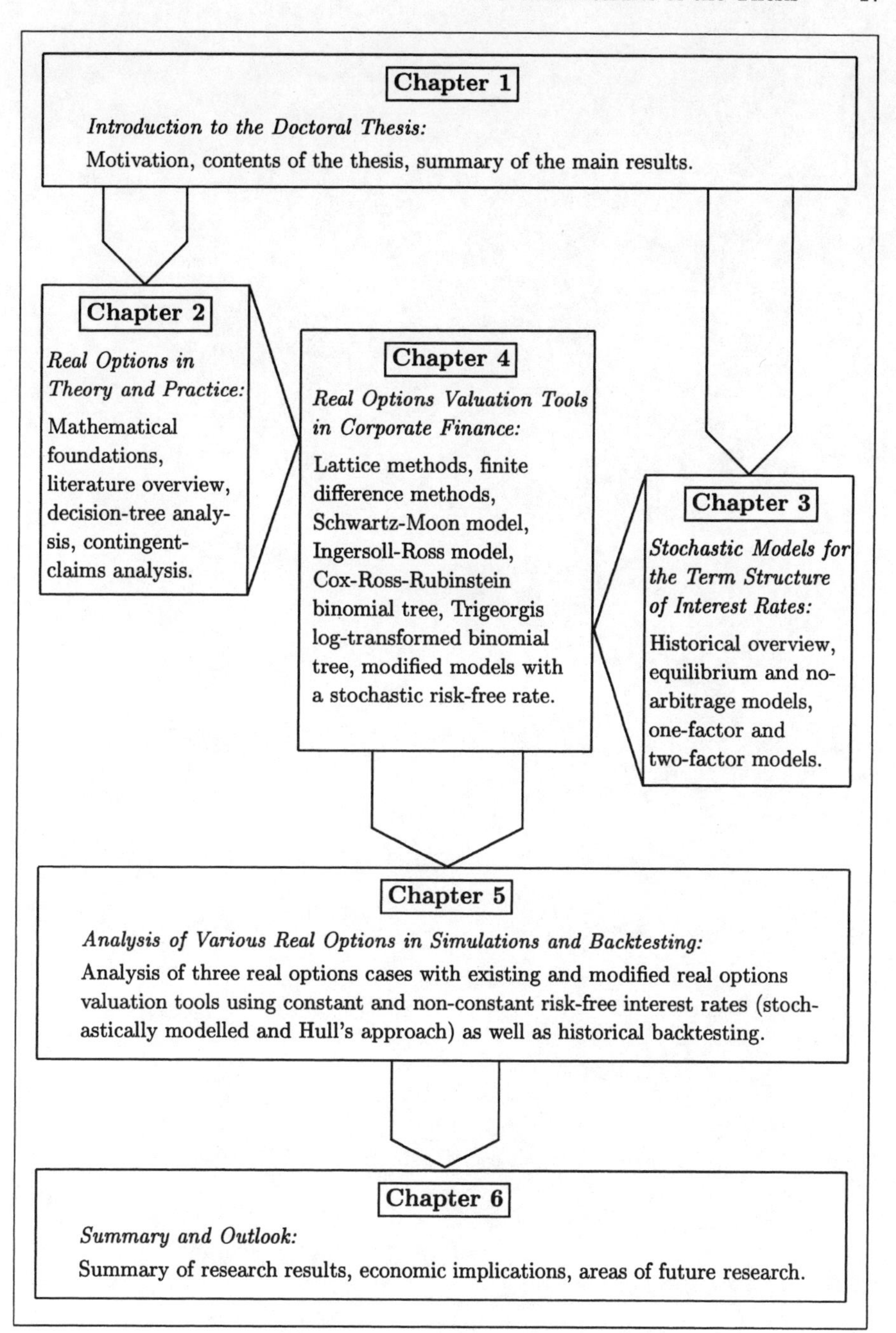

Fig. 1.9. Contents and structure of the thesis: a flow chart.

Fig. [illegible] structure of the thesis as a flow chart

2

Real Options in Theory and Practice

2.1 Introduction

In this chapter, the theory and practice of real options will be introduced. Although their application belongs to the field of Corporate Finance, the valuation methods origined in the option pricing theory for financial securities. In 1977, Stewart C. Myers introduced real options as a new area of financial research with the publication of his famous article[1] on this subject. He also coined the term *real option* and, as a result, is often referred to as the *father of real options theory*.

The introduction to the concept of real options is the topic of Section 2.2. Sections 2.3, 2.4, and 2.5 are devoted to Financial Engineering. The language of Financial Engineering and hence that of real options is stochastic calculus, which will be used throughout this thesis when quantifying real options. Section 2.4 introduces the basics of this theory in order to provide an understanding of the mathematical language used to describe real options valuation and stochastic term structure models. The term *diffusion process* will take center stage in the following chapters. However, since the meaning of diffusion processes can be best explained by applying methods of classical probability theory, a mathematical detour will be taken in Section 2.3 about classical probability theory for diffusion processes. Section 2.5 is devoted to numerical simulation methods of continuous-time stochastic processes as needed for the computer simulation program. The available literature on the quantitative description of real options models is the topic of Section 2.6. This section comprises four subsections: The first section (2.6.1) deals with decision-tree analysis. Subsequently, Section 2.6.2 introduces the idea of contingent-claims analysis and Section 2.6.3 categorizes the various real options valuation methods, distinguishing between analytical and numerical methods. The flexibility of an investment project due to interest rate uncertainty is the topic of Section 2.6.4. The final section (2.7) gives a summary of Chapter 2.

[1] See Myers [99].

2.2 Basics of Real Options

This section provides an introduction to the world of real options. It outlines what they are and how they have developed in the past 27 years. The origin of real options theory is the theory of financial options, options that are written on an exchange-traded underlying. The valuation of financial options has been an area of research for more than three decades. The important breakthrough was achieved by Myron C. Scholes[2] and Fisher Black[3] in the early 70s at the MIT Sloan School of Management with their famous Black-Scholes formula[4] for European options. An important contribution to this formula was made by Robert C. Merton[5] with his no-arbitrage argument. A summary of the historical development of the Black-Scholes formula can be found in Black [7].

Although the Black-Scholes pricing formula was applied immediately at Wall Street, it was not obvious at first how to apply this theory to Corporate Finance. The transition from Capital Markets theory to Corporate Finance theory took place in 1977 when Stewart C. Myers published his idea of perceiving discretionary investment opportunities as *growth options*[6]. The term *real option* was coined by Stewart C. Myers as well: *Strategic planning needs finance. Present value calculations are needed as a check on strategic analysis and vice versa. However, standard discounted cash flow techniques will tend to understate the option value attached to growing profitable lines of business. Corporate Finance theory requires extension to deal with "real options".*[7]

Since 1977 academic researchers started to publish a number of articles[8] on this subject. However, it was not before 1996 that the first successful attempt to make the real options theory accessible to financial practitioners was made when Lenos Trigeorgis published his famous book *Real Options*[9]. Trigeorgis presented the theory and the practice of real options, explained various valuation methods and presented several case studies where the real options approach creates value. This book is well suited for both practitioners and Corporate Financial Engineers. From a more applied point of view, Amram

[2] Myron C. Scholes, Professor of Finance, Emeritus, at the Graduate School of Business at Stanford University. Professor at the MIT Sloan School of Management 1968-1973.

[3] Fisher Black (1938-1995), Professor at the MIT Sloan School of Management 1971 - 1984.

[4] See Black & Scholes [9].

[5] Robert C. Merton, Professor of Finance at the Harvard Business School. Professor at the MIT Sloan School of Management 1970-1988.

[6] See Myers [99] and Trigeorgis [132], page 15.

[7] See Myers [99], page 147, or Myers [100], page 137.

[8] See Section 2.6 for an overview.

[9] See Trigeorgis [132].

and Kulatilaka approached this area in their book *Real Options*[10], published in 1999. Copeland and Antikarov offered an applied, less quantitative, approach to real options as well[11].

Real options practitioners and academics have also contributed to the diffusion of the real options concept. A book published by Hommel, Scholich, and Vollrath[12] in 2001 on the real options approach in Corporate Finance practice and theory is one of the most successful books in Germany on this subject. Another book on real options was published in 2003 by Hommel, Scholich, and Baecker[13]. Both books contain articles from various authors, from the corporate sector as well as from the academic world.

Although all these books are a valid introduction to real options for a broad audience (and this list is by no means complete), the main valuation tool in capital budgeting is still the DCF method. Until recently, capital budgeting problems, i.e. the valuation of investment projects, were almost exclusively done via the DCF method. This method calculates the project's net present value NPV given a deterministic cash flow structure of the project and a known discount factor k as

$$NPV = \sum_{i=0}^{N} \frac{C_{t_i}}{(1+k)^{t_i}} .$$

A positive cash flow $C_{t_i} > 0$ indicates a cash inflow, a negative cash flow $C_{t_i} < 0$ indicates a cash outflow. In total the project is assumed to have $N+1$ discrete cash in- and outflows. In order to apply the formula above very strong restrictions have to be fulfilled. Especially, the cash flow structure of the whole project needs to be known at the beginning of the project and the discount factor k needs to be constant over the lifetime of the project. However, many projects do not fulfill these criteria.

The theory of the DCF method excludes management from making decisions and capitalizing on emerging opportunities during the lifetime of the project. However, in practice those decisions are made and change the project's cash flow structure and the discount factor that should be applied for the project valuation. Unfortunately, neither the DCF approach nor any other traditional approach of capital budgeting is apt to integrate these changes and capture the asymmetric information embedded in these investment opportunities. To quote Dixit and Pindyck: *The simple NPV rule is not just wrong; it is often very wrong*[14].

[10] See Amram & Kulatilaka [2].
[11] See Copeland & Antikarov [33].
[12] See Hommel, Scholich & Vollrath [61].
[13] See Hommel, Scholich & Baecker [60].
[14] See Dixit & Pindyck [40], page 136.

However, it has to be stressed that a real options approach is not necessary in each and every investment situation. Amram and Kulatilaka developed a list of criteria that show under which circumstances the real options approach is fruitful[15]: *A real options analysis is needed in the following situations:*

1. *When there is a contingent investment decision. No other approach can correctly value this type of opportunity.*

2. *When uncertainty is large enough that it is sensible to wait for more information, avoiding regret for irreversible investment.*

3. *When the value seems to be captured in possibilities for future growth options rather than current cash flow.*

4. *When uncertainty is large enough to make flexibility a consideration. Only the real options approach can correctly value investments in flexibility.*

5. *When there will be project updates and mid-course strategy corrections.*

Both authors also categorize strategic investments and put them into the perspective of real options[16]. The various types of strategic investments are as follows:

1. **Irreversible investments:**
   Once these investments are in place, they cannot be reversed without losing much of their value. The value of an irreversible investment with its associated options is greater than recognized by traditional tools because the options truncate the loss.
2. **Flexibility investments:**
   These are investments that incorporate flexibility in the form of options into the initial design. This is hardly achievable with traditional valuation tools.
3. **Insurance investments:**
   These investments are investments that reduce the exposure to uncertainty, i.e., a put option.
4. **Modular investments:**
   Modular investments create options through product design. Each module has a tightly specified interface to the other, allowing modules to be independently changed and upgraded. A modular product can be viewed as a portfolio of options to upgrade.

[15] See Amram & Kulatilaka [2], page 24.
[16] See Amram & Kulatilaka [2], pape 25-27.

5. **Platform investments:**
   These investments create valuable follow-on contingent investment opportunities. This is particularly important in R&D, *the* classic platform investment. The value of such an investment results from products it releases for further development that may in turn lead to marketable products. Traditional tools greatly undervalue these investments, whereas the real options approach is ideally suited to value such an investment.
6. **Learning investments:**
   Learning investments are investments that are made to obtain information that is otherwise unavailable. A typical learning investment is the exploration work undertaken on a site prior to exploiting it.

Based on this list of different investment types, a categorization was developed and is currently widely used in Finance. This list is based on Trigeorgis [132], pages 1-14. Trigeorgis groups real options into the following classes:

- **Option to defer (learning option):**
  This includes options where the time point of an investment is not determined but flexible allowing this time point to be optimized. Those options can also arise from changes in the term structure of interest rates over time even if the future cash flow is deterministic. Given the sharp interest rates decline in the U.S. in 2001 as seen in Section 1.1 (Figure 1.3), this option is of special use for firms in volatile capital markets: The NPV increases as interest rates decrease and it may be optimal to invest at a later point of time.
- **Time-to-build option:**
  This is an option that allows to stop a step-by-step investment within a project if market conditions turn unfavorable. Such an option is particularly valuable in R&D.
- **Option to alter the operating scale:**
  This is the option to react upon a changing market and to expand operations (favorable market conditions) or to scale down operations (unfavorable market conditions). Such an option can be implemented when a firm wants to introduce a new product or would like to enter a new market, for example in the consumer goods industry.
- **Option to abandon (put option, insurance):**
  The option to sell a project is called *option to abandon*. The value that can be regained by selling the project (*salvage value*) is included in the pricing of the project and can substantially alter the project's NPV calculation.
- **Option to switch:**
  This option comprises the possibility to react upon changed market conditions by changing the input and output factors via input shifts and/or output shifts. This is *the* classical real option.

- **Growth option:**
  Growth options are strategic options. They are particularly revelant for projects which are not advantageous in themselves but may generate lucrative opportunities in the future. This type of option can especially be found in R&D[17]. In the pharmaceutical industry, e.g., it takes more than ten years for a product to develop from the original idea to the final product with a success probability of only a few percent[18]. However, during the course of a project the original investment may generate various other applications which are profitable and generate a positive NPV for the whole investment.
- **Multiple interacting options:**
  Multiple interacting options are combinations of real options of the types described above. In practice, of course, they are the most frequent ones.

To price (real) options different methods can be applied. An overview is given in Figure 2.1. A broad classification and qualitative discussion of these can be found in Hommel & Lehmann [59], Chapter 3. A similar overview with an in-depth discussion and mathematical descriptions of some specific methods can be found in Schulmerich [118], pages 64-67.

Although the real options theory has been a topic of research for over 25 years it has been made broadly accessible to practitioners in Corporate Finance only since the mid 1990s. Therefore, the diffusion of this theory in capital budgeting has still a long way to go. These are the main results of an article by Pritsch and Weber[19] who found that real options pricing models are not well known among senior managers accustomed to the NPV method. The importance of the real options approach in the academic field has yet to be gained in corporate financial practice[20]. As Pritsch and Weber point out, this will not be easily achieved. However, they also emphasize that the same held true for the NPV method in the last decades[21]. In Figure 2.2 they developed an overview of the diffusion of the NPV method in U.S. companies in the last decades[22].

[17] For a detailed analysis of R&D and the valuation approach with multi-dimensional models on American compound options in that area see Pojezny [107].
[18] See Schwartz & Moon [121], pages 87-89.
[19] See Pritsch & Weber [108].
[20] See Pritsch & Weber [108], page 38.
[21] See Pritsch & Weber [108], page 31.
[22] Figure 2.2 is based on the articles Klammer & Walker [76] (for 1955 and 1984), Klammer [75] (for 1959, 1964, and 1970), Gitman & Forrester [47] (for 1970 and 1977) as well as Reichert, Moore & Byler [110] (for 1992).

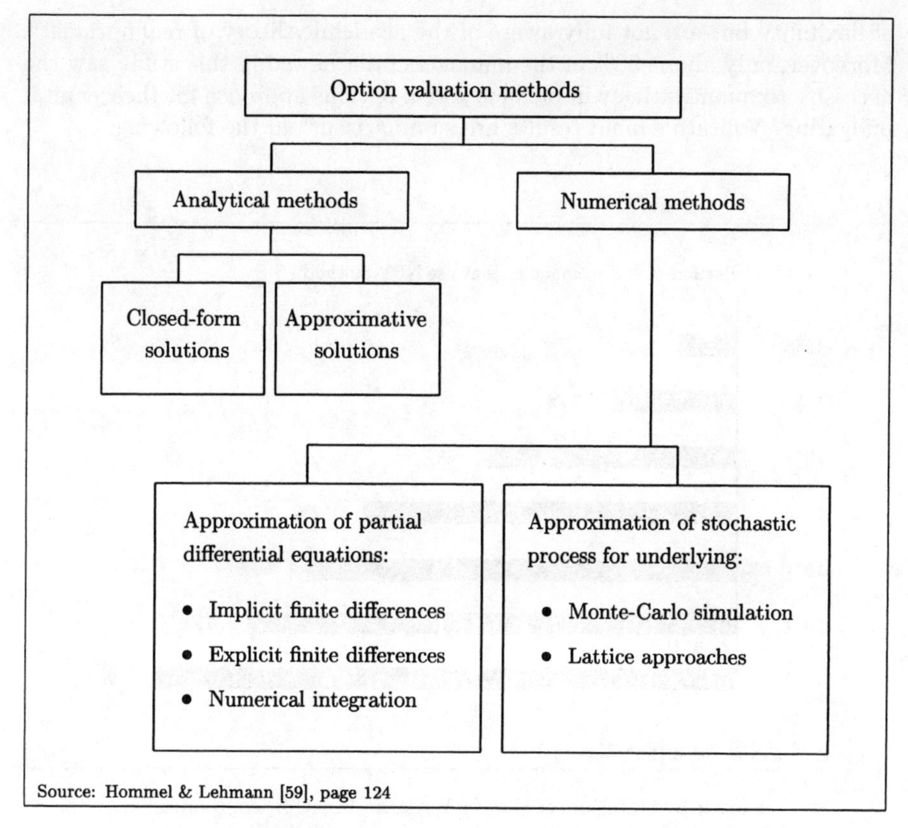

Fig. 2.1. Classification of valuation methods for real options.

A thorough analysis of the importance of the real options approach in capital budgeting practice was undertaken only by a few others[23]. One of the latest and best empirical studies was done by Vollrath[24] who in 2000 investigated a sample of companies with headquarters in Germany. He found that although the real options approach is theoretically superior to traditional capital budgeting tools, it is not widespread in companies.

A detailed study that reveals similar findings was undertaken by Bubsy and Pitts[25] in 1997 for all firms in the FTSE-100 index in the U.K.[26]. This study shows that decision-makers in those firms intuitively realize the importance

[23] See Vollrath [134], pages 46-52.
[24] See Vollrath [134].
[25] See Bubsy & Pitts [24].
[26] Although all FTSE-100 firms were asked to participate, only 44% finally took part and answered the questionnaire.

of flexibility but are not fully aware of the academic theory of real options[27]. Moreover, only about 50% of the managers interviewed in this study saw the necessity to quantitatively implement a real options approach for their capital budgeting. Vollrath's main results are summarized[28] in the following:

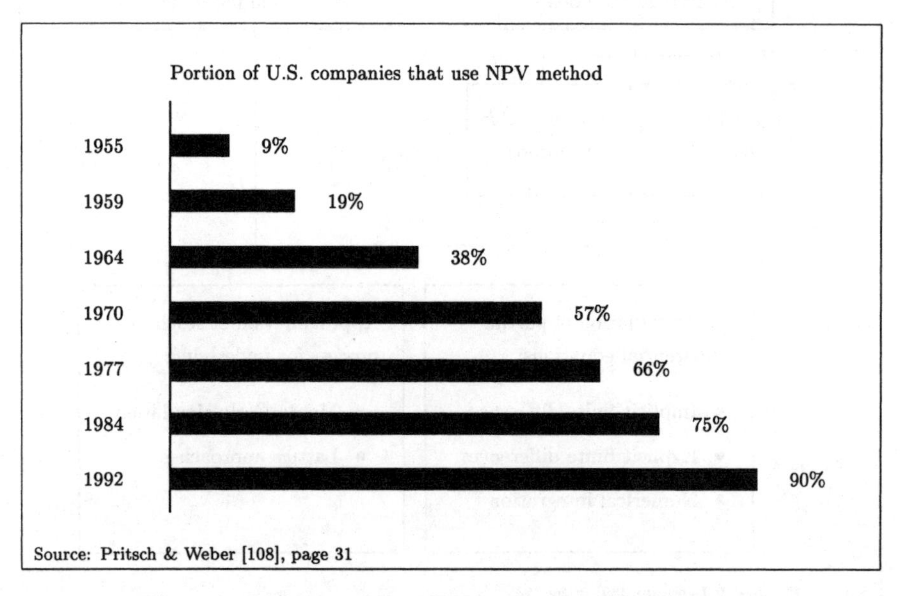

Fig. 2.2. Diffusion of the NPV method in U.S. companies.

1. **Implemented methods for capital budgeting:**
   The major investment decisions are made at the corporate headquarters with several valuation tools used in parallel. However, the real options approach is not applied.

2. **Implementation of methods:**
   Investment projects will be undertaken due to a strategic reason even if the NPV is negative.

3. **Intuitive realization of real options values:**
   Flexibility is intuitively incorporated and assigned a positive value. However, there is no formal option valuation due to the complexity of options pricing. Furthermore, the influence of various parameters on the real option cannot be evaluated sufficiently.

[27] See Vollrath [134], page 50.
[28] See Vollrath [134], pages 72-73, for the complete analysis.

This thesis intends to bridge the gap between the theoretical and the practical importance of real options by providing various real options valuation tools that can readily be used in practice[29]. Moreover, by applying a stochastic term structure instead of a constant risk-free interest rate, the thesis intends to improve these numerical methods in order to model reality more accurately.

2.3 Diffusion Processes in Classical Probability Theory

When dealing with real options pricing and term structure models the mathematical tool called Ito calculus or stochastic calculus has to be used. Term structure models are just one example that uses diffusion processes to describe the movement of the yield curve over time. Diffusion processes are also used to model the underlying of options or to model cost processes as done in the Schwartz-Moon model that will be described in Section 4.3. Diffusion processes are best described with stochastic calculus since classical probability theory can only give limited insight into this type of stochastic process. Nevertheless, it is advisable to introduce diffusion processes using classical probability theory since the parameters of a diffusion process can be best explained with classical probability theory. Therefore, this section is devoted to the introduction of a diffusion process using classical probability theory.

This section closely follows Schulmerich [117], pages 3-14, and assumes that the theory of continuous-time Markov processes is known. The state space of a process is always an interval I with left interval point l and right interval point r. The interval I has one of the four forms $(l,r), (l,r], [l,r)$ or $[l,r]$ with $l,r \in \mathbb{R} \cup \{\pm\infty\}$.

2.3.1 Definition: Diffusion Process - 1st Version

A continuous-time stochastic process $(X_t)_{t\geq 0}$, which has the strong Markov property[30], is called (one-dimensional) *diffusion process*, if all paths of the process are smooth functions in t in the almost sure sense.

This definition is called *1st version* since another definition of a diffusion process (called *2nd version*) will be given in connection with *Ito calculus*. The relationship between these two definitions will then be explained in detail.

[29] This has been pointed out as an important point of critique and as an area of future research by Trigeorgis, see Trigeorgis [132], page 375, Section 12.3 - Future Research Directions: *Developing generic options-based user-friendly software packages with simulation capabilities that can handle multiple real options as a practical aid to corporate planners*. This is a major purpose of this thesis which is realized with the computer simulation program that was developed for all analyses in Chapter 5 based on the models presented in Chapters 3 and 4.

[30] See Karlin & Taylor [73], page 149.

2.3.2 Definitions: Infinitesimal Expectation and Variance

Let $(X_t)_{t \geq 0}$ be a diffusion process with real limits

$$\lim_{h \downarrow 0} \frac{1}{h} E(\Delta_h X_t \,|\, X_t = x) \quad =: \quad \mu(x,t), \tag{2.1}$$

$$\lim_{h \downarrow 0} \frac{1}{h} E([\Delta_h X_t]^2 \,|\, X_t = x) \quad =: \quad \sigma^2(x,t) \tag{2.2}$$

for each $l < x < r$ and $t \geq 0$, where l and r are the left and right interval points of the state space I and $\Delta_h X_t := X_{t+h} - X_t$. The functions $\mu(x,t)$ and $\sigma^2(x,t)$ are the infinitesimal parameters of the diffusion process; $\mu(x,t)$ is called *infinitesimal expectation, expected infinitesimal change* or *(infinitesimal) drift*; $\sigma^2(x,t)$ is called *infinitesimal variance* or *diffusion parameter.* Functions μ and σ^2 are called *infinitesimal expectation* and *infinitesimal variance*, respectively, since from (2.1) and (2.2) it is known[31]:

$$E(\Delta_h X_t \,|\, X_t = x) = \mu(x,t)\, h \; + \; o_1(h), \qquad \lim_{h \downarrow 0} \frac{o_1(h)}{h} = 0,$$

$$Var(\Delta_h X_t \,|\, X_t = x) = \sigma^2(x,t)\, h \; + \; o_2(h), \qquad \lim_{h \downarrow 0} \frac{o_2(h)}{h} = 0.$$

Here, $o_i, i = 1, 2$, is the Landau symbol. Generally, $\mu(x,t)$ and $\sigma^2(x,t)$ are smooth functions of x and t.

In this thesis various types of diffusion processes will be used. Some of them are *time-homogeneous* diffusion processes[32], i.e., $\mu(x,t) = \mu(x)$ and $\sigma^2(x,t) = \sigma^2(x)$ are independent of t (see Karlin & Taylor [73], page 160). Other diffusion processes have parameters that depend only on time t but not on the state of the process, such as the short-rate process in the Ho-Lee model. Other processes depend both on time t and state x like in the Hull-White one-factor model. In some cases (see the Hull-White two-factor model) the infinitesimal drift even contains a random component that is modelled via a second diffusion process, an Ornstein-Uhlenbeck process.

In the remainder of Section 2.3 the focus is on time-homogeneous processes, i.e., diffusion processes with $\mu(x,t) = \mu(x)$ and $\sigma^2(x,t) = \sigma^2(x)$ independent of t. It is assumed that μ and σ are smooth in x, and $I = (l, r)$ is assumed to be the state space of the diffusion process. This provides the mathematical tools to handle the Ornstein-Uhlenbeck process, the Vasicek model, and the Cox-Ingersoll-Ross model in Chapter 3. The goal of this section is to derive the stationary distribution of those diffusion processes (see 2.3.4).

[31] See Karlin & Taylor [73], page 160.

[32] The short-rate processes in the Vasicek model and in the Cox-Ingersoll-Ross model are examples of time-homogeneous diffusion processes as will be seen later.

2.3.3 Denominations for Diffusion Processes

Let $(X_t)_{t\geq 0}$ be a diffusion process with the infinitesimal parameters $\mu(x)$ and $\sigma^2(x)$, $l < x < r$, where $I = (l, r)$ is the state space of the process. Via

$$S(x) := \int_c^x \exp\left\{ -\int_{\eta_0}^{\eta} \frac{2\mu(\xi)}{\sigma^2(\xi)} d\xi \right\} d\eta, \quad l < x < r,$$

the function $S : I \longrightarrow \mathbb{R}$ is defined and referred to as the *scale function* of the diffusion process. Parameters c and η_0 are flexible constants[33] from the state space I. The first derivative of S,

$$S'(\eta) = \exp\left\{ -\int_{\eta_0}^{\eta} \frac{2\mu(\xi)}{\sigma^2(\xi)} d\xi \right\}, \quad l < \eta < r,$$

will be denoted with $s(\eta)$. Via

$$M(x) := \int_{\tilde{c}}^{x} \frac{2}{\sigma^2(\eta)\, s(\eta)} d\eta, \quad l < x < r,$$

the function $M : I \longrightarrow \mathbb{R}$ is defined and referred to as the *speed function* of the diffusion process. Here, $\tilde{c}$ is a flexible constant from the state space I. In a natural manner a measure on I can be defined, which will be (as common in literature) denoted with M:

$$M((a, b]) := M(b) - M(a), \quad l < a < b < r.$$

The measure M on I is called *speed measure* of the diffusion process[34]. Via

$$m(x) := M'(x) = \frac{2}{\sigma^2(x)s(x)}, \quad l < x < r,$$

the function $m : I \longrightarrow \mathbb{R}$ is defined, which is called *speed density* of the diffusion process. This density is a *Radon-Nikodym density.*

The functions S and s are unique only up to an affine transformation with a monotonically increasing transformation function in the case of S and up to a positive multiplicative factor in the case of s. For a given S, the

[33] The term *flexible constant* refers to the mathematical idea that the choice of c and η does not matter. However, if certain values for c and η_0 are chosen, these values cannot be changed any more.

[34] It is common in literature to use the M for the speed measure and for the speed function. This does not create confusion since both functions operate on different sets.

speed density m is unique. A process with scale function S in the form of $S(x) = C_1 + C_2 x$, $x \in I$ with $C_1, C_2 \in \mathbb{R}$ constant, is called *in natural scale/canonical scale.*

According to Karlin & Taylor [73], page 194, the speed measure M is a measure $M : I \longrightarrow \mathbb{R}$, such that for each $a < x < b$ with $a, x, b \in I$ it holds:

$$E(T_{a \wedge b} | X_0 = x) = \int_{(a,b)} G_{S(a),S(b)}(S(x), S(y))\, dM(y) \quad \text{with} \tag{2.3}$$

$$G_{q_1,q_2}(u,v) := \left\{ \begin{array}{ll} \dfrac{[u - q_1]\,[q_2 - v]}{q_2 - q_1}, & q_1 < u \le v < q_2 \\ \dfrac{[q_2 - u]\,[v - q_1]}{q_2 - q_1}, & q_1 < v \le u < q_2 \end{array} \right\} \tag{2.4}$$

as the *Green function* on $(q_1, q_2) \subseteq \mathbb{R}$, $q_1 < q_2$.

Knowing this relationship helps to understand the meaning of the speed density of a diffusion process $(X_t)_{t \geq 0}$, which is in natural scale, with state space I. For this let $x \in I$ be flexible but constant and let $\varepsilon > 0$ be so small that $[x - \varepsilon, x + \varepsilon]$ lies in interval I. Then:

$$\begin{aligned}
& E(T_{x-\varepsilon \wedge x+\varepsilon} | X_0 = x) \\
\overset{(2.3)}{=}\ & \int_{(x-\varepsilon,x+\varepsilon)} G_{S(x-\varepsilon),S(x+\varepsilon)}(S(x), S(y))\, dM(y) \\
=\ & \int_{(x-\varepsilon,x+\varepsilon)} G_{x-\varepsilon,x+\varepsilon}(x,y)\, dM(y) \\
=\ & \int_{(x-\varepsilon,x)} G_{x-\varepsilon,x+\varepsilon}(x,y)\, dM(y) + \int_{[x,x+\varepsilon)} G_{x-\varepsilon,x+\varepsilon}(x,y)\, dM(y) \\
\overset{(2.4)}{=}\ & \int_{x-\varepsilon}^{x} \frac{([x+\varepsilon] - x)(y - [x-\varepsilon])}{[x+\varepsilon] - [x-\varepsilon]}\, m(y)\, dy \ + \\
& \qquad \int_{x}^{x+\varepsilon} \frac{(x - [x-\varepsilon])([x+\varepsilon] - y)}{[x+\varepsilon] - [x-\varepsilon]}\, m(y)\, dy \\
=\ & \frac{1}{2} \left\{ \int_{x-\varepsilon}^{x} (y - x + \varepsilon)\, m(y)\, dy + \int_{x}^{x+\varepsilon} (x + \varepsilon - y)\, m(y)\, dy \right\}.
\end{aligned}$$

Therefore:

$$
\begin{aligned}
&\frac{1}{\varepsilon^2} E(T_{x-\varepsilon \wedge x+\varepsilon} | X_0 = x) \\
= \quad & \frac{1}{2} \left\{ \frac{1}{\varepsilon^2} \int_{x-\varepsilon}^{x} (y - x + \varepsilon)\, m(y)\, dy \; + \; \frac{1}{\varepsilon^2} \int_{x}^{x+\varepsilon} (x + \varepsilon - y)\, m(y)\, dy \right\} \\
= \quad & \frac{1}{2} \left\{ \frac{1}{\varepsilon^2} (x - \delta_1(\varepsilon)\varepsilon - x + \varepsilon)\, m(x - \delta_1(\varepsilon)\varepsilon)\, (x - [x - \varepsilon]) \right. \\
& \quad \left. + \; \frac{1}{\varepsilon^2} (x + \varepsilon - (x + \delta_2(\varepsilon)\varepsilon))\, m(x + \delta_2(\varepsilon)\varepsilon)\, ([x + \varepsilon] - x) \right\}, \\
& \quad \delta_1(\varepsilon) \text{ and } \delta_2(\varepsilon) \in [0,1] \text{ appropriate with} \\
& \quad \delta_i \xrightarrow[\varepsilon \to 0]{} 0 \; (i = 1, 2) \\
= \quad & \frac{1}{2} \{ (1 - \delta_1(\varepsilon))\, m(x - \delta_1(\varepsilon)\varepsilon) \; + \; (1 - \delta_2(\varepsilon))\, m(x + \delta_2(\varepsilon)\varepsilon) \} \\
\xrightarrow[\varepsilon \to 0]{} \; & m(x), \quad (m \text{ is smooth, see 2.3.2 and definition of } m).
\end{aligned}
$$

This means that the speed density $m(x)$ is the speed with which the process $(X_t)_{t \geq 0}$ moves forward if it is in state $x \in I$.

2.3.4 Stationary Distribution of a Diffusion Process

Let $(X_t)_{t \geq 0}$ be a diffusion process with state space $I = (l, r)$, whereby the process is *in natural scale*. Let

$$
|m| \quad := \quad \int_{l}^{r} m(x)\, dx
$$

be the total mass of the speed density m on I. If $|m| < \infty$ then there exists a stationary distribution of the process $(X_t)_{t \geq 0}$ which has a Lebesgue measure with density

$$
\psi(x) \quad = \quad \frac{m(x)}{|m|}, \quad x \in I.
$$

Since m and $|m|$ are unique up to the same multiplicative constant (see 2.3.3), ψ is uniquely defined. The proof can be found in Rogers & Williams [113], page 303, theorem 54.5. According to Schulmerich [117], Section 1.16, the restriction *in natural scale* on the diffusion process $(X_t)_{t \geq 0}$ above can be omitted.

This is the main result of 2.3. In the following an example is given of a special diffusion process, the one-dimensional Brownian Motion.

2.3.5 Example: One-Dimensional Brownian Motion

A stochastic process $(B_t)_{t\geq 0}$ with $B_0 = 0$ is called *one-dimensional Brownian Motion* (or *Wiener process*) with drift $\mu \in \mathbb{R}$, if for each choice of $0 \leq t_0 < t_1 < \ldots < t_{n-1} < t_n$, $n \in \mathbb{N}$, the joint distribution of $B_{t_0}, B_{t_1}, \ldots, B_{t_n}$ is determined by two conditions:

1. the $(B_{t_k} - B_{t_{k-1}})$, $k = 1, \ldots, n$, are independent and
2. $(B_{t_k} - B_{t_{k-1}}) \stackrel{d}{=} N([t_k - t_{k-1}]\mu, [t_k - t_{k-1}]\sigma^2)$, $k = 1, \ldots, n$, for suitable constants $\mu \in \mathbb{R}$ and $\sigma \in \mathbb{R}^+$.

If $\mu = 0$ and $\sigma^2 = 1$, this process is called a *Standard Brownian Motion.* A Standard Brownian Motion $(B_t)_{t\geq 0}$ is a time-homogeneous diffusion process with state space $I = (-\infty, \infty)$.

If a Standard Brownian Motion $(B_t)_{t\geq 0}$ is given, a Brownian Motion $Y := (Y_t)_{t\geq 0}$ with drift $\mu \in \mathbb{R}$ and volatility $\sigma \in \mathbb{R}^+$ can be derived via $Y_t := \sigma B_t + \mu t, t \geq 0$. By doing so the state space $I = (-\infty, \infty)$ remains the same. $(Y_t)_{t\geq 0}$ is a time-homogeneous diffusion process as well and has the infinitesimal parameters $\mu_Y(x) = \mu$ and $\sigma_Y^2(x) = \sigma^2 \in \mathbb{R}^+, x \in I$. According to 2.3.3 it holds for each $x \in \mathbb{R}$:

Case I with $\mu \neq 0$:

$$
\begin{aligned}
s(x) &= \exp\left\{-\frac{2\mu x}{\sigma^2}\right\}, \quad \text{where } \eta_0 := 0 \in I, \\
S(x) &= \frac{\sigma^2}{2\mu}\left(1 - \exp\left\{-\frac{2\mu x}{\sigma^2}\right\}\right), \quad \text{where } c := 0, \\
M(x) &= \frac{1}{\mu}\left(\exp\left\{\frac{2\mu x}{\sigma^2}\right\} - 1\right), \quad \text{where } \tilde{c} := 0, \\
m(x) &= \frac{2}{\sigma^2}\exp\left\{\frac{2\mu x}{\sigma^2}\right\}.
\end{aligned}
$$

Case II with $\mu = 0$:

$$
\begin{aligned}
s(x) &= 1, \quad \text{where } \eta_0 := 0 \in I, \\
S(x) &= x, \quad \text{where } c := 0, \\
M(x) &= \frac{2}{\sigma^2}x, \quad \text{where } \tilde{c} := 0, \\
m(x) &= \frac{2}{\sigma^2}.
\end{aligned}
$$

The Geometric Brownian Motion $Z = (Z_t)_{t\geq 0}$ can be derived via $Z_t := \exp\{Y_t\}$, $t \geq 0$. The resulting state space I is then $\mathbb{R}^+$ and, according to the transformation formula in Karlin & Taylor [73], pages 173-175, it holds for the Geometric Brownian Motion:

$$\mu_Z(z) = \left(\mu + \frac{\sigma^2}{2}\right) z \qquad \text{and} \qquad \sigma_Z^2(z) = \sigma^2 z^2, \quad z \in \mathbb{R}^+.$$

2.4 Introduction to the Ito Calculus

This section introduces the language of Financial Engineering, the *stochastic calculus*, referred to as *Ito calculus.* While traditional probability theory is well suited to introduce the notion of a diffusion process as done in the previous section, the main tool for describing and especially for calculating such a process is stochastic calculus. This introduction concentrates on presenting results without giving a detailed proof. However, literature will always be cited where a detailed proof and further information about any of the presented topics can be found. The following introduction closely follows Schulmerich [117], pages 15-28, but focuses on those aspects of the theory that will be needed for this thesis.

2.4.1 Definition: Ito Integral

Let $(\Omega, \mathcal{F}, P)$ be the probability space and $\mathcal{B}$ the *Borel σ-Field* on $\mathbb{R}_0^+$.

1. Let $\{\mathcal{N}_t | t \geq 0\}$ be a family of increasing σ-fields over Ω. A process $h : \mathbb{R}_0^+ \times \Omega \longrightarrow \mathbb{R}$ is called $(\mathcal{N}_t)_{t \geq 0}$-*adapted* if for each $t \geq 0$ the function $\omega \longmapsto h(t, \omega)$ is $\mathcal{N}_t$-measurable, $\omega \in \Omega$.
2. Let $(B_s)_{s \geq 0}$ be a one-dimensional Standard Brownian Motion. Let $\mathcal{V} := \mathcal{V}(S, T)$, $S, T \in \mathbb{R}_0^+$ with $S < T$ be a class of functions $f : \mathbb{R}_0^+ \times \Omega \longrightarrow \mathbb{R}$ with the following properties:

   (a): $(t, \omega) \longmapsto f(t, \omega)$ is $\mathcal{B} \times \mathcal{F}$-measurable, $t \geq 0$, $\omega \in \Omega$, i.e.: function f is *progressively measurable.*

   (b): f is $\mathcal{F}_t$-adapted, where $\mathcal{F}_t := \sigma(B_s | 0 \leq s \leq t)$, $t \geq 0$.

   (c): $E\left(\int_S^T f(t, \omega)^2 dt\right) < \infty.$
3. For S and T from point 2 above and for fixed $n \in \mathbb{N}$ let $t_j^{(n)}, j \in \mathbb{N}_0$, be defined by

$$t_j^{(n)} := \begin{cases} S, & \text{if } j\,2^{-n} < S, \\ j\,2^{-n}, & \text{if } S \leq j\,2^{-n} \leq T, \\ T, & \text{if } T < j\,2^{-n}. \end{cases}$$

   A function $\lambda_n : \mathbb{R}_0^+ \times \Omega \longrightarrow \mathbb{R}$ from $\mathcal{V}$ is called *elementary function,* if it has the form

$$\lambda_n(t,\omega) = \sum_{j\in\mathbb{N}_0} e_j(\omega)\,1_{[t_j^{(n)},t_{j+1}^{(n)})}(t), \quad t \geq 0,\ \omega \in \Omega.$$

e_j is a random variable on $(\Omega, \mathcal{F}, P)$ that needs to be $\mathcal{F}_{t_j}$-measurable because of $\lambda_n \in \mathcal{V}$. $1_{[t_j^{(n)},t_{j+1}^{(n)})}$ is the indicator function with respect to the set $[t_j^{(n)}, t_{j+1}^{(n)})$, $j \in \mathbb{N}_0$.

For an elementary function λ_n the *Ito integral* $\int_S^T \lambda_n(t,\omega)\,dB_t(\omega)$ is defined for each $\omega \in \Omega$ as

$$\int_S^T \lambda_n(t,\omega)\,dB_t(\omega) := \sum_{j\in\mathbb{N}_0} e_j(\omega)[B_{t_{j+1}^{(n)}} - B_{t_j^{(n)}}](\omega).$$

According to Øksendal [103], pages 23-25, it holds:
For each $f \in \mathcal{V}$ there exist elementary functions $\lambda_n \in \mathcal{V}, n \in \mathbb{N}$, with

$$E\left(\int_S^T |f-\lambda_n|^2\,dt\right) \xrightarrow[n\to\infty]{} 0.$$

For such a function f the *Ito integral* $\int_S^T f(t,\omega)dB_t(\omega), \omega \in \Omega$, is defined as

$$\int_S^T f(t,\omega)\,dB_t(\omega) := \lim_{n\to\infty}\int_S^T \lambda_n(t,\omega)\,dB_t(\omega).$$

This limit exists as an element of the set $L_2(\Omega, P)$. The Ito integral is well-defined, i.e., independent from the choice of the sequence of elementary functions $(\lambda_n)_{n\in\mathbb{N}}$.

2.4.2 Properties of an Ito Integral

The following rules hold for an Ito integral:

1. $E\left(\left[\int_S^T f\,dB_t\right]^2\right) = E\left(\int_S^T f^2\,dt\right) \quad \forall f \in \mathcal{V}(S,T)$ *(Ito isometry).*

2. For $f, g \in \mathcal{V}(0,T)$ it holds with $0 \leq S < U < T$:

$$(a): \int_S^T f\,dB_t = \int_S^U f\,dB_t + \int_U^T f\,dB_t \quad \text{almost sure.}$$

$$(b): \quad \int_S^T (cf+g)\,dB_t = c\int_S^T f\,dB_t + \int_S^T g\,dB_t \quad \text{almost sure, } c \in \mathbb{R}.$$

$$(c): \quad E\left(\int_S^T f\,dB_t\right) = 0.$$

$$(d): \quad \left(\int_S^T f\,dB_t\right)_{T\geq S} \quad \text{is a martingale.}$$

$$(e): \quad \int_S^T f\,dB_t \quad \text{is } \mathcal{F}_T\text{-measurable.}$$

3. For each $f \in \mathcal{V}(0,T)$ there exists an in t smooth version of the Ito integral

$$\int_0^t f\,dB_s, \quad 0 \leq t \leq T,$$

i.e., there exists an in t smooth process $(X_t)_{t\geq 0}$ on $(\Omega, \mathcal{F}, P)$ with

$$P\left(X_t = \int_0^t f\,dB_s\right) = 1 \quad \text{for each} \quad 0 \leq t \leq T.$$

In the following only in t smooth Ito integrals will be considered. The proof of these properties can be found in Øksendal [103], pages 26-30.

2.4.3 Definition: A Broader Class of Ito Integrals

1. The Ito integral can also be defined for a broader class of integrands than $\mathcal{V}$: Let $\mathcal{W} \supset \mathcal{V}$ be the class of functions $g : \mathbb{R}_0^+ \times \Omega \longrightarrow \mathbb{R}$ with the following properties:
   a) g is progressively measurable.
   b) There exists a family $(\mathcal{H}_t)_{t\geq 0}$ of increasing σ-fields, such that $(B_t)_{t\geq 0}$ is a martingale with respect to $(\mathcal{H}_t)_{t\geq 0}$ and such that $g(t,\cdot)$ is adapted on $\mathcal{H}_t$, $t \geq 0$.
   c) $P\left(\left\{\omega \in \Omega \;\middle|\; \int_0^t g(s,\omega)^2\,ds < \infty \quad \forall t \geq 0\right\}\right) = 1.$

Analogous to the description in 2.4.1, the Ito integral can also be defined for $g \in \mathcal{W}$. Yet this will not be done here. For details see Øksendal [103], pages 31-32, whose definition will be used in the following. All of the properties in 2.4.2 are fullfilled for each $g \in \mathcal{W}$ as well (see Karatzas & Shreve [72], Chapter 3.2).

2. It is also possible to define the Ito integral from 2.4.1 not only with respect to the Standard Brownian Motion $(B_t)_{t\geq 0}$ as done in the original construction from Ito, but also, under suitable conditions, with respect to each stochastic process $(M_t)_{t\geq 0}$ that is a smooth, quadratic integrable martingale. This broader definition of the stochastic integral can be found in Karatzas & Shreve [72], pages 128-139.

2.4.4 Definition: One-Dimensional Ito Process

Let $(B_t)_{t\geq 0}$ be a one-dimensional Standard Brownian Motion on a probability space $(\Omega, \mathcal{F}, P)$. A *one-dimensional Ito process* or *stochastic integral* is a stochastic process $(X_t)_{t\geq 0}$ on $(\Omega, \mathcal{F}, P)$ of the form

$$X_t(\omega) = X_0(\omega) + \int_0^t a(s,\omega)ds + \int_0^t b(s,\omega)dB_s(\omega), \quad \omega \in \Omega \quad \text{and} \quad t \geq 0. \quad (2.5)$$

Here, it is assumed that $b \in \mathcal{W}$ holds and that for a the following equation holds true:

$$P\left(\left\{\omega \in \Omega \;\middle|\; \int_0^t |a(s,\omega)|\, ds \;<\; \infty \quad \forall t \geq 0\right\}\right) \;=\; 1.$$

Moreover, let a be adapted to $(\mathcal{H}_t)_{t\geq 0}$ where $(\mathcal{H}_t)_{t\geq 0}$ is a family of increasing σ-fields from $\mathcal{W}$. The differential notation for (2.5) is:

$$dX_t(\omega) \;=\; a(t,\omega)\, dt \;+\; b(t,\omega)\, dB_t(\omega), \quad t \geq 0,$$

or in short notation:

$$dX_t \;=\; a\, dt \;+\; b\, dB_t, \quad t \geq 0. \quad (2.6)$$

(2.6) is called *stochastic differential equation* (*SDE*).

The usage of $f \in \mathcal{V}$ in 2.4.1 as integrand in the Ito integral is more specific than the usage of random variables $X_t, t \geq 0$. In particular, each element $f \in \mathcal{V}$ is a stochastic process. The reverse of this statement is not true since not each stochastic process fulfills the conditions (a) and (b) of point 2 in 2.4.1. However, if the notation from definition 2.4.4 is applied, it yields: For each stochastic process $(X_t)_{t\geq 0}$ the parameters $a(t, X_t)$ and $b(t, X_t)$ fulfill the conditions (a) and (b) of point 2 in 2.4.1.

2.4.5 Theorem: Existence and Uniqueness of the Solution of a Stochastic Differential Equation

With $T > 0$ let $a : [0,T] \times \mathbb{R} \longrightarrow \mathbb{R}$ and $b : [0,T] \times \mathbb{R} \longrightarrow \mathbb{R}$ be Lebesgue-measurable functions. It is further assumed that the following two conditions hold:

(*i*) $|a(t,x)| + |b(t,x)| \;\leq\; C\,(1+|x|),\;\; x \in \mathbb{R},\;\; t \in [0,T],$

$C \in \mathbb{R}^+$ appropriate (*growth condition*).

(*ii*) $|a(t,x) - a(t,y)| + |b(t,x) - b(t,y)| \;\leq\; D|x-y|,\;\; x,y \in \mathbb{R},$

$t \in [0,T],\;\; D \in \mathbb{R}^+$ appropriate (*Lipschitz condition*).

Let $(B_t)_{t\geq 0}$ be a one-dimensional Standard Brownian Motion and let Z be a random variable with $E(|Z|^2) < \infty$ that is independent from $\mathcal{F}_\infty := \sigma(B_s | s \geq 0)$. Then the SDE

$$dX_t \;=\; a(t,X_t)\,dt \;+\; b(t,X_t)\,dB_t,\;\; 0 \leq t \leq T,\;\; X_0 = Z, \tag{2.7}$$

has a unique and in t smooth solution $(X_t)_{t\geq 0}$ that is an element of $\mathcal{V}(0,T)$. More precisely, the process $(X_t)_{t\geq 0}$ is referred to as a *strong solution* since $(B_t)_{t\geq 0}$ is given and the solution process is $\mathcal{F}_t$-adapted, $t \geq 0$. A solution is called a *weak solution* if the Standard Brownian Motion is not given but a pair $((B_t)_{t\geq 0}, (X_t)_{t\geq 0})$ is searched for that fulfills (2.7).

A strong solution $(X_t)_{t\geq 0}$ of (2.7) possesses the strong Markov property and fulfills

$$E\left(|X_t|^2\right) \;<\; \infty \qquad \forall\; t \geq 0.$$

If μ and σ do not depend on t, i.e., if the SDE

$$dX_t \;=\; a(X_t)\,dt \;+\; b(X_t)\,dB_t,\;\;\; t \geq 0,\;\; X_0 = Z,$$

is considered, (*i*) and (*ii*) are simplified to

(*iii*) $|a(x) - a(y)| + |b(x) - b(y)| \;\leq\; D\,|x-y|,\quad x,y \in \mathbb{R}.$

SDE (2.7) is the type of an SDE that describes most of the term structure models in Chapter 3, e.g., the Vasicek model, the Cox-Ingersoll-Ross (CIR) model, the Ho-Lee model, and the Hull-White one-factor model.

The goal for the remainder of this section is to give explicit solutions for some specific SDEs. These formulas will be then further used in Chapter 3 to describe various term structure models.

2.4.6 Coefficients of an Ito Process

According to Karlin & Taylor [73], page 376, it can be shown: If an Ito process from 2.4.4 fulfills the criteria of definition 2.3.1 and also fulfills the conditions 2.4.5 (i), (ii), then $a \equiv \mu$ and $b \equiv \sigma$, using the notation μ and σ from definition 2.3.2. In the following only the notation μ and σ is used. σ is also called *volatility*.

2.4.7 Definition: Diffusion Process - 2nd Version

A one-dimensional diffusion process $(X_t)_{t\geq 0}$ is a stochastic process that fulfills the SDE

$$dX_t \;=\; \mu(t, X_t)\, dt \;+\; \sigma(t, X_t)\, dB_t, \;\; t \geq 0, \;\; X_0 = x_0 \in \mathbb{R} \text{ constant},$$

where $\mu : \mathbb{R}_0^+ \times \mathbb{R} \longrightarrow \mathbb{R}$ and $\sigma : \mathbb{R}_0^+ \times \mathbb{R} \longrightarrow \mathbb{R}$ fulfill the conditions 2.4.5 (i), (ii), and $(B_t)_{t\geq 0}$ is a one-dimensional Standard Brownian Motion. $(X_t)_{t\geq 0}$ is also called *Ito diffusion* or *diffusion.*

According to 2.4.5 a diffusion process is smooth and possesses the strong Markov property. If functions μ and σ are independent from t the diffusion process is a time-homogeneous diffusion process (see Øksendal [103], page 104). Moreover, this definition can be generalized to a multi-dimensional diffusion process, see Schulmerich [117], page 21.

2.4.8 Remark: Relationship of the two Definitions of a Diffusion Process

The two definitions of a diffusion process given in 2.3.1 (version 1) and 2.4.7 (version 2), respectively, are not equivalent. In particular, a diffusion process described in 2.4.7 is smooth and possesses the strong Markov property. This yields:

$(X_t)_{t\geq 0}$ is a diffusion according to version 2 $\Longrightarrow$ $(X_t)_{t\geq 0}$ is a diffusion according to version 1.

The opposite direction does generally not hold true as the *Feller-McKean process* demonstrates (see Rogers & Williams [113], page 271). This process is a diffusion process according to version 1 but not according to version 2. Therefore, there exist *more general* diffusion processes (version 1) than diffusion processes which are solutions of an SDE (version 2).

In the following, a process is called a diffusion process if it complies with the more general definition 2.3.1 (1st version of a diffusion process). This is common in literature.

2.4.9 Theorem: Solution of a One-Dimensional, Linear Stochastic Differential Equation

A one-dimensional, linear SDE is given via

$$dX_t = [A(t)X_t + a(t)]\,dt + [C(t)X_t + c(t)]\,dB_t, \; t \geq 0, \qquad (2.8)$$

where X_0 is a one-dimensional random variable with $E(X_0^2) < \infty$ that is independent from $\mathcal{F}_\infty$ (see 2.4.5). The functions $A, a, C,$ and c are assumed to have real values[35], and $(B_t)_{t\geq 0}$ is a one-dimensional Standard Brownian Motion. Functions $A, a, C,$ and c are further assumed to be smooth in t. Then the stochastic differential equation (2.8) has the solution[36]:

$$X_t = \Phi(t)\left[X_0 + \int_0^t \Phi^{-1}(u)\,[a(u) - C(u)c(u)]\,du + \int_0^t \Phi^{-1}(u)\,c(u)\,dB_u\right],$$

where

$$\Phi(t) = \exp\left\{\int_0^t \left[A(u) - \frac{1}{2}C^2(u)\right] du + \int_0^t C(u)\,dB_u\right\}, \quad t \geq 0.$$

A specific case is given for $C \equiv 0$ in (2.8). Under the assumption

$$E\left(\int_0^t \left[\Phi^{-1}(u)\sigma(u)\right]^2 du\right) < \infty \quad \forall t \geq 0$$

this yields:

$$(i) \;\; E(X_t) = \Phi(t)\left[E(X_0) + \int_0^t \Phi^{-1}(u)a(u)\,du\right], \quad t \geq 0.$$

$$(ii) \;\; Cov(X_s, X_t) = \Phi(s)\left[Var(X_0) + \int_0^{s\wedge t} \left(\Phi^{-1}(u)\sigma(u)\right)^2 du\right]\Phi(t), \quad s, t \geq 0.$$

For a proof of this theorem see Schulmerich [117], pages 26-28.

---

[35] The theorem holds also true for a multi-dimensional linear SDE, see Schulmerich [117], Sections 2.16, 2.19, and 2.20. In this case, A and C are real matrices and a and c are real vectors.

[36] Φ is a so-called *fundamental solution* in the one-dimensional case. For a multi-dimensional stochastic process Φ is a *fundamental matrix*. For more details see Schulmerich [117], pages 22-26. The proof of the solution is done with the Ito formula and is explained in detail in Göing [48], pages 60-62.

As mentioned earlier, stochastic calculus is also the language to describe the term structure models that will be explained in Chapter 3. In that context the term *mean reversion* will play a crucial role.

2.4.10 Definition: Mean Reversion Process

Let $(X_t)_{t\geq 0}$ be a diffusion process that is given as the solution of the SDE

$$\left.\begin{array}{l} dX_t = \mu(X_t)dt + \sigma(X_t)dB_t, \;\; t \geq 0, \;\; X_0 = x_0 \in I, \quad \text{where} \\ \mu(x) = \beta - \alpha x \; = \; \alpha\left(\frac{\beta}{\alpha} - x\right), \;\; x \in I, \;\; \alpha \in \mathbb{R}^+, \;\; \beta \in \mathbb{R}, \\ \sigma(x) > 0 \;\; \forall x \in I. \end{array}\right\} \quad (2.9)$$

$I = (l, r)$ denotes the state space of the process (see Section 2.3). Such a process is called a *mean reverting* or *mean reversion process*; parameter α is called *mean reversion force* and $\frac{\beta}{\alpha}$ is called *mean reversion level*. These names are based on the notion that α is the "force" that pulls the process back to its "mean" $\frac{\beta}{\alpha}$. The model is called a *mean reversion model.*

According to this definition only a constant mean reversion level is allowed. However, this can be extended to a time-dependent mean reversion level or even a mean reversion level that contains a stochastic process, e.g., an Ornstein-Uhlenbeck process[37]. This process will be introduced in the following.

2.4.11 Example: Ornstein-Uhlenbeck Process

The Ornstein-Uhlenbeck process is a one-dimensional Ito diffusion that is given as the solution of the following linear SDE in the stronger sense:

$$dX_t = -\alpha X_t \, dt + \sigma \, dB_t, \;\; t \geq 0, \;\; X_0 = x_0 \in \mathbb{R} \text{ constant}, \;\; \alpha \in \mathbb{R}^+, \;\; \sigma \in \mathbb{R}^+.$$

This process mean reverts around zero[38]. According to 2.4.9 it holds

$$\Phi(t) = e^{-\alpha t}, \; t \geq 0,$$

and

$$X_t = x_0 \, e^{-\alpha t} + \sigma \, e^{-\alpha t} \int_0^t e^{\alpha s} \, dB_s, \quad t \geq 0.$$

[37] An example is the Hull-White two-factor model which includes a drift term with a stochastic part which is modelled via an Ornstein-Uhlenbeck process.

[38] This property follows directly from 2.4.9 (i) for $t \to \infty$. See also Schulmerich [117], page 30.

2.5 Discretization of Continuous-Time Stochastic Processes

After having described thoroughly the underlying mathematics for real options and term structure modelling in the previous two sections, the next question to be considered is how to implement these continuous-time models in the discrete, numerical world of a computer.

Section 2.5 answers this question by providing three algorithms to simulate a discrete path of a continuous-time stochastic process. These algorithms have different degrees of "accuracy", a term that will be explained as well. Moreover, it will be shown how to generate normally distributed and correlated realizations of random variables as the basis for the three algorithms presented. The principles of this simulation can be found in Kloeden & Platen [77], Chapters 5.5, 9.6, 10.2, 10.3, and 10.4. A similar introduction to this topic can be found in Schulmerich [117], pages 143-146.

For a constant $T > 0$ consider a one-dimensional Ito process $(X_t)_{0 \leq t \leq T}$ that is given as the solution of the SDE

$$dX_t \;=\; \mu(t, X_t)\, dt \;+\; \sigma(t, X_t)\, dB_t, \;\; 0 < t \leq T, \;\; X_0 = x_0 \in \mathbb{R}.$$

$(B_t)_{t \geq 0}$ is a one-dimensional Standard Brownian Motion. To simulate this continuous-time stochastic process $(X_t)_{0 \leq t \leq T}$ the *simulation time points*

$$0 \;=\; \tau_0 \;<\; \tau_1 \;<\; \ldots \;<\; \tau_N \;=\; T, \;\; N \in \mathbb{N},$$

of the *observation period* $[0, T]$ have to be specified. T is the *time horizon* of the observation period. Moreover, let $\Delta_s^i := \tau_{i+1} - \tau_i$, $0 \leq i \leq N-1$. Δ_s^i is the i-th *simulation step size.* In the computer simulation program for the thesis only equidistant simulation time points are used, i.e., $\tau_i = i\Delta_s, i = 0, \ldots, N$, with $\Delta_s := \frac{T}{N}$. Therefore, the following three simulation methods are specified for equidistant simulation time points. The number of simulated points is then $N = \frac{T}{\Delta_s}$, i.e., the total number of data points for one single simulated path is $N+1$ (including the non-random and given starting point x_0 of the process).

The basic tool for the following simulation methods is the Ito-Taylor expansion of a stochastic process $(X_t)_{0 \leq t \leq T}$ that can be found in Kloeden & Platen [77], Chapter 5.5. This expansion is the stochastic counterpart to the Taylor expansion. Depending on how many terms are used in the Ito-Taylor expansion for the simulation of the process $(X_t)_{0 \leq t \leq T}$, a different simulation method results. The Ito-Taylor expansion will not be explained here.

2.5.1 Mathematical Methods to Generate Random Variables

Let U_1 and U_2 be two on $(0, 1)$ uniformly distributed and independent random variables. Then

$$\left.\begin{aligned} N_1^{(0,1)} &:= \sqrt{-2\ln(U_1)}\ \cos(2\pi U_2), \\ N_2^{(0,1)} &:= \sqrt{-2\ln(U_1)}\ \sin(2\pi U_2) \end{aligned}\right\} \tag{2.10}$$

are two independent and standard-normally distributed random variables. The superscriptions $(0,1)$ indicate that these random variables are standard-normal (mean zero and variance one). (2.10) is called *Box-Muller-method.* With $\mu_1, \mu_2 \in \mathbb{R}$ and $\sigma_1, \sigma_2 \in \mathbb{R}^+$ the definitions

$$N_1^{(\mu_1,\sigma_1^2)} := \sigma_1 N_1^{(0,1)} + \mu_1 \quad \text{and} \quad N_2^{(\mu_2,\sigma_2^2)} := \sigma_2 N_2^{(0,1)} + \mu_2 \tag{2.11}$$

give two independent, normally distributed random variables with

$$N_1^{(\mu_1,\sigma_1^2)} \overset{d}{=} N(\mu_1,\sigma_1^2) \qquad \text{and} \qquad N_2^{(\mu_2,\sigma_2^2)} \overset{d}{=} N(\mu_2,\sigma_2^2).$$

To create realizations of these random variables within a computer simulation program, the random number generator of a programming language can be used. Since this random number generator usually produces only realizations with a uniform distribution on interval $(0,1)$ (like in C++), formulas (2.10) and (2.11) can be used to get the appropriate realization of a normally distributed random variable. This is needed for the Euler scheme in 2.5.2 and the Milstein scheme in 2.5.3.

In the Taylor 1.5 scheme that will be presented in 2.5.4 two correlated normally distributed random variables are needed with correlation coefficient $\frac{\sqrt{3}}{2}$. The definitions

$$\Delta B^{(1)} := N_1^{(0,1)}\sigma_1 \quad \text{and} \quad \Delta B^{(2)} := \frac{1}{2}\sigma_1^3\left(N_1^{(0,1)} + \frac{1}{\sqrt{3}}N_2^{(0,1)}\right) \tag{2.12}$$

give two correlated normally distributed random variables with

$$\left.\begin{aligned} &\Delta B^{(1)} \overset{d}{=} N(0,\sigma_1^2), \quad \Delta B^{(2)} \overset{d}{=} N(0,\tfrac{1}{3}\sigma_1^6), \quad \text{and} \\ &Corr(\Delta B^{(1)},\Delta B^{(2)}) = \tfrac{\sqrt{3}}{2}. \end{aligned}\right\} \tag{2.13}$$

2.5.2 Euler Scheme

The sequence $(Y_i)_{i=0}^N \equiv (Y_{\tau_i})_{i=0}^N$ is defined iteratively via the first three terms of the Ito-Taylor expansion of the continuous-time process $(X_t)_{0\leq t\leq T}$:

$$\left.\begin{aligned} Y_0 &:= x_0, \\ Y_{i+1} &:= Y_i + \mu(\tau_i,Y_i)\,\Delta_s + \sigma(\tau_i,Y_i)\,\Delta B_i^{(1)}, \\ &\qquad i = 0,\ldots,N-1, \end{aligned}\right\} \tag{2.14}$$

where $\Delta B_i^{(1)} \stackrel{d}{=} N(0, \Delta_s)$ for $i = 0, \ldots, N-1$. $(Y_i)_{i=0}^N$ is the *Euler approximation* of the continuous-time process $(X_t)_{0 \le t \le T}$ in the observation period $[0, T]$. (2.14) is called *Euler scheme.* In the computer simulation program a realization of $\Delta B_i^{(1)}$ is calculated via the method presented in (2.10) and (2.11) with $\mu = 0$ and $\sigma_1 = \sqrt{\Delta_s}$.

2.5.3 Milstein Scheme

As opposed to the Euler scheme, the Milstein scheme uses one more term of the Ito-Taylor formula. In the following, a partial derivative with respect to the local component (=second component) is denoted with a stroke next to the variable. The sequence $(Y_i)_{i=0}^N \equiv (Y_{\tau_i})_{i=0}^N$ can now iteratively be defined via:

$$\left.\begin{aligned} Y_0 \quad &:= \quad x_0, \\ Y_{i+1} \quad &:= \quad Y_i \; + \; \mu(\tau_i, Y_i)\,\Delta_s \; + \; \sigma(\tau_i, Y_i)\,\Delta B_i^{(1)} \\ &\qquad + \; \tfrac{1}{2}\sigma(\tau_i, Y_i)\sigma'(\tau_i, Y_i)\left\{(\Delta B_i^{(1)})^2 - \Delta_s\right\}, \\ &\qquad i = 0, \ldots, N-1, \end{aligned}\right\} \tag{2.15}$$

where $\Delta B_i^{(1)} \stackrel{d}{=} N(0, \Delta_s)$ for $i = 0, \ldots, N-1$. $(Y_i)_{i=0}^N$ is the *Milstein approximation* of the continuous-time process $(X_t)_{0 \le t \le T}$ in the observation period $[0, T]$. (2.15) is called *Milstein scheme.* In the computer simulation program a realization of $\Delta B_i^{(1)}$ is calculated via the method presented in (2.10) and (2.11) with $\mu = 0$ and $\sigma_1 = \sqrt{\Delta_s}$.

2.5.4 Taylor 1.5 Scheme

As opposed to the Milstein scheme, the Taylor 1.5 scheme uses several more terms of the Ito-Taylor formula. The sequence $(Y_i)_{i=0}^N \equiv (Y_{\tau_i})_{i=0}^N$ can iteratively be defined, see (2.16). Here, $\Delta B_i^{(1)} \stackrel{d}{=} N(0, \Delta_s)$ and $\Delta B_i^{(2)} \stackrel{d}{=} N(0, \frac{1}{3}\Delta^{1.5})$ for $i = 0, \ldots, N-1$. In the computer simulation program realizations of $\Delta B_i^{(1)}$ and $\Delta B_i^{(2)}$ are calculated via the method presented in (2.12), which according to (2.13) yields the appropriate distributions with $\mu = 0$ and $\sigma_1 = \sqrt{\Delta_s}$. $(Y_i)_{i=0}^N$ is the *Taylor 1.5 approximation* of the continuous-time process $(X_t)_{0 \le t \le T}$ in the observation period $[0, T]$. (2.16) is the *Taylor 1.5 scheme.* The denomination of the simulation method (2.16) in the thesis as *Taylor 1.5 scheme* stems from the following introduction of the term *strong convergence of order* κ with $\kappa \in \mathbb{R}^+$ which will be used for comparing the three simulation methods introduced above with each other.

$$
\left.
\begin{aligned}
Y_0 &:= x_0, \\
Y_{i+1} &:= Y_i + \mu(\tau_i, Y_i)\,\Delta_s + \sigma(\tau_i, Y_i)\,\Delta B_i^{(1)} \\
&\quad + \frac{1}{2}\sigma(\tau_i, Y_i)\sigma'(\tau_i, Y_i)\ \left\{(\Delta B_i^{(1)})^2 - \Delta_s\right\} \\
&\quad + \mu'(\tau_i, Y_i)\sigma(\tau_i, Y_i)\,\Delta B_i^{(2)} \\
&\quad + \frac{1}{2}\left(\mu(\tau_i, Y_i)\mu'(\tau_i, Y_i) + \frac{1}{2}\sigma^2(\tau_i, Y_i)\mu''(\tau_i, Y_i)\right)\Delta_s^2 \\
&\quad + \left(\mu(\tau_i, Y_i)\sigma'(\tau_i, Y_i) + \frac{1}{2}\sigma^2(\tau_i, Y_i)\sigma''(\tau_i, Y_i)\right) \\
&\quad\ \left\{\Delta B_i^{(1)}\,\Delta_s - \Delta B_i^{(2)}\right\} \\
&\quad + \frac{1}{2}\sigma(\tau_i, Y_i)\ (\sigma(\tau_i, Y_i)\sigma''(\tau_i, Y_i) \\
&\quad + [\sigma'(\tau_i, Y_i)]^2)\ \left\{\frac{1}{3}[\Delta B_i^{(1)}]^2 - \Delta_s\right\}\Delta B_i^{(1)}, \\
&\quad i = 0, \ldots, N-1.
\end{aligned}
\right\}
\tag{2.16}
$$

2.5.5 Strong Convergence of Order κ

Let $(X_t)_{t\geq 0}$ be an Ito process that is given as the solution of the SDE

$$dX_t = \mu(t, X_t)\,dt + \sigma(t, X_t)\,dB_t, \ t \geq 0, \ X_0 = x_0.$$

Let $0 = \tau_0 < \tau_1 < \ldots < \tau_N = T$ be the simulation time points (not necessarily equidistant any more) in the time interval $[0, T]$ with the maximal step size

$$\delta := \max\{\tau_{i+1} - \tau_i \,|\, i = 0, \ldots, N-1\}$$

and $(Y_i^\delta)_{i=0}^N$ be a discrete-time approximation of $(X_t)_{t\geq 0}$ on $[0, T]$ with respect to the simulation time points $\{\tau_0, \ldots, \tau_N\}$. The approximation $(Y_i^\delta)_{i=0}^N$ converges strongly with order $\kappa > 0$ towards $(X_t)_{t\geq 0}$ at time T, if there exists a constant $C > 0$ that is independent from δ, and if there exists a constant $\delta_0 > 0$ with

$$E\left(|X_T - Y_T^\delta|\right) \leq C\,\delta^\kappa \quad \forall\, \delta \in (0, \delta_0).$$

Under appropriate assumptions[39] it can be shown:

[39] See Kloeden & Platen [77], page 323, page 345, and page 351.

1. The Euler scheme is strongly convergent of order $\kappa = 0.5$.
2. The Milstein scheme is strongly convergent of order $\kappa = 1$.
3. The Taylor 1.5 scheme is strongly convergent of order $\kappa = 1.5$.

In this sense the Taylor 1.5 scheme is better suited to simulate an Ito process than the other two schemes. However, all three simulation methods are implemented for the simulation of the short-rate process in the computer simulation program whenever possible. In this thesis, the approximation scheme that is best suited to simulate the short-rate process is always applied in the test situations 2-5 of Chapter 5.

2.6 Evolution of the Real Options Theory and Models in the Literature

Real options have already been introduced in Section 2.2. An introduction to the mathematics of Financial Engineering was also provided in Sections 2.3, 2.4, and 2.5. The developed theory will now be used to give an overview of the evolution of the real options field in the literature but without focusing on the mathematical derivation of the models in this section. However, it is not the goal to give a plain literature review since this is already done is great detail in, e.g., Trigeorgis [132], pages 14-21. Rather, the goal is to provide a broad overview with classifications of the available articles, theories, and methods and, thereby, set the context for what is done analytically and numerically in Chapters 4 and 5 of this thesis.

The study of real options *arose in part as a response to the dissatisfaction of corporate practitioners, strategists, and some academics with traditional techniques of capital budgeting*[40]. This refers especially to the DCF method introduced in Section 2.2. Although some of the unsatisfactory valuation results from the DCF method stemmed from its being misapplied in practice, its inappropriateness to the valuation of investments with operating or strategic options was clear[41]. In particular, the DCF method becomes more inappropriate as the number of options in an investment project rises.

Even before the idea of real options was born, economists like Roberts and Weitzman found that in sequential decision-making, even in the case of a negative NPV, it may be worthwhile to undertake an investment[42]. Several authors stressed the limitations of the DCF method for evaluating investment decisions with strategic considerations[43], i.e., undervaluing those investment

[40] See Trigeorgis [132], pages 14-15.
[41] See Myers [100].
[42] See Roberts & Weitzman [112].
[43] See Dean [38], Hayes & Abernathy [49], and Hayes & Garvin [50].

projects. In the words of Trigeorgis[44]: *The basic inadequacy of the NPV approach and other DCF approaches to capital budgeting is that they ignore, or cannot properly capture, management's flexibility to adapt and revise later decisions (i.e., review its implicit operating strategy). The traditional NPV approach, in particular, makes implicit assumptions concerning an "expected scenario" of cash flows and presumes management's commitment to a certain "operating strategy".*

One of the results of this criticism was the development of the decision-tree analysis by Hertz and Magee[45] in 1964. Interestingly enough, the idea of the decision-tree analysis was published by these authors in the Harvard Business Review, which is more focused on application and is not a classical academic journal. In the following the basic idea of decision-tree analysis will be explained and elaborated to show its advantanges and disadvantages.

2.6.1 Decision-Tree Analysis

Decision-tree analysis (DTA) picks up at the point of uncertain future cash flows in describing the project within a tree structure that allows different paths during the life of the project. The decisions that need to be made within the tree are marked by quadratic nodes; those decisions do not need to be made at the time of the valuation but later when more information has arrived. If the project contains several realizations at one point during the project's life that cannot be influenced by management, this is represented by a circular node in the tree. Real probabilities q have to be assigned to all possible realizations (i.e., branches). The superscripts ^{H}, ^{M}, and ^{L} describe the economic condition (high, medium, and low market, respectively). By this method management can visualize the project's inherent options and price them into the project's NPV. Within the tree each alternative must be represented as a branch. Moreover, each final branch in the tree must be assigned a numerical value in case of its realization.

Example 1. A platform investment in R&D is a typical example where DTA can be used. It is graphically explained in Figure 2.3. The numbers on the right next to the final point (denoted with c_t) give the NPVs at the end of year 3 in the respective market stage of the expected discounted cash flows of the years 3 and 4 - 10. It is assumed that at the end of year 10, the last year of production within the project, the plant is closed down at no cost. The cash flows are assumed to occur at the end of the respective year, although the investments have to be made at the beginning of that year.

[44] See Trigeorgis [132], page 121.
[45] See Hertz [55] and Magee [85].

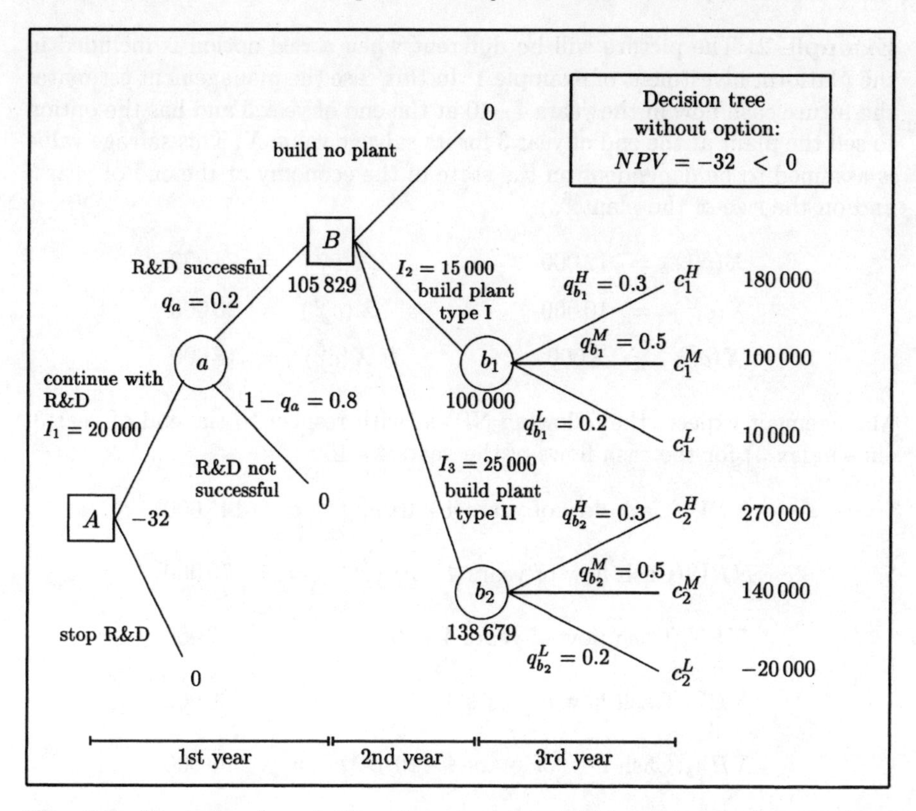

Fig. 2.3. Example of a platform investment in R&D within the framework of decision-tree analysis.

The calculations of the values in the tree are done by discounting with the cost of capital k, which is assumed to be 6%, and taking into consideration the investment cost and the management decision at point B:

$$\text{Point } b_1: \quad 100\,000 = \frac{0.3 \cdot 180\,000 + 0.5 \cdot 100\,000 + 0.2 \cdot 10\,000}{1 + 0.06}$$

$$\text{Point } b_2: \quad 138\,679 = \frac{0.3 \cdot 270\,000 + 0.5 \cdot 140\,000 + 0.2 \cdot (-20\,000)}{1 + 0.06}$$

$$\text{Point } B: \quad 105\,829 = max\left(0, \frac{100\,000}{1 + 0.06} - 15\,000, \frac{138\,679}{1 + 0.06} - 25\,000\right)$$

$$\text{Point } A: \quad -32 = \frac{0.2 \cdot 105\,829 + 0.8 \cdot 0}{1 + 0.06} - 20\,000$$

Applying DTA yields a negative NPV of −32, which means that the platform investment would not be undertaken: stop R&D.

Example 2. The picture will be different when a real option is included in the platform investment of example 1. In this case the management estimates the future cash flow in the years 4 - 10 at the end of year 3 and has the option to sell the plant at the end of year 3 for its salvage value X. This salvage value is assumed to be dependent on the state of the economy at the end of year 3 and on the size of the plant[46]:

$$\begin{aligned} X(c_1^H) &= 12\,000 & \qquad X(c_2^H) &= 24\,000 \\ X(c_1^M) &= 10\,000 & \qquad X(c_2^M) &= 20\,000 \\ X(c_1^L) &= 8\,000 & \qquad X(c_2^L) &= 18\,000 \end{aligned}$$

Management expects the following NPVs (with respect to the end of year 3, thus index 3) for the cash flows of the years 4 - 10:

$$\begin{aligned} NPV_3(\text{Cash flow of years 4 - 10}, c_1^H) &= 145\,000 \\ NPV_3(\text{Cash flow of years 4 - 10}, c_1^M) &= 70\,000 \\ NPV_3(\text{Cash flow of years 4 - 10}, c_1^L) &= 7\,000 \\ NPV_3(\text{Cash flow of years 4 - 10}, c_2^H) &= 200\,000 \\ NPV_3(\text{Cash flow of years 4 - 10}, c_2^M) &= 90\,000 \\ NPV_3(\text{Cash flow of years 4 - 10}, c_2^L) &= -14\,000 \end{aligned}$$

This also gives the following cash flows at the end of year 3 for year 3 as $E(c_i^j) = c_i^j - NPV_3(c_i^j)$:

$$\begin{aligned} E(\text{Cash flow in year } 3, c_1^H) &= 180\,000 - 145\,000 &&= 35\,000 \\ E(\text{Cash flow in year } 3, c_1^M) &= 100\,000 - 70\,000 &&= 30\,000 \\ E(\text{Cash flow in year } 3, c_1^L) &= 10\,000 - 7\,000 &&= 3\,000 \\ E(\text{Cash flow in year } 3, c_2^H) &= 270\,000 - 200\,000 &&= 70\,000 \\ E(\text{Cash flow in year } 3, c_2^M) &= 140\,000 - 90\,000 &&= 50\,000 \\ E(\text{Cash flow in year } 3, c_2^L) &= (-20\,000) - (-14\,000) &&= -6\,000 \end{aligned}$$

[46] Alternative methods are presented in Trigeorgis [132], pages 61-65.

At the end of year 3 management will only sell the plant if NPV_3 is smaller than the salvage value. For both plant types this is only the case for a low market when plant type I sells for 8 000 and plant type II sells for 18 000. Taking this salvage value into consideration at the end of year 3, gives the following situation (compared with the right side of Figure 2.3):

- Don't change 180 000 at c_1^H since $35\,000+\max(145\,000,\ 12\,000) = 180\,000$
- Don't change 100 000 at c_1^M since $30\,000+\max(70\,000,\ 10\,000) = 100\,000$
- Replace 10 000 at c_1^L with $3\,000+\max(7\,000,\ 8\,000) = \underline{\mathbf{11\,000}}$
- Don't change 270 000 at c_2^H since $70\,000+\max(200\,000,\ 24\,000) = 270\,000$
- Don't change 140 000 at c_2^M since $50\,000+\max(90\,000,\ 20\,000) = 140\,000$
- Replace −20 000 at c_2^L with $-6\,000+\max(-14\,000,\ 18\,000) = \underline{\mathbf{12\,000}}$

The remaining calculations are the same as in example 1. They produce the decision tree of Figure 2.4 in the case of an option to abandon for a salvage value in this example of a platform investment in R&D. It finally yields an NPV of 1 043, so that the investment would be undertaken. This NPV is called a *strategic NPV* since it includes a real option. The NPV of −32 previously calculated in the absence of a real option is called *static NPV*. The value of the real option can therefore be calculated as:

$$\begin{aligned}\text{Value of the option to abandon} &= \text{option premium}\\ &= \text{strategic NPV} - \text{static NPV}\\ &= 1\,043 - (-32)\\ &= 1\,075.\end{aligned}$$

In summary, DTA is well suited to price some types of real options. It can price sequential investment decisions in which management decisions are made at discrete points in the future and uncertainty is resolved at discrete points in time as well. DTA is able to handle this embedded flexibility but the practical application has serious limitations: the number of discrete points in time can get large, thus, creating an extremely complex tree. Trigeorgis refers to this as *decision-bush analysis*[47] instead of decision-tree analysis. Realistic corporate budgeting situations cannot be properly handled that way. Second, the applied discount rate poses a problem since it is usually assumed to be constant. So, when uncertainty gets resolved at decision nodes, the DTA method does not use a changed discount factor. DTA cannot therefore reflect this change in the riskiness of the project expressed in the discount rate.

[47] See Trigeorgis [132], page 66.

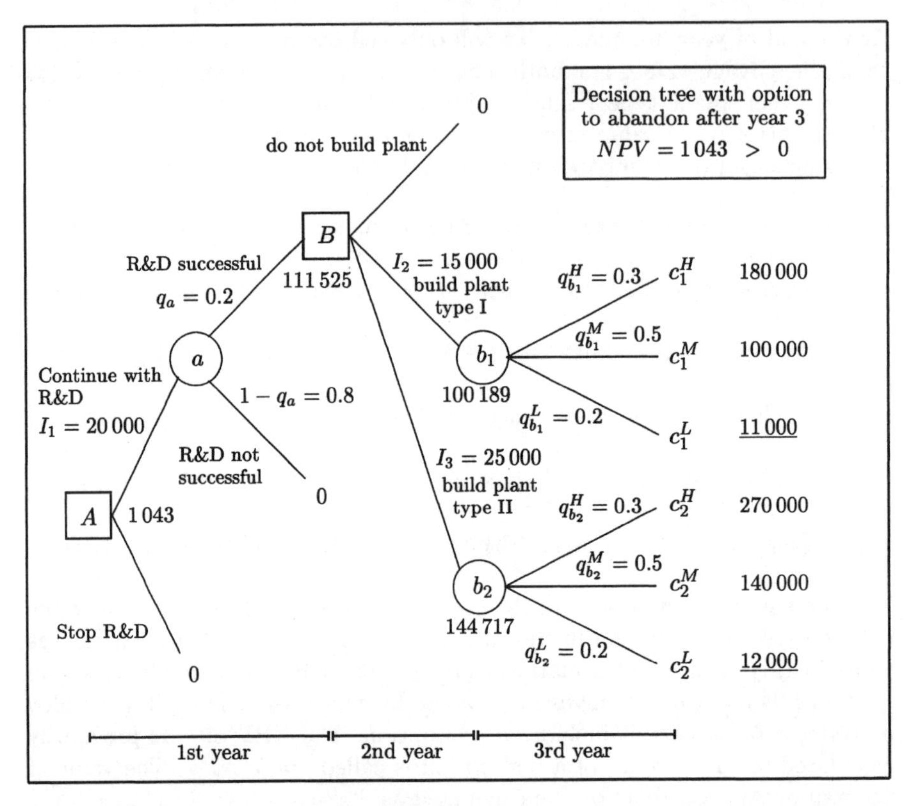

Fig. 2.4. Example of a platform investment in R&D with option to abandon within the framework of decision-tree analysis.

2.6.2 Contingent-Claims Analysis

A comparison of the DTA method and the DCF method shows that DTA improves on the DCF method by including real options. However, the DTA has two major drawbacks. On the one hand it applies a constant discount rate. On the other hand the decision tree gets very complicated even in simple cases. Both reasons limit its application in practice. Contingent-claims analysis (CCA) is a method that gets rid of the discount rate problem which the DTA method exhibits. CCA seeks to replicate the pay-off structure of the project and its real options via financial transactions in order to determine the NPV of the project. This section builds on the ideas of Trigeorgis[48] in illustrating the CCA method with an option to defer.

---

[48] See Trigeorgis [132], pages 153-161.

The characteristic of a real option is that it alters the risk structure of the project. In example 2 of the previous Section 2.6.1, the option to abandon after year 3 gets rid of negative cash flows in a low-case market. This should result in a new risk-adjusted interest rate as discount factor below 6%. Since, however, the discount factor stays the same, the true NPV is understated. Therefore, the true NPV is above the calculated 1 043.

Trigeorgis characterizes DTA and CCA in a very precise way[49]: *DTA can actually be seen as an advanced version of DCF or NPV - one that correctly computes unconditional expected cash flows by properly taking account of their conditional probabilities given each state of nature. As such, DTA is correct in principle and is particularly useful for analyzing complex sequential investment decisions. Its main shortcoming is the problem of determining the appropriate discount rate to be used in working back through the decision tree. [...] The fundamental problem with traditional approaches to capital budgeting lies in the valuation of investment opportunities whose claims are not symmetrical or proportional. The asymmetry resulting from operating flexibility options and other strategic aspects of various projects can nevertheless be properly analyzed by thinking of discretionary investment opportunities as options on real assets (or as real options) through the technique of contingent-claims analysis.*

The CCA method builds upon the DTA method and eliminates the weak point of "constant interest rate". With DTA the basic idea is to calculate with real probabilities and a constant, risk-adjusted interest rate as the discount rate. With CCA the basic idea is to transform the real probabilities into risk-adjusted probabilities such that the algorithm can use a constant, risk-free interest rate that is independent of the project's risk structure. To explain the idea of CCA some definitions are necessary:

V := total value of the project,

S := price of the twin security that is almost perfectly correlated with V,

E := equity value of the project for the shareholder,

k := return of the twin security,

r_f := risk-free interest rate,

p := risk-neutral probability for up-movements of V and S per period,

q := real probability for up-movements of V and S per period,

u := multiplicative factor for up-movements of V and S per period,

d := multiplicative factor for down-movements of V and S per period.

[49] See Trigeorgis [132], page 155.

As the starting point the current values for V and S are assumed to be $V = 100$ and $S = 20$. S is the price of the so-called *twin security* that exhibits the same risk profile as the project with value V and can be traded on the capital markets. Furthermore, the real probability q is assumed to be 0.5. This gives a development of S and V with $u = 1.8$ and $d = 0.6$ as shown in Figure 2.5.

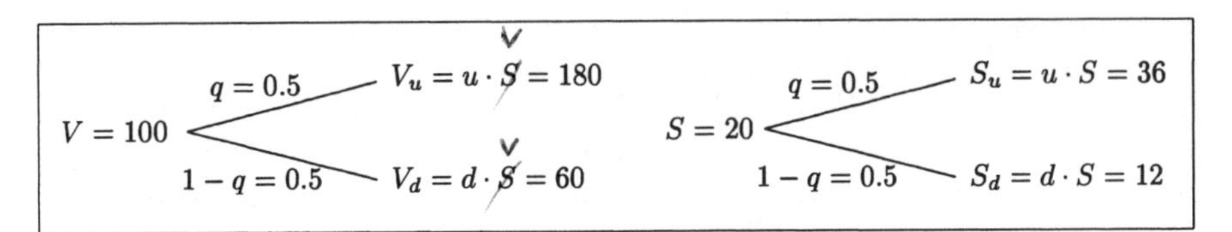

Fig. 2.5. Development of the total value V of a project and of the price S of a twin security.

The goal is to replicate the pay-off structure of the project after one year through the purchase of n shares of the twin security and through issuing of a 1-year Zero bond with nominal value B and risk-free return r_f. This is illustrated in Figure 2.6. The issue of the Zero is equivalent to getting a loan of value B for the interest rate r_f. Therefore, a positive B indicates a bond issue.

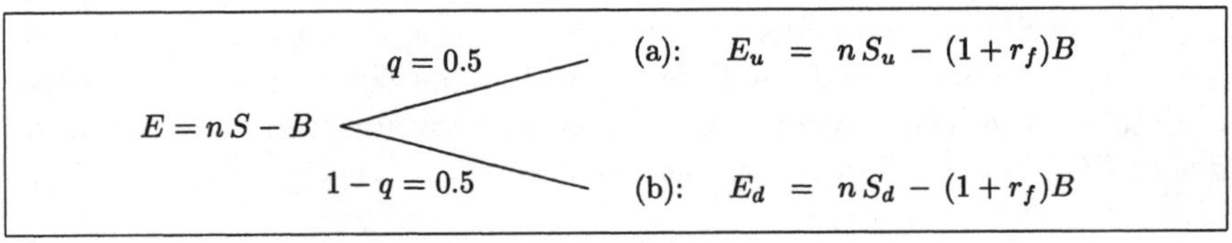

$E = nS - B$, with $q = 0.5$ and $1 - q = 0.5$:

$$\text{(a)}: \quad E_u = n\,S_u - (1+r_f)B$$

$$\text{(b)}: \quad E_d = n\,S_d - (1+r_f)B$$

Fig. 2.6. Idea of the replicating portfolio.

$$\text{n} = \frac{E_u - E_d}{S_u - S_d} \qquad \text{E} = \frac{pE_u + (1-p)E_d}{1+r_f}$$

$$\text{B} = \frac{E_u S_d - E_d S_u}{(S_u - S_d)(1+r_f)} \qquad \text{p} = \frac{(1+r_f) - d}{u - d}$$

Fig. 2.7. Solutions of the replicating portfolio.

By choosing the replicating portfolio approach $E = nS - B$ (i.e., E is replicated via n shares of stocks and the issue of a Zero worth nominal B) equations

(a) and (b) in Figure 2.6 have to be solved simultaneously to arrive at n and B. The solutions are summarized in Figure 2.7.

When dealing with a project that starts immediately, it is obvious that $V = E, V_u = E_u$, and $V_d = E_d$ have to hold since the value of the project has to be exactly the same as the value the shareholder gets out of the project. Putting the given data into the formulas, with a risk-free rate of $r_f = 8\%$, yields the following result:

$$p = \frac{(1+r_f)-d}{u-d} = \frac{(1+0.08)-0.6}{1.8-0.6} = 0.4$$

$$V = \frac{pV_u+(1-p)V_d}{1+r_f} = \frac{0.4\cdot 180+(1-0.4)\cdot 60}{1+0.08} = 100$$

$$B = \frac{V_uS_d-V_dS_u}{(S_u-S_d)(1+r_f)} = \frac{180\cdot 12-60\cdot 36}{(36-12)(1+0.08)} = 0$$

$$n = \frac{V_u-V_d}{S_u-S_d} = \frac{180-60}{36-12} = 5$$

With an assumed investment cost of $I_0 = 104$, the NPV is $V - I_0 = 100-104 = -4 < 0$. Thus, in case of an immediate start, this project would not be undertaken.

In the above formulas, the risk-adjusted probability $p = 0.4$ is applied (and not the real probability $q = 0.5$) as well as a risk-free interest rate r_f. This is the crucial difference between CCA and DTA since the latter uses real probabilities and a risk-adjusted interest rate that cannot be easily determined (if at all). If the DTA is used to value the project the return k of the security is needed:

$$k = \frac{qS_u+(1-q)S_d}{S} - 1 = \frac{0.5\cdot 36+(1-0.5)\cdot 12}{20} - 1 = 20\%.$$

This gives:

$$V^{DTA} = \frac{qV_u+(1-q)V_d}{1+k} = \frac{0.5\cdot 180+(1-0.5)\cdot 60}{1+0.2} = 100.$$

In this case, therefore, the same value of V is yielded by DTA and CCA. This will change when the project contains a real option. Only the CCA with its risk-free interest rate gives the correct value for projects with real options, unlike the DTA with its risk-adjusted interest rate. This will be illustrated in the example of an option to defer by calculating the values for V and E if the project starts immediately or one year from today. For the NPV calculation the development of V, S, and E over two years with $S_{uu} = u^2S$, $S_{ud} = udS = S_{du}$, and $S_{dd} = d^2S$ needs to be determined. $S_{ud} = S_{du}$ leads to a recombining tree, i.e., the branches of the tree merge in the next stage and each additional stage creates one new node.

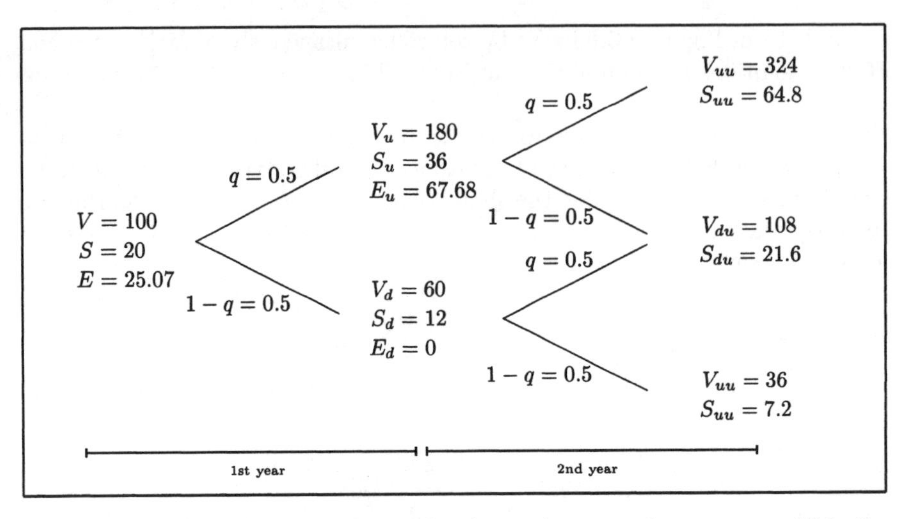

Fig. 2.8. Example of an option to defer the project start by one year within the framework of contingent-claims analysis.

An immediate start of the project has so far been assumed with an investment cost of $I_0 = 104$. If the start of the project is in one year from today, an investment cost of $I_1 = (1 + r_f)I_0 = 112.32$ can be assumed[50]. The valuation of the real option is explained in Figure 2.8. The shareholder values in Figure 2.8 can be calculated as follows:

$$
\begin{aligned}
E_u &= max(V_u - I_1, 0) &&= max(180 - 112.32, 0) &&= 67.68 \\
E_d &= max(V_d - I_1, 0) &&= max(60 - 112.32, 0) &&= 0 \\
E &= \frac{pE_u + (1-p)E_d}{1 + r_f} &&= \frac{0.4 \cdot 67.68 + (1 - 0.4) \cdot 0}{1 + 0.08} &&= 25.07
\end{aligned}
$$

Thus, in the case of a 1-year deferred project start, different values for V and E at $t = 0$ are calculated: $V = 100$ but $E = 25.07$. In this situation the NPV as the equity value to the shareholder is 25.07. The question that arises here is: What is an option worth that allows the start of the project to be deferred by one year? It has to be the difference of both NPVs, i.e., $25.07 - (-4) = 29.07$. This means that management would be willing to pay up to 29.07 to have the option to defer the project start by exactly one year.

Applying the DTA method yields a different value for E and, therefore, for the NPV:

$$
NPV^{DTA} = E^{DTA} = \frac{qE_u + (1-q)E_d}{1+k} = \frac{0.5 \cdot 67.68 + 0.5 \cdot 0}{1 + 0.2} = 28.20\,.
$$

[50] See Trigeorgis [132], page 158.

So the question is which NPV is the correct one? This question can be easily answered by applying the theory of the replicating portfolio, see Figure 2.7.

$$n = \frac{E_u - E_d}{S_u - S_d} = \frac{67.68 - 0}{36 - 12} = 2.82$$

shares of stocks and the issue of a 1-year Zero with nominal value

$$B = \frac{E_u\, S_d \; - \; E_d\, S_u}{(S_u - S_d)\,(1 + r_f)} = \frac{67.68 \cdot 12 - 0 \cdot 36}{(36 - 12)\,(1 + 0.08)} = 31.33$$

are needed to replicate the equity value in the up- and down-cases for a risk-free interest rate of $r_f = 8\%$. The cost of a replicating portfolio is $2.82 \cdot 20 - 31.33 = 25.07$, so that $E = 25.07$ is the correct answer. DTA, therefore, overestimates the equity value. The reason for this is that the DTA approach uses a constant interest rate, which is not appropriate if real options are present.

In summary, asymmetrical cash flows change the risk structure of the project and, in so doing, the discount rate - something the DTA does not consider. The flaw of the DCF is nicely described in Trigeorgis [132], page 152: *The presence of flexibility embedded in future decision nodes, however, changes the payoff structure and the risk characteristics of an actively managed asset in a way that invalidates the use of a constant discount rate. Unfortunatelly, classic DTA is in no better position than DCF techniques to provide any recommendation concerning the appropriate discount rate.*

This error is corrected in the CCA method since CCA transforms the real probabilities into risk-adjusted probabilities and the risk-adjusted interest rates into risk-free interest rates to correctly mirror the asymmetry within a real option. In the words of Trigeorgis[51]: *Traditional DTA is on the right track, but although mathematically elegant it is economically flawed because of the discount rate problem. An options approach can remedy these problems.*

Therefore, only a CCA approach can capture the flexibility that is inherent in investment projects. *Flexibility*, according to Trigeorgis [132], page 151, *is nothing more than the collection of options associated with an investment opportunity, financial or real.* However, the general question of the justification of an option based approach with replicating portfolios and its limits has to be answered. A good justification for this approach is given in Trigeorgis [132], pages 124-129. While his full justification will not be restated here, a summary of Trigeorgis' most important points is needed[52]: *Can the standard technique of valuing options on the basis of a no-arbitrage equilibrium, using portfolios*

[51] See Trigeorgis [132], page 68.
[52] See Trigeorgis [132], page 127.

of traded securities to replicate the payoff to options, be justifiably applied to capital budgeting where projects may not be traded? As Mason and Merton[53] *in 1985 pointed out, the answer is affirmative if the same assumptions are adopted that are used by standard DCF approaches - including NPV - which attempt to determine what an asset or project would be worth if it were to be traded.* [...] *Given the prices of the project's twin security, management can, in principle, replicate the returns to a real option by purchasing a certain number of shares of its twin security while financing the purchase partly by borrowing at the riskless rate.* [...] *Risk-neutral valuation can be applied, whether the asset is traded or not, by replacing the actual growth rate, α, with a "certainty-equivalent" or risk-neutral growth rate, $\hat{\alpha}$, after substracting an appropriate risk premium ($\hat{\alpha} = \alpha -$ risk premium).*

To summarize the analogies of an investment opportunity (= real option on a project) and a call option on a stock, the comparison of financial and real options given by Trigeorgis[54] will be provided here in Table 2.1.

Table 2.1. Comparison between a real option on a project and a call option.

Call option on stock	Real option on project
current value on stock	(gross) present value of expected future cash flows
exercise price	investment cost
time to expiration	time until opportunity disappears
stock value uncertainty	project value uncertainty
risk-free interest rate	risk-free interest rate

Source: Trigeorgis [132], page 127

In the words of Trigeorgis[55]: *Option valuation can be seen operationally as a special, economically corrected version of decision-tree analysis that is better suited in valuing a variety of corporate operating and strategic options.*

However, the limitation of the analogy of a real option to a call option needs to be considered as well. In 1993, Kester analyzed the main differences in this analogy[56]. Besides the fact that real assets are non-traded and financial securities are traded (such that dividend-like adjustment are necessary[57]) the main differences are:

---

[53] See Mason & Merton [88].
[54] See Trigeorgis [132], page 125.
[55] See Trigeorgis [132], page 15. Compare also with Mason & Trigeorgis [89].
[56] See Kester [74].
[57] See Trigeorgis [132], page 127.

1. (Non) Exclusiveness of ownership and competitive interaction
2. Non-tradability and preemption
3. Across-time (strategic) interdependence and option compoundness

For a more detailed analysis on this subject see Trigeorgis [132], page 127.

2.6.3 Categorization of Real Options Valuation Methods

The valuation of investment projects with strategic and operating options started in 1977 by Stewart C. Myers. His pioneering idea was to perceive discretionary investment opportunities as *growth options*, giving corporate valuation a totally new direction, the *real options direction.* Since 1977 many articles have been published that deal with the valuation of real options. Two main avenues of real options valuation can be distinguished.

1. **Analytical methods:** approximative analytical solutions and closed-form solutions, including the Schwartz-Moon model[58], which will be introduced in Section 4.3.
2. **Numerical methods:** approximation of the partial differential equation that describes the option and approximation of the underlying stochastic process of the real option.

The following provides an overview of these two methods.

1. **Analytical methods.** Analytical methods can be divided into closed-form solutions and approximative analytical solutions. For the former, according to Trigeorgis [132], page 17, *a series of papers gave a boost to the real options literature by focusing on valuing quantitatively - in many cases deriving analytic, closed-form solutions.* Much work in that area was characterized by these analytical solutions that offer a nice (since closed-form) solution to simplified problems that seldom reflect reality. Trigeorgis offers a good overview of the academic articles taking this approach for various types of real options (see Trigeorgis [132], pages 2-3). In Chapter 6 of Trigeorgis [132] he also discusses in more detail some of the continuous-time models (including the martingale approach in Section 6.6), which are briefly summarized in the following:
   a) Option to defer (McDonald & Siegel [92], 1986; Paddock, Siegel & Smith [104], 1988; Majd & Pindyck [86], 1987): For the gross project value $(V_t)_{t\geq 0}$ McDonald and Siegel use a diffusion process given via the SDE

$$dV_t \;=\; \alpha\, V_t\, dt \;+\; \sigma\, V_t\, dB_t,\;\; t \geq 0, \alpha \in \mathbb{R}^+, \sigma \in \mathbb{R}^+.$$

[58] See Schwartz & Moon [121].

α is the instantaneous expected return on the project and σ its instantaneous standard deviation. Paddock, Siegel, and Smith use a similar process given via the SDE

$$dV_t = (\alpha - D)\, V_t\, dt + \sigma\, V_t\, dB_t, \quad t \geq 0, \alpha \in \mathbb{R}^+, \sigma \in \mathbb{R}^+,$$

with D as the payout rate for the valuation of off-shore petroleum leases. Majd and Pindyck use the same process for the market value of the completed project in valuing the option to defer irreversible construction in a project where a series of outlays must be made sequentially but construction has an upper speed boundary.

b) Option to shut down or abandon (McDonald & Siegel [91], 1985; Myers & Majd [101], 1990): McDonald and Siegel assume that the diffusion process $(P_t)_{t \geq 0}$ for the output price is given by the SDE

$$dP_t = \alpha\, P_t\, dt + \sigma\, P_t\, dB_t, \quad t \geq 0, \alpha \in \mathbb{R}^+, \sigma \in \mathbb{R}^+, \tag{2.17}$$

i.e., the same process as in McDonald & Siegel [92]. Myers and Majd use a similar process, given by

$$dP_t = (\alpha - D)\, P_t\, dt + \sigma\, P_t\, dB_t, \quad t \geq 0, \alpha \in \mathbb{R}^+, \sigma \in \mathbb{R}^+, \tag{2.18}$$

where D represents the instantaneous cash payout (calculated as cash flow divided by the project value).

c) Option to switch (Margrabe [87], 1978; Stulz [126], 1982): Margrabe analyzes the value of an option to exchange one non-dividend paying risky asset for another where the prices V and S of the risky assets are modelled via the same processes as in (2.17), however with different coefficients for the V process and the S process. Stulz analyzes European options on the minimum or maximum of two risky assets, assuming the same two processes as in (2.17) and (2.18).

d) (Simple) Compound options (Geske [44], 1979): Geske calculates the value of an option on a stock where the stock can be seen as a European call option on the value of the firm's assets. Trigeorgis stresses this application for real options as to value a compound growth option where earlier investments have to be undertaken as prerequisites[59]. The value of the underlying is assumed to follow the same process as in (2.17).

e) Compound options to switch (Carr [27], 1988; Carr [28], 1995): Carr uses the same processes to describe the value of two risky assets to analyze European compound (or sequential) exchange options as in (2.17), however with different coefficients for the two processes.

[59] See Trigeorgis [132], page 213.

In addition to these articles several books were published in the past presenting specific real options models. 17 articles on real options by various authors were compiled in Brennan & Trigeorgis [21], published in 2000. The models presented are mathematically very complex and specific. The analytical analysis is often accompanied by empirical valuation results for the situation analyzed. Important to notice is that all articles assume a constant risk-free rate apart from Miltersen [96]. Another (less quantitative) work on real options was published in 1995 by Trigeorgis[60]. This book also contains 17 separate articles on real options, all assuming a constant risk-free rate.

Two main disadvantages are inherent in all analytical solutions. First, the capital budgeting problem needs to be "nice" in order to be analytically trackable. This means that for a capital budgeting problem one must be in the position to write down the describing partial differential equation with the underlying stochastic process. In practice this is almost never the case. Second, even if a solution to the capital budgeting problem can be found for each single real option, this does not say anything about the value of the total option package. The valuation for complex options, i.e., option packages with many different real options and real option types[61], need to take option interactions into consideration, something almost always impossible to handle in an analytical approach.

2. **Numerical methods.** One criticism of analytical approaches is that they are not suited to valuing complex real options. The ability to value such complex options has been enhanced through various numerical techniques[62]. These can be divided into methods that approximate the partial differential equation and methods that approximate the underlying stochastic process[63]:
   a) **Approximation of the partial differential equation:** This includes numerical integration methods as well as the explicit and implicit finite difference methods.
   b) **Approximation of the underlying stochastic process:** This includes Monte Carlo simulation and lattice approaches like the Cox-Ross-Rubinstein binomial tree method and the Trigeorgis log-transformed binomial tree method.

   In the following the basic characteristics of both numerical approximation methods will be described:

[60] See Trigeorgis [131].
[61] See Section 2.2.
[62] See Trigeorgis [132], pages 20-21.
[63] See Trigeorgis [132], page 21.

a) **Approximation of the partial differential equation.** This type of approximation includes numerical integration methods as well as the explicit and implicit finite difference methods. The methods of finite differences can only be applied if the time development of the option value can be described via a partial differential equation. This means it is sufficient to build a partial differential equation; a closed-form solution does not need to exist. The finite difference approach discretizes this partial differential equation. The name of the specific difference method derives from the way the resulting grid is solved: the implicit finite difference method, the explicit finite difference method or hybrid methods like the Crank-Nicholson finite difference method. Moreover, the underlying can be log-transformed before discretizing the partial differential equation, which yields better mathematical properties for the finite difference method. The log-transformed explicit and implicit methods are both presented in Chapter 4 and numerically analyzed in Chapter 5. Due to the properties *consistency*, *stability*, and *efficiency*[64] this thesis will primarily focus on the implicit method.

All finite difference methods can be used to value European options and American options, and they can handle several state variables (multi-dimensional grids). An important point is that the method gives option values for many different start values for the underlying, something lattice approaches do not provide. Moreover, Trigeorgis characterizes the finite difference methods as more mechanical, requiring less intuition than lattice approaches[65].

Various authors have contributed to this type of approximation. Trigeorgis cites many of the following publications in his book entitled *Real Options*[66]:

- **Numerical integration:**
  Parkinson [105], 1977.
- **Finite difference schemes:**
  Brennan & Schwartz [16], 1977; Brennan & Schwartz [17], 1978; Brennan[15], 1979; Majd & Pindyck [86], 1987.

---

[64] *Consistency*, also called *accuracy*, refers to the idea that the discrete-time process used for calculation has the same mean and variance for every time-step size as the underlying continuous process. *Stability*, or *numerical stability*, means that the approximation error in the computations will be dampened rather than amplified. *Efficiency* refers to the number of operations, i.e., the computational time needed for a given approximation accuracy. For more information see Trigeorgis [132], page 308. These terms are addressed in Chapter 4 where the various methods are introduced in detail.

[65] See Trigeorgis [132], page 305.

[66] See Trigeorgis [132], page 306.

- **Analytic approximation:**
  Johnson [71], 1983; Geske & Johnson [45], 1984; MacMillan [84], 1986; Blomeyer [10], 1986; Barone-Adesi & Whaley [4], 1987; Ho, Stapleton & Subrahmanyam [58], 1997.

b) **Approximation of the underlying stochastic process.** When pricing financial options the Monte Carlo simulation plays an important role for the approximation of the underlying stochastic process. However, this type of simulation is mainly used for the pricing of European options since in the past it could not be used to price American options. Recently, however, several publications have addressed the issue of pricing American type options with Monte Carlo simulation. A good overview is given by Pojezny[67]. He classifies Monte Carlo simulation for American type options into four groups:

- Combination procedures
- Parametrisation of early excercise boundary
- Estimation of bounds
- Approximation of value function

These groups of Monte Carlo simulation are not a focus of this thesis. For more information on the subject see Pojezny [107], Section 3.4.3. However, the basic idea of Monte Carlo simulation (as used for European option valuation) is described.

The starting point of a Monte Carlo simulation is the stochastic differential equation that describes the underlying. The underlying $(S_t)_{t\geq 0}$ is often described via

$$dS_t = \alpha S_t\, dt + \sigma S_t\, dB_t, \quad t \geq 0, \tag{2.19}$$

where $\alpha \in \mathbb{R}^+$ is the infinitesimal or instantaneous return of the underlying, $\sigma \in \mathbb{R}^+$ is the infinitesimal or instantaneous standard deviation of the underlying and dB_t is a normal distributed random variable with variance dt. The parameters in this equation can be estimated from financial data and have already been introduced in the various methods presented in the analytical methods part. One aspect to consider is that α has to be chosen in such a way that the generated path represents the underlying in a risk-neutral world. If the underlying is an exchange traded stock, α has to be replaced by $k - \delta$ with k as the return of the stock and δ as its dividend yield[68].

[67] See Pojezny [107], Section 3.4.3.
[68] See Lieskovsky, Onkey, Schulmerich, Teng & Wee [81], pages 13-14, and Trigeorgis [132], page 310.

The stochastic differential equation (2.19) describes the paths of the underlying for $t \geq 0$. A discretization of the stochastic differential equation allows the simulation of such a path with a computer simulation program. These discretization methods were already described in detail in Section 2.5. The idea is to divide the time interval $[0, T], T > 0$, in $N \in \mathbb{N}$ subintervals with equal length $\Delta_s := \frac{T}{N}$. Δ_s is called *stepsize* of the simulation. The goal is to simulate a path value $S_i := S(\tau_i)$ for each of the time points $\tau_i := i\Delta_s,\ i = 0, 1, \ldots, N$. To do this a start value S_0 of the process is needed. If this start value is given, the path values can be calculated iteratively, e.g., via

$$S_{i+1} \quad := \quad S_i \ + \alpha\, S_i\, \Delta_s \ + \ \sigma\, S_i\, \Delta B_i, \quad i = 0, 1, \ldots, N,$$

where $\Delta B_i \stackrel{d}{=} N(0, \Delta_s)$. This is the so-called *Euler scheme* as presented in 2.5.2. Having simulated such a path, the further steps depend on the type of option that needs to be priced. In the following a European call option with strike price X will be valued. To do this the final value S_T of the path is needed and has to be compared with the strike price of the option. Let $P_j := max(S_T - X, 0)$ be the price of a European call at maturity date T where index j indicates that this is the price for the jth simulated path. This simulation of the paths has to be done many times with each simulation independent of each other. Let A denote the number of simulations (e.g., $A := 10\,000$). Each j gives a different value P_j. The mean

$$B \quad := \quad \sum_{j=1}^{A} P_j \qquad (2.20)$$

then gives the value of the option at time T, and $B\, e^{-r_f T}$ gives the current discounted value of the option with r_f as the current risk-free interest rate. An interesting application of path simulations and option pricing can be found in Lieskovsky, Onkey, Schulmerich, Teng & Wee [81] for pricing and hedging Asian options on copper.

As already mentioned, the latest publications also show how Monte Carlo simulation can be used to price American type options. If the underlying does not pay any dividends, the exercise of an American call option prior to expiration is never optimal[69]. However, if the underlying pays dividends, it can be optimal to exercise the call option prior to maturity. For an American put option it can be optimal to exercise early even in the absence of dividend payments[70].

---

[69] See Merton [94].

[70] See Schweser [123], pages 609-611.

Another way of approximating the underlying is done in lattices, e.g., by using binomial or trinomial trees that start with the current underlying value as the start value. This approach can also accommodate American options. One of the building blocks of the lattice approach was the work of Cox, Ross, and Rubinstein in 1979 for the pricing of financial options (see Cox, Ross & Rubinstein [34]). Here, the construction of a binomial tree was a breakthrough for option valuation in discrete time, achieved by building on the earlier work (1976) of Cox and Ross[71] who introduced the notion of a *replicating portfolio* to create a *synthetic option* from an equivalent portfolio comprising the underlying of the option and a bond. This risk-neutral valuation allows to value an option using risk-adjusted probabilities and risk-free interest rates. This idea is framed as a contingent-claims analysis, an option-based approach that enables management to quantify the additional value of a project's operating flexibility[72]. This approach was already introduced in Section 2.6.2 and will be elaborated in Chapter 4. The positive characteristics of lattice methods are best summarized by Trigeorgis[73]: *Lattice approaches are generally more intuitive, simpler, and more flexible in handling different stochastic processes, options payoffs, early exercise of other intermediate decisions, several underlying variables, etc.*

The main disadvantage of lattices is that they only give the option value for one single underlying start value. Therefore, the whole procedure has to be run many times with different start values. This is time consuming, especially if the goal is to calculate an option value distribution depending on the start value of the underlying. On the other hand, lattice methods can handle real options and real options packages, compounded options, dividend payments on the underlying and interaction among multiple options within an option package.

Again, various authors contributed to the development of this type of approximation. Trigeorgis cites the following publications[74]:

- **Monte Carlo simulation:**
  Boyle [13], 1977.
- **Lattice methods:**
  Cox, Ross & Rubinstein [36], 1979; Boyle [14], 1988; Hull & White [63], 1988; Trigeorgis [130], 1991.

---

[71] See Cox & Ross [36].
[72] See Trigeorgis [132], page 155.
[73] See Trigeorgis [132], page 306.
[74] See Trigeorgis [132], page 306.

A final word has to be addressed to the valuation of multiple options. Interactions among multiple options are the reason why option value additivity can usually not be assumed when valuing multiple options. Valuing each option of such an option package separately and then adding up the values does not take into consideration the fact that options usually interact. Consider, for example, an option to abandon and an option to switch. The value of the option to switch is zero if the option to abandon has been exercised. Therefore, the value of the option to switch depends on whether the project is still valid (i.e., option to abandon has not been exercised and is still active) or not. In such a case the values of each single option cannot simply be added together; rather, both options have to be valued together. This gives a multiple option value less than the sum of the single option values, since the sum is less the more interactions there are among the single options. Trigeorgis elaborates on the aspect that the incremental value of an additional option, in the presence of another option, is generally less than its value in isolation (in Chapter 7 of Trigeorgis [132]). An example of value additivity is an option package comprising an option to expand and an option to contract, each one year after the project was started. The underlying value, the present value of the project, can be split into exercise ranges for each of the options. These ranges do not overlap in the case of an option to expand or an option to abandon, which is obvious.

2.6.4 Flexibility due to Interest Rate Uncertainty

So far the focus has been on real options that were valued using a constant risk-free interest rate. The existence of a non-deterministic future cash flow structure and a deterministic (since constant) future risk-free interest rate was sufficient to create a real option. This situation can also be turned around: Even the case of deterministic future cash flows but uncertain future interest rates creates a real option. As Trigeorgis points out[75], management has the flexibility to wait with the investment (option to defer) or to abandon (option to abandon) the project completely.

An option to defer created by a stochastic interest rate even in the case of a deterministic future cash flow structure was first analyzed by Ingersoll and Ross[76] in their publication *Waiting to Invest: Investment and Uncertainty* in 1992 with the simple case of a 1-year Zero bond that pays 1 $ and was issued at time $t \geq 0$ with maturity $T = t+1$. This can be seen as a project that starts at time t for an investment cost of P and that guarantees a 1 $ payment at time $t+1$. The authors show how, with stochastically modelled interest rates

[75] See Trigeorgis [132], page 197.
[76] See Ingersoll & Ross [69].

of a special type[77], the optimal time t to invest can be determined. These pioneering ideas will be introduced in detail in 4.4.1. In 2001, when the U.S. Federal Reserve Bank reduced the risk-free interest rate several times, this idea became especially important since a decreasing interest rate increases a project's present value and might lead to earlier initiation of a project.

Trigeorgis also mentions, if only briefly, the notion of uncertain interest rates creating real options in his book *Real Options*[78]. He addresses this idea in the simple case that Ingersoll & Ross dealt with in their publication mentioned above[79]: *In reality, of course, situations do not present themselves in such simple scenarios. For example, a parallel upward shift in the level of interest rates (favoring early investment) in the presence of rising interest rate uncertainty (favoring project delay) would result in an unclear mixed effect. Moreover, if the cash flows of a project are growing and/or the project's life expires at a specified time (e.g., upon expiration of a patent or upon competitive entry), then delaying the project would involve an opportunity cost analogous to a "dividend" effect, reducing the value of the option to wait and favoring earlier initiation. Furthermore, if the cash flows are uncertain and correlated with the interest rate (and the market return), the cost of capital will differ from the risk-free interest rate. A risk-neutral valuation approach will then be necessary, especially if other real options are also present. The risk-neutral probabilities will again be derived from market price and interest rate information, but now they may vary over time and across states (since they will depend on changing interest rates).*

Other authors who dealt with stochastic interest rate modelling in (real) options valuation are, e.g., Sandmann[80] (1993, European options only, see Section 4.4), Ho, Stapleton, and Subrahmanyam[81] (1997, valuation of American options with a stochastic interest rate by generalizing the Geske-Johnson model), Miltersen[82] (2000) as well as Alvarez and Koskela[83] (2002). Miltersen used a model developed by Miltersen and Schwartz[84] to value natural resource investment projects. He assumed that the multiple product decisions are independent, and hence can be valued as the sum of European options[85]. His work is mathematically very complex, but he also provides various examples.

[77] Ingersoll and Ross used the Cox-Ingersoll-Ross model to describe the term structure of interest rates whereby the parameters α and β are omitted, i.e., the short-rate process has a zero-expected change.

[78] See Trigeorgis [132].

[79] See Trigeorgis [132], page 199.

[80] See Sandmann [115].

[81] See Ho, Stapleton & Subrahmanyam [58].

[82] See Miltersen [96].

[83] See Alvarez & Koskela [1].

[84] See Miltersen & Schwartz [97].

[85] See Miltersen [96], page 196.

The recent article of Alvarez and Koskela builds[86] on Ingersoll & Ross [69] and analyzes irreversible investments under interest rate variability. According to Alvarez and Koskela[87], *the current extensive literature on irreversible investment decisions usually makes the assumption of constant interest rate.*

As Alvarez and Koskela further state[88]: *In these studies dealing with the impact of irreversibility in a variety of problems and different types of frameworks the constancy of the discount rates has usually been one of the most predominant assumptions. The basic motivation of this argument is that interest rates are typically more stable (and consequently, less significant) than revenue dynamics. [...] If the exercise of such investment opportunities takes a long time, the assumed constancy of the interest rate is questionable. This observation raises several questions: Does interest rate variability matter and, if so, in what way and how much?*

Consequently, their studies are motivated by the importance of interest rate variability[89]: *It is known from empirical research that interest rates fluctuate a lot over time and that in the long run these follow a more general mean reverting process (for an up-to-date theoretical and empirical survey in the field see e.g. Björk [6] and Cochrane [32]). Since variability may be deterministic and/or stochastic, we immediately observe that interest rate variability in general can be important from the point of view of exercising real investment opportunities.*

The article of Alvarez and Koskela is very thorough and focuses on the mathematical aspects but does not provide detailed real world examples or historical backtesting. In this thesis, the focus is on numerical real options pricing with interest rate uncertainty. Although articles about option pricing in general are plentiful in the literature, thorough numerical analyses using simulations and historical backtesting of complex real options situations with stochastic interest rates are rare at best. Therefore, the goal of this thesis is to fill this gap by providing in-depth insight into how a stochastically modelled interest rate influences the real options value in various real options cases common in practice and to derive rules for application in Corporate Finance.

[86] See Alvarez & Koskela [1], page 2: *[...] we generalize the important findings by Ingersoll and Ross both by allowing for stochastic interest rate of a mean reverting type and by exploring the impact of combined interest rate and revenue variability on the value and the optimal exercise policy of irreversible real investment opportunities.*

[87] See the abstract in Alvarez & Koskela [1].

[88] See Alvarez & Koskela [1], page 1.

[89] See Alvarez & Koskela [1], page 2.

2.7 Summary

Chapter 2 introduced the idea of real options: qualitatively, quantitatively, and with respect to the historical development of this research field. Various real options approaches have been categorized and explained in terms of their development over time and use in financial practice. While Section 2.2, *Basics of Real Options*, introduced the concept of real options in general, Sections 2.3, 2.4, and 2.5 introduced the necessary financial mathematics needed in the following chapters. Section 2.3, *Diffusion Processes in Classical Probability Theory*, presented the idea of the diffusion process that is necessary to model real options and the term structure of interest rates. Section 2.4, *Introduction to Ito Calculus*, introduced Ito calculus in its basic form, especially with a special focus on diffusion processes. Finally, Section 2.5, *Discretization of Continuous-Time Stochastic Processes*, provided the necessary algorithms to numerically calculate a realized discrete path of a continuous-time stochastic process in a computer simulation program. The *Evolution of the Real Options Theory and Models in the Literature* was the topic of Section 2.6, a section divided into four subsections: *Decision-Tree Analysis* in 2.6.1, *Contingent-Claims Analysis* in 2.6.2, *Categorization of Real Options Valuation Methods* in 2.6.3, and *Flexibility due to Interest Rate Uncertainty* in 2.6.4.

In the remaining chapters of this thesis the focus is on real options pricing under interest rate uncertainty, which was introduced in 2.6.4. The central thrust will be to thoroughly analyze complex real options situations with non-constant (especially, stochastically modelled) risk-free interest rates using numerical simulations and historical backtesting. The primary valuation tools will be methods used for valuing financial options but modified to value real options problems (i.e. the investment opportunities) since the similarities between these two are manifold. To quote Dixit and Pindyck[90]: *Opportunities are options - rights but no obligations to take some action in the future. Capital investments, then, are essentially about options.*

The stochastic interest rate movement in these models will be described via term structure models. Each term structure model is specified by a diffusion process that will be introduced in the following chapter. The real options valuation methods will be explained in Chapter 4 and numerically analyzed in Chapter 5.

[90] See Dixit & Pindyck [40], page 105.

3

Stochastic Models for the Term Structure of Interest Rates

3.1 Introduction

This chapter is devoted to stochastic interest rate models. Since the 1970s stochastic interest rate models have played a central role in Finance, not only in research but also in practical applications. These models were immediately implemented by practitioners to accurately price financial securities, like interest rate derivatives. Their pricing, however, was mathematically far more demanding than equity pricing, due to the fact that the whole term structure of interest rates and its development over time was required.

Chapter 3 aims to give an introduction to the term structure of interest rates, to the historical development of term structure models and to their classification. Section 3.2, *The Term Structure of Interest Rates*, introduces term structure models and short-rate models (3.2.1). It presents the idea of cubic splines for yield curve interpolation (3.2.2), and gives a general overview of stochastic interest rate models (3.2.3). Section 3.2.1 defines the basic terms such as *no-arbitrage model*, *equilibrium model*, *mean reversion model*, *one-factor model* and *multi-factor model*. Moreover, it gives a summary of the notation for term structure models which is used in Chapter 3 and the following chapters. In 3.2.3 stochastic interest rate models are described from a general point of view. The models are categorized, the differences between them are pointed out and their historical development is explained.

Section 3.3, *Stochastic Interest Rate Models*, explains the models in mathematical detail, taking into account their application in this thesis. After providing more information on the mean reversion feature of short-rate models (3.3.1), the one-factor models will be explained (3.3.2), followed by the two-factor models (3.3.3). Section 3.3 draws heavily on the stochastic calculus introduced in Section 2.4. While all results for the models used in this thesis are presented, only some derivations of formulas are given, the main focus being on their application rather than on providing their proof. Finally, Section 3.4 gives a summary of this chapter and contains a table that provides an easy comparison of the interest rate models presented.

3.2 The Term Structure of Interest Rates

3.2.1 Introduction to the Term Structure of Interest Rates and Short Rate Models

The *term structure of interest rates* is the yield curve that describes the interest rate for different times to maturity. The term structure models describe the movement of the term structure over time according to a stochastic model.

Figure 1.3 in Chapter 1 showed the yield curve movement of U.S. Zero bonds between April 2000 and March 2003 in a three-dimensional plot. Figure 3.1 now gives a two-dimensional plot for the term structure provided in Figure 1.3 for five different time points starting on December 1, 2000. It takes some information from Figure 1.3 and displays it in a way that starting on December 1, 2000 the development of the term structure of interest rates over four quarters becomes visible.

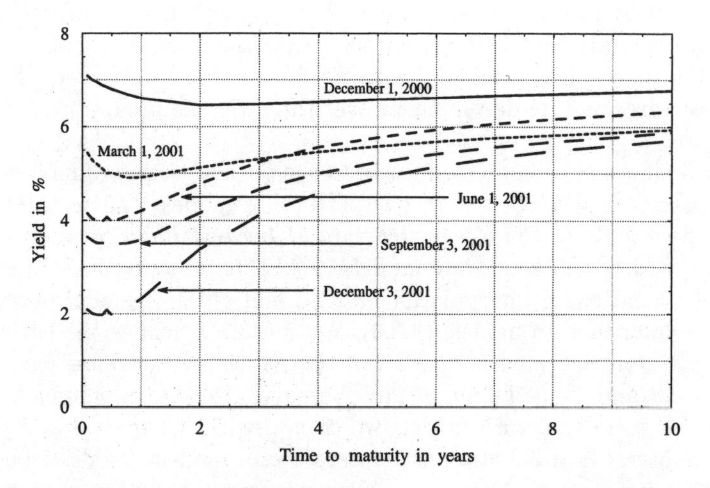

Fig. 3.1. Development of the term structure of interest rates for U.S. Zero bonds over four quarters starting December 1, 2000.

It took until the mid 1970s for the first stochastic term structure model to be developed[1]. This section introduces the *term structure of interest rates* and the theory of *short-rate models* that form the core of each interest rate model. Common to all models are the same basic terminology and notation that will be consistently applied throughout the thesis.

[1] The first of those models was the Vasicek model which was developed by Oldrich Vasicek in 1977, see Vasicek [133].

In order to introduce the basic terminology, a Zero bond is considered that pays 1 \$ at maturity. Today the time point is $t = 0$ and the Zero is assumed to mature at time point $t = T \geq 0$. The unit for the time axis is always *years* throughout the thesis. This means that today the Zero is a T-year Zero. The question is how much would someone be willing to pay for this bond at time t, $0 \leq t \leq T$? Figure 3.2 explains the situation of the Zero graphically.

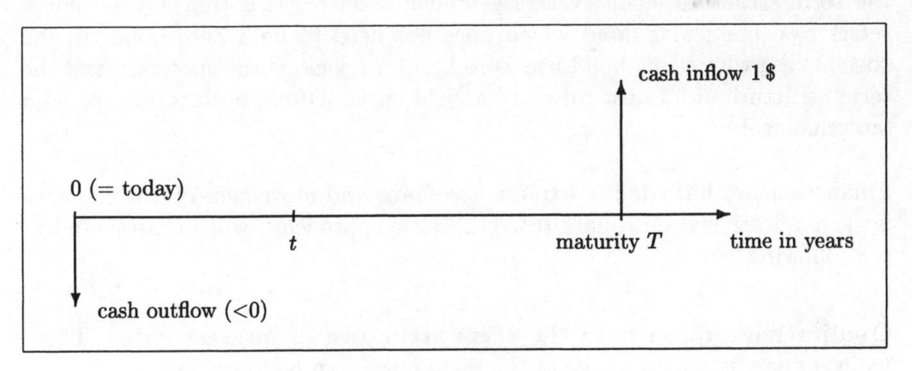

Fig. 3.2. Cash flows of a Zero bond.

Summary of Notation[2]:

$$\begin{aligned}
P(t,T) &:= \text{price at time } t \text{ of a Zero bond that matures at time } T,\\
R(t,T) &:= \text{continuously compounded yield at time } t \text{ on the}\\
&\quad\ \text{Zero bond that matures at time } T \text{ (also called spot rate)},\\
r_t &:= \text{short-term interest rate at time } t \text{ (also called short rate)},\\
f(t,T) &:= \text{instantaneous forward rate at time } t \text{ for time } T,\\
\sigma(r_t) &:= \text{volatility of the short rate } r_t,\\
\sigma_R(t,T) &:= \text{volatility of the spot yield } R(t,T).
\end{aligned}$$

The pricing relationship for a Zero bond is:

$$P(t,T) = e^{-R(t,T)(T-t)}. \tag{3.1}$$

If the bond price is given, the spot rate $R(t,T)$ can be calculated as

$$R(t,T) = -\frac{1}{T-t} \ln P(t,T). \tag{3.2}$$

The instantaneous forward rate $f(t,T)$ is defined as

$$f(t,T) := -\frac{\partial}{\partial T} \ln P(t,T). \tag{3.3}$$

[2] The notation follows Clewlow & Strickland [31], page 184.

Combining (3.1), (3.2), and (3.3) yields

$$R(t,T) = \frac{1}{T-t}\left(\int_t^T f(t,\tau)\,d\tau\right) \text{ and } P(t,T) = \exp\left\{-\int_t^T f(t,\tau)\,d\tau\right\}.$$

The function R describes the *term structure of interest rates.* Function σ_R describes the *term structure of interest rate volatilities.* The difference between the term structure of interest rates and a *yield curve* is that a yield curve refers to a particular bond which does not need to be a Zero bond. In the special case where this bond is a Zero bond, its yield is the spot rate and the term structure of interest rates is its yield curve. Often, both terms are used interchangeably.

Finance theory has tried to explain the shape and movement of a yield curve both qualitatively and quantitatively. Both approaches will be presented in the following.

Qualitative approach to the term structure of interest rates. Three main influences on the shape of the yield curve can be identified[3]:

1. Interest rate expectations
2. Bond risk premia
3. Convexity bias

The first influence, interest rate expectations, is based on the *pure expectations theory.* This theory has already been briefly introduced in Section 1.1. Pure expectations theory states that the shape of the yield curve is purely driven by the market's collective expectations regarding future interest rates. This has three basic implications for the shape of the yield curve:

- If the market expects no changes in future interest rates, today's spot curve will be flat.
- If the market expects a parallel shift in the yield curve, the spot curve will be linear.
- If the market expects increasing interest rates and a flattening of the yield curve, today's spot curve will be upward sloping and concave. If the market expects decreasing interest rates and a flattening of the yield curve, today's spot curve will be downward sloping and convex.

The second influence, the bond risk premia[4], can be explained by several theories. The *liquidity theory* states that investors demand a positive risk premium

[3] The description is based on a summary presented in Schweser [122], pages 80-82.
[4] The bond risk premium is defined as a long-term bond's expected one-period return that is in excess of a one-period bond's risk-free return. The bond risk premium can be positive or negative.

to compensate them for bearing risk that arises from the fact that long-term bonds have greater price volatility (and, therefore, greater interest rate risk). The *preferred habitat hypothesis* states that some investors will have a preferred maturity habitat. They may accept lower yields to be in that maturity range, resulting in a negative risk premium.

The third and last influence, the convexity bias, is based on the observation that investors will pay more (thereby accepting a lower yield) for a more convex bond. An approximative relationship between bond price changes and interest rate changes is

$$\frac{dP}{P} = -D_{mod} \cdot dy + \frac{1}{2} C \cdot (dy)^2.$$

D_{mod} is the modified duration of a bond, C its convexity, P its price, and dp its absolute price change because of the absolute yield change of dy. The higher the convexity, the less the bond will lose in value if interest rates rise, and the more the bond will gain in value if interest rates fall. The greater the expected yield volatility, the greater the value of convexity. Therefore, the relative yields of more convex bonds will be lower in times of greater yield volatility. Short-term bonds have lower convexity and long-term bonds have greater convexity. This can explain why the long end of the yield curve is flatter than the short end.

Quantitative approach to the term structure of interest rates. The quantitative approach to the term structure of interest rates is one of the major fields of financial research. After the first model was proposed in 1977 by Oldrich Vasicek[5], many researchers developed various types of interest rate models. In this thesis, seven models of the term structure of interest rates will be presented in detail:

- Vasicek model
- Cox-Ingersoll-Ross model
- Ho-Lee model
- Hull-White one-factor model and two-factor model
- Heath-Jarrow-Morton one-factor model and two-factor model

Models of the term structure of interest rates can be categorized in several ways. A general classification is:

- **Equilibrium models versus no-arbitrage models:**
  Equilibrium models are models that start with an equilibrium condition for deriving the term structure of interest rates. They attempt to explain the market bond prices rather than taking them as an exogenous input

[5] See Vasicek [133].

parameter. These models contain specific assumptions about the dynamics of the driving factors. These factors also *determine the attainable shapes of the yield curve, and, at the same time, apportion in different ways the contributions to the curvature of the yield curve to future expectations of rates, and to future expectations of volatility*[6]. A no-arbitrage model is a model that starts with a no-arbitrage condition. Such a no-arbitrage condition is implied in the current term structure of interest rates since otherwise the arbitrage would be quickly eliminated in the capital markets. Therefore, no-arbitrage models usually take the current term structure of interest rates as an input parameter. As a consequence, a no-arbitrage model gives the current term structure of interest rates as a solution for time $t = 0$. This, however, does not need to be the case for an equilibrium model.

- **One-factor models versus multi-factor models:**
  A one-factor model is a model for the term structure of interest rates with one single factor that influences the shape and development of the term structure over time. Consequently, a one-factor model can only model a parallel shift in the term structure but no twist or butterfly movements[7]. A one-factor model always assumes that the short and the long end of the yield curve are perfectly correlated, i.e., move in the same direction (parallel shift). However, in reality it can be observed that the short and the long end of the yield curve are influenced by different risk factors that are not perfectly correlated. Therefore, at least two factors are needed for a model that quantifies non-parallel shifts of the term structure of interest rates. The minimum number of risk factors an interest rate model needs to have are two, in order to allow for uncorrelated movements of the short and the long end of the curve. Models with two or more factors are called multi-factor models.
- **Mean reversion models versus no mean reversion models:**
  A mean reversion model is a model where the interest rate gets pulled back to a mean interest rate value. Mean reversion models are not only used for interest rate modelling but also in the modelling of commodity price movements[8]. A mathematical definition for a mean reversion model with constant mean reversion level was already given in 2.4.10.

The core of any model of the term structure of interest rates is the *short-rate process*. The short rate is the instantaneous interest rate. This means that $r_\tau, \tau \geq 0$, is the annualized interest rate prevailing at time τ for the infinitesimal time interval $[\tau, \tau + dt]$ with dt as an infinitesimal short time

[6] See Rebonato [109], page 252.
[7] See Rudolf [114], page 135.
[8] See Lieskowsky, Onkey, Schulmerich, Teng & Wee [81], pages 10-13.

period[9]. Therefore, in time period $[0, t], t \geq 0$, a 1 \$ bank deposit earns

$$exp\left\{\int_0^t r_\tau d\tau\right\} - 1$$

dollars as interest. For any of the term structure models the model is described via the short-rate process[10]. Various formulas for, e.g., Zero prices and Zero yields will be derived from the short-rate process in the course of this thesis.

3.2.2 Yield Curve Interpolation with Cubic Splines

Interest rates are given for special maturity buckets when downloading them from data providers like *Bloomberg* or *Datastream.* Those buckets usually are 1 mos., 2 mos., ..., 12 mos., 2 yrs., 3 yrs., 4 yrs., 5 yrs., 6 yrs., 7 yrs., 8 yrs., 9 yrs., 10 yrs., 15 yrs., 20 yrs. and 30 yrs. In order to get interest rates for time points in between these buckets, an interpolation method has to be applied. Mathematical software packages like *Mathematica* which was used to generate all the plots in this thesis have these interpolation methods already built in. However, in order to generate a stochastic term structure with a no-arbitrage model that needs the current yield curve as an input parameter, it was necessary to include such an interpolation scheme in the computer simulation program itself. If for example, *Bloomberg* only quotes the 25 time points above, while many more interest rate points are needed for the term structure calculation, both the interpolation of the data between the smallest and largest maturity bucket as well as the extrapolation between zero and the smallest maturity bucket are necessary. This means that for the case of a time step size of 1 day, distinct interest rates are needed for $1 \cdot 360 \cdot 30 = 10\,800$ time points to interpolate the whole yield curve over 30 years (assuming a 30/360 day count method).

The cubic spline method is used for the interpolation procedure due to its favorable properties. A cubic spline is, as will be shown below, twice differentiable (with a smooth second derivative), a feature needed for the simulation of short-rate paths in the Ho-Lee and the two Hull-White models that will be presented in Section 3.3. For the extrapolation part an ordinary linear extension will be used.

The cubic spline interpolation presented here will allow for two different boundary conditions: the Hermite and the natural conditions. The meaning of both conditions will be explained but only the natural condition will be used

---

[9] The short rate will be described by a diffusion process. Since in Chapter 2 all stochastic processes were denoted with X_t, this notation with t as an index will be used to ensure consistency throughout the thesis.

[10] For more information on the short rate see Rudolf [114], page 22.

in the computer simulation program for reasons explained later. This condition leads to the linear extrapolation method below the lowest given maturity bucket.

Definition: Cubic Spline Function
A cubic spline function of smoothness order 2 for a node set $\Delta = \{a = x_0 < x_1 < x_2 < \ldots < x_N = b\}$ with $N \in \mathbb{N}$ and $a, b \in \mathbb{R}$ with $a < b$ is a function $s : [a, b] \longrightarrow \mathbb{R}$ with the following properties:

(*i*) $s\Big|_{[x_{i-1}, x_i]} \in \Pi_3, \; i = 1, 2, \ldots, N,$

(*ii*) $s \in C^2([a, b])$.

The set of all cubic spline functions of smoothness order 2 for the node set Δ is noted by S_Δ. Note that this parameter N has nothing to do with other parameters in this thesis called N which are applied to discretize a time interval $[0, T]$.

Theorem: Basis of the Set S_Δ
The set S_Δ of all cubic spline functions of smoothness order 2 for the node set Δ is a linear space with dimension $N + 3$ and the basis

$$\{g_i, \, h_j \mid i = 0, 1, 2, 3 \text{ and } j = 1, 2, \ldots, N - 1\}.$$

Thereby:

$$g_i := (x - x_0)^i, \qquad i = 0, 1, 2, 3,$$

$$h_j := ((x - x_j)_+)^j, \quad j = 1, 2, \ldots, N - 1, \quad \text{with}$$

$$(x - x_j)_+ := \begin{cases} 0 & \text{for } x \leq x_j, \\ x - x_j & \text{for } x \geq x_j. \end{cases}$$

Graphically, the functions g_i and h_j are plotted below.

Proof: It will be shown that s can be expressed as

$$s(x) = \sum_{i=0}^{3} a_i \, (x - x_0)^i + \sum_{j=1}^{N-1} b_j \, ((x - x_j)_+)^3 \tag{3.4}$$

with appropriate coefficients $a_i, b_j \in \mathbb{R}$. To show this, function s has to be constructed (i.e., the coefficients $a_i, b_j \in \mathbb{R}$ have to be determined) and the *linear independence* of the basis functions has to be shown.

Obviously, condition $s \in S_\Delta$ is fulfilled. To show the linear independence, given that s can be expressed in form (i), it has to be shown that $s \equiv 0$ on $[a,b]$ implies that all coefficients a_i and b_j are zero. Let $x \in [x_0, x_1]$. Then:

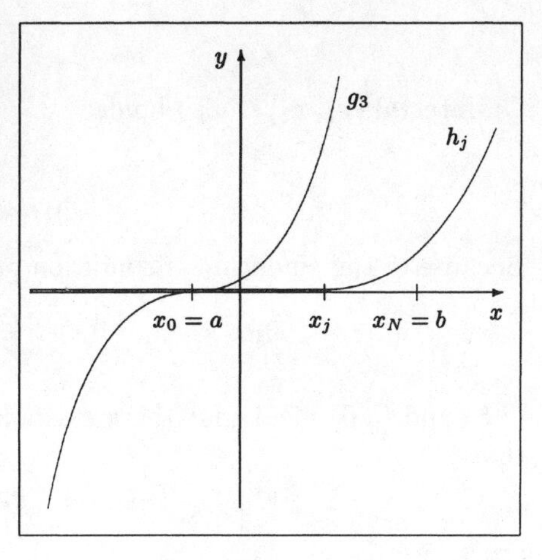

$$\sum_{i=1}^{N-1} a_i(x-x_0)^i + \sum_{j=1}^{N-1} b_j(x-x_j)_+^3$$

$$\stackrel{a \le x_1}{=} \sum_{i=1}^{N-1} a_i \underbrace{(x-x_0)}_{\ge 0} a^i = 0.$$

This implies $a_i = 0$ for $i = 0, 1, 2, 3$. On interval $[x_1, x_2]$ this yields:

$$\sum_{i=1}^{N-1} a_i(x-x_0)^i + \sum_{j=1}^{N-1} b_j(x-x_j)_+^3 = b_1 \underbrace{(x-x_1)_+^3}_{\ge 0} = s(x) = 0.$$

This implies $b_1 = 0$. Via induction it is now assumed that for a flexible, but constant $k \in \{2, 3, \ldots, N-1\}$ the coefficients $b_1 = b_2 = \ldots = b_{k-1} = 0$ have already been determined. This then gives on interval $[x_k, x_{k+1}]$:

$$\sum_{i=1}^{N-1} a_i(x-x_0)^i + \sum_{j=1}^{N-1} b_j(x-x_j)_+^3 \stackrel{assumption}{=} b_k \underbrace{(x-x_k)_+^3}_{\ge 0} = s(x) = 0.$$

This implies $b_k = 0$. Then by induction the value of all coefficients a_i and b_j is zero if assuming $s \equiv 0$ on $[a,b]$. This is the property of linear independence.

To construct s, i.e., to show that coefficients a_i and b_j exist such that spline s can be written in form (3.4), s has first to be restricted on interval $[x_0, x_1]$. It has to be shown that on $[x_0, x_1]$ coefficients a_i exist with

$$s\Big|_{[x_0,x_1]}(x) = \sum_{i=0}^{3} a_i(x-x_0)^i.$$

However, such coefficients can be found because of property (i) of a spline function. With

$$s_1(x) := \sum_{i=0}^{3} a_i(x-x_0)^i, \quad x \in [a,b],$$

it holds that

$$s\Big|_{[x_0,x_1]} = s_1\Big|_{[x_0,x_1]}.$$

On interval $[x_1, x_2]$ it also holds:

$$(s - s_1)\Big|_{[x_1,x_2]} \in \Pi_3. \tag{3.5}$$

Because of the smoothness condition of order 2 on s it holds

$$\lim_{x \downarrow x_1} (s - s_1)^{(i)}(x) = 0 \quad \text{for } i = 0, 1, 2. \tag{3.6}$$

(3.5) and (3.6) yield that there exists an appropriate coefficient $b_1 \in \mathbb{R}$ such that

$$(s - s_1)\Big|_{[x_1,x_2]}(x) = b_1\,(x - x_1)_+^3, \quad x \in [x_1, x_2].$$

With

$$s_2(x) := \sum_{i=0}^{3} a_i(x - x_0)^i + b_1(x - x_1)_+^3, \quad x \in [a, b],$$

it can be shown that

$$s\Big|_{[x_0,x_2]} = s_2\Big|_{[x_0,x_2]}.$$

Iteratively, this idea can be continued until reaching interval $[x_{N-1}, x_N]$ that finally specifies the last coefficient b_{N-1}. Therefore, function s is completely determined, q.e.d.

The *interpolation question* is if for a given set of nodes $\{x_0, x_1, x_2, \ldots, x_N\}$ and corresponding values $\{y_0, y_1, y_2, \ldots, y_N\}$ there exists a cubic spline function $s \in S_\Delta$ with the following properties:

$$s(x_j) = y_j \quad \text{for } j = 0, 1, 2, \ldots, N.$$

It will be shown that solutions always exist to this question. However, these solutions are not unique, since there are two degrees of freedom left. To show this, the construction process from the proof above has to be repeated. This will provide the general idea of the method required to determine the cubic spline. Thereafter, the numerical algorithm used in the computer simulation program for interpolating the yield curve (spot and forward curves) is presented.

First, a function h_j has to be defined as

$$h_j := x_j - x_{j-1}, \quad j = 1, 2, \ldots, N.$$

Define further parameter a_0 as $a_0 := s(x_0)(= y_0)$ and define parameter a_1, a_2 and a_3 via the equation

$$s(x_1) \quad = \quad a_0 + a_1 h_1 + a_2 h_1^2 + a_3 h_1^3 \quad = \quad y_1.$$

However, this means that two parameters cannot be uniquely specified. Therefore, the following condition will be introduced:

$$s(x_2) \;=\; s_1(x_2) + b_1\left((x_2 - x_1)_+\right)^3 \;=\; s_1(x_2) + b_1 h_2^3 \;=\; y_2.$$

This now gives b_1 as an affine function of the two unspecified parameters above. In the next step the condition

$$s(x_3) \quad = \quad s_2(x_3) + b_2 h_3^3 \quad = \quad y_3$$

is introduced. This defines coefficient b_2 implicitely. This procedure continues until the final condition

$$s(x_N) \quad = \quad s_{N-1}(x_N) + b_{N-1} h_N^3 \quad = \quad y_N$$

is reached that determines b_{N-1} implicitely. After having applied all conditions on s, two parameters remain unspecified, i.e., the problem has two degrees of freedom.

Now the question arises, what additional conditions should be defined in order to get a unique cubic spline function s. These conditions are called *boundary conditions.* The following introduces the two most common boundary conditions among the many that exist:

(*i*) *Hermite boundary condition:* $s'(a) = y'_0$ and $s'(b) = y'_N$.

(*ii*) *Natural boundary condition:* $s''(a) = 0$ and $s''(b) = 0$.

y'_0 and y'_N are the first derivatives in the two end points x_0 and x_N, respectively. The values of y'_0 and y'_N can be chosen freely. Depending on the specification of the boundary condition a cubic spline is called a *natural cubic spline* or a *Hermite cubic spline.*

In the computer simulation program the natural boundary condition is used since it says that at the two end time points the yield curve has no curvature. This assumption directly leads to an ordinary linear extrapolation below the lowest given time point x_0 down to zero.

In the following a numerical algorithm will be described to construct a cubic spline function for a given set of nodes $\Delta = \{a = x_0 < x_1 < x_2 < \ldots < x_N = b\}$ and the corresponding set of values $\{y_0, y_1, \ldots, y_N\}$. Note that the values $x_j - x_{j-1}, j = 1, \ldots, N$, do not need to be the same. This construction process is done in five steps:

Algorithm: Cubic Spline Construction

Step 1:

According to the spline conditions, the cubic spline function s is stepwise cubic and an element of C^2. This means that s'' is stepwise linear and smooth. Function s'' can be expressed for $x \in [x_{j-1}, x_j], j = 1, 2, \ldots, N$, as

$$s''(x) = M_{j-1} \frac{x_j - x}{x_j - x_{j-1}} + M_j \frac{x - x_{j-1}}{x_j - x_{j-1}} = M_{j-1}(1-t) + M_j t$$

with

$$x = x_{j-1} + th_j, \quad t \in [0,1], \quad \text{and} \quad M_j := s''(x_j).$$

The coefficients $M_j, j = 1, 2, \ldots, N$, are unknown so far and will be specified in the following step. They are generally called *moments*.

Step 2:

On interval $[x_{j-1}, x_j], j \in \{1, 2, \ldots, N\}$, function s can be derived by integrating s'' (from step 1) twice. Obviously, s then includes two integration constants that both depend on j. It can easily be shown[11] that s has the following form:

$$\left.\begin{aligned} s(x) = y_{j-1}(1-t) &+ y_j t + \\ \frac{M_{j-1}h_j^2}{6}(2-t)(1-t)(-t) &+ \frac{M_j h_j^2}{6}(t-1)t(t+1). \end{aligned}\right\} \quad (3.7)$$

Here, $x \in [x_{j-1}, x_j], j \in \{1, 2, \ldots, N\}$, with

$$t := \frac{x - x_{j-1}}{x_j - x_{j-1}} \in [0,1].$$

Step 3:

For each $j \in \{1, 2, \ldots, N\}$ function s from (3.7) can be differentiated on $[x_{j-1}, x_j]$. This yields:

$$s'(x) = \frac{y_j - y_{j-1}}{h_j} + \frac{M_{j-1}\, h_j}{6}[1 - 3(1-t)^2] + \frac{M_j\, h_j}{6}[3t^3 - 1].$$

Moreover, it holds:

$$\lim_{x \downarrow x_j} s'(x) = \frac{y_{j+1} - y_j}{h_{j+1}} - \frac{h_{j+1}}{6}[2M_j + M_{j+1}], \quad (3.8)$$

$$\lim_{x \uparrow x_j} s'(x) = \frac{y_j - y_{j-1}}{h_j} + \frac{h_j}{6}[M_{j-1} + 2M_j]. \quad (3.9)$$

[11] For a proof see Börsch-Supan [11], Chapter IV A, § 1.

The smoothness condition on s' requires

$$\lim_{x \downarrow x_j} s'(x) \quad = \quad \lim_{x \uparrow x_j} s'(x).$$

Subtract (3.8) - (3.9) and multiply the result by $\frac{6}{h_j + h_{j+1}}$. With the definitions

$$\mu_j \ := \ \frac{h_j}{h_j + h_{j+1}} \qquad \text{and} \qquad \lambda_j \ := \ \frac{h_{j+1}}{h_j + h_{j+1}}, \quad j \in \{1, \ldots, N-1\},$$

this yields the condition

$$\mu_j M_{j-1} \ + \ 2M_j \ + \ \lambda_j M_{j+1} \quad = \quad 6\,[x_{j-1}, x_j, x_{j+1}].$$

Here, the so-called *Gauss notation of the divided differences* is used[12].

Step 4:

In step 4, the missing two conditions from the $(N+1)$ moments of the boundary conditions will be determined:

Hermite Boundary Condition (i):

The Hermite boundary condition (i) can be rewritten using moments as

$$2M_0 + M_1 = \frac{6}{h_1} \left([x_0, x_1] - y_0'\right),$$
$$M_{N-1} + 2M_n = \frac{6}{h_N} \left(y_N' - [x_{N-1}, x_N]\right).$$

But as this condition forces the interpolating cubic spline to assume a predefined shape at the end of the interpolating interval, it will not be applied in the computer simulation program.

Natural Boundary Condition (ii):

The natural boundary condition (ii) can be rewritten as $M_0 \ = \ 0 \ = \ M_n$. This condition says that there is no curvature at the end points a and b of the interpolating interval $[a, b]$. This means that at these points the interpolating cubic spline could be extended by two straight lines and that at these points this extended cubic spline would have no curvature. The term *natural* is associated with an elastic ruler fit to each of the N points (x_i, y_i) and at the ends no conditions are posed on this ruler. Mathematically, this is the natural boundary condition.

[12] A detailed explanation of this standard notation in numerical mathematics can be found in Stoer [125], Section 2.1.3.

Step 5:

The previous steps lead to a linear equation system of order $N-1$ for the moments which has to be solved now. This linear equation system can be written in matrix notation as:

$$\begin{pmatrix} 2 & \lambda_1 & & & & \mu_1 \\ \mu_2 & 2 & \lambda_2 & & & \\ & \mu_3 & 2 & \lambda_3 & & \\ & & \ddots & \ddots & \ddots & \\ & & & \mu_{N-1} & 2 & \lambda_{N-1} \\ \lambda_N & & & & \mu_N & 2 \end{pmatrix} \cdot \begin{pmatrix} M_0 \\ M_1 \\ M_2 \\ \vdots \\ M_{N-2} \\ M_{N-1} \end{pmatrix} = \begin{pmatrix} 6\,[M_0, M_1, M_2] \\ 6\,[M_1, M_2, M_3] \\ 6\,[M_2, M_3, M_4] \\ \vdots \\ 6\,[M_{N-3}, M_{N-2}, M_{N-1}] \\ 6\,[M_{N-2}, M_{N-1}, M_N] \end{pmatrix}$$

The question whether this linear equation system can always be solved and whether the solution is unique can be answered by using the following lemma.

Lemma:

Let $A := (a_{ij})_{1 \le i,j \le N} \in \mathbb{R}^{N,N}$ with the property

$$|a_{ii}| > \sum_{j=1;j\neq i}^{N} |a_{ij}| \quad \text{for } i = 1, 2, \ldots, N. \tag{3.10}$$

Then matrix A is regular[13] and an LR decomposition[14] exists for it. A matrix with the property (3.10) is called *strictly diagonally dominant.* A proof of this lemma can be found in Börsch-Supan [11], Chapter IV A, 1.2.

Since

$$|a_{ii}| = 2 > \sum_{j=1;j\neq i}^{N} |a_{ij}| \quad \forall i \in \{1, 2, \ldots, N\}$$

holds true for both the normal and the Hermite boundary conditions, the linear equation system can always be solved and the solution is unique. To solve it, a Gauss algorithm can be chosen[15], see, e.g., Stoer [125], page 149.

---

[13] A matrix is called *regular* if the determinant of the matrix is different from zero.

[14] An *LR decomposition* of a matrix is a special decomposition of the matrix in a product of two matrices. For detailed information on this see Stoer [124], Section 6.6.4.

[15] This Gauss algorithm is used in the computer simulation program accompanying this thesis.

3.2.3 General Overview of Stochastic Interest Rate Models

This section provides a general overview and discussion of stochastic interest rate models. The purpose is to show the differences and similarities of the most popular term structure models and how these models developed over time. The section also provides a short empirical comparison for some of the models[16].

The mathematical derivation of term structure models shows aspects similar to the derivation of the famous Black-Scholes formula. For financial derivatives on equities, the core of the Black-Scholes formula is the Black-Scholes partial differential equation. This differential equation can be solved for all Ito processes. The core of the theory of term structure models is the term structure equation, a differential equation that needs to hold true for interest rate derivatives. But in contrast to the Black-Scholes partial differential equation, the term structure equation cannot be solved for every Ito process, but only for the affine processes of term structure models. For more details on the derivation of the term structure equation see Rudolf [114], pages 35-40.

All term structure models are distinguished by the choice of the short-rate process. The first model for the term structure of interest rates was the famous Vasicek model, developed by Oldrich Vasicek in 1977. This model was an equilibrium model with a mean reversion feature. The mean reversion assumption can be traced to an economic assumption[17]: In phases with high interest rates, economic activity is restricted. This in turn will drive the monetary policy to arrange for decreasing interest rates. If lower levels of interest rates are reached these low rates drive up economic activity again and interest rates will rise. The Vasicek model is a generalization of an Ornstein-Uhlenbeck process that was suggested by Merton in 1974 for short-rate modelling[18].

The next important step in stochastic term structure modelling was done by Cox, Ingersoll, and Ross in 1985 through their acclaimed Cox-Ingersoll-Ross model (CIR)[19]. This equilibrium model exhibits the same mean reversion feature as the Vasicek model but it no longer contains a constant volatility term. The volatility in the CIR model is a square-rooted process that depends on the current level of the short rate: the lower the short rate the lower the volatility. This feature ensures that the short rate never gets negative (which is possible in the Vasicek model).

---

[16] For the most thorough analysis of term structure models and their empirical validation see Rebonato [109]. This book is well suited for mathematicians and Financial Engineering practitioners.

[17] See Rudolf [114], page 6, and Hull & White [65].

[18] See Merton [94].

[19] See Cox, Ingersoll & Ross [34].

The Vasicek and CIR models described above are equilibrium models as opposed to no-arbitrage models. Both concepts were introduced in Section 3.2.1: Equilibrium models start with an equilibrium condition for deriving the term structure of interest rates[20]. The main feature is that they attempt to explain the market bond prices rather than taking them as exogenous input parameters. This is also their weak point because the bond prices they produce can be far off from the current market values. On the other hand, no-arbitrage models start with a no-arbitrage condition which is, e.g., implied in the current term structure of interest rates. Therefore, no-arbitrage models take the current term structure of interest rates (and/or volatilities, forward rates) as an input parameter and calculate them for the special case $t = 0$. These models are hence called *term structure consistent.* This is the reason why no-arbitrage models are used primarily in the world of Finance today.

The first no-arbitrage model was the Ho-Lee model, published in 1986 in the Journal of Finance[21]. It was the first model to pioneer the use of arbitrage-free computational lattices[22]. This one-factor model takes the initial instantaneous forward rate as the input parameter to get term structure consistency. This model is fully analytically trackable but does not exhibit mean reversion and can also produce negative interest rates. On the other hand, the Hull-White one-factor model[23], published in 1993, has a mean reversion feature, but can also produce negative interest rates[24]. This model is currently very popular among practitioners[25] since it exhibits several desirable features: It is a fully analytically trackable, easy to use mean reversion model.

The main advantage of the Hull-White one-factor model over the Ho-Lee model, which is a special case of the Hull-White one-factor model, is its mean reversion feature which allows for a more realistic modelling of the term structure. Another important shortcoming[26] of the Ho-Lee model is its inability to fit an arbitrary term structure of volatilities and a pre-assigned future behaviour for the volatility of the short rate at the same time[27]. The Ho-Lee model requires that the volatility of the short rates diminishes over time from today's level so that mean reversion takes place. Therefore, the Hull-White one-factor model is more suited to model term structure movements[28]. On

[20] See Rebonato [109], page 252.

[21] See Ho & Lee [57].

[22] See Rebonato [109], page 187. The lattice model was later extended into a continuous-time model, see, e.g., Clewlow & Strickland [31], page 208.

[23] See Hull & White [66].

[24] See Rebonato [109], page 282.

[25] See Clewlow & Strickland [31], page 215.

[26] See Rebonato [109], page 281.

[27] The same holds true for the Black-Derman-Toy model that will be introduced later.

[28] See Rudolf [114], page 130.

the negative side, it only has one single risk factor to model the interest rate movements over time. This limits the use of the model since movements of the term structure need at least two risk factors to provide realistic term structure movements over time that include twists and butterflies.

One-factor and multi-factor models can be distinguished with respect to the number of factors that drive the term structure movement over time. Among the latter, the two-factor models are popular in practice since they can incorporate twists and butterfly movements of the term structure and not only shifts, to which the one-factor models are limited. Although the two-factor models give much better empirical results, they are more difficult to implement than one-factor models[29]. While one-factor models are insufficient to model the reality of term structure movements (i.e., no accomodation of twists and butterflies), two-factor models (and even worse multi-factor models) provide an additional challenge besides being more complex: the economic interpretation of the factors[30]. The factors for two-factor models are mostly specified, i.e., in Brennan & Schwartz [19], Hull & White [68], and Heath, Jarrow & Morton [52] the short-end and the long-end interest rates are the two factors. The importance of the risk factor specification is obvious since its economic interpretation determines the specification of the volatility functions[31].

To quote Rudolf[32]: *The most important advantage for the use of two-factor models over one-factor models is the better opportunity to adjust the modelled term structure to empirical data. One-factor models only allow to model perfectly correlated changes of the interest rates over all maturities. In reality, however, the interest rates between various maturity ranges are not perfectly correlated, i.e., the term structure often exhibits twist and butterfly movements. Those can only be modelled with two-factor models.* Therefore, two-factor models play a more important role in academia and practice as Rudolf also stresses[33]: *In the practical application and in academic papers the two-factor models, which usually model the short-end and the long-end interest rates, play a dominant role.*

Only a few studies have been undertaken that deal with the question of the number of risk factors needed. One of these was done by Bühler and Zimmermann in 1996[34]; they found the three factors *interest rate level*, *slope*, and *curvature* to be the three main components that determine the term structure of interest rates. However, this study was restricted to interest rates in Switzerland and Germany.

[29] See Rudolf [114], page 2 and page 66.
[30] See Rudolf [114], page 67.
[31] See Rudolf [114], page 67.
[32] See Rudolf [114], pages 3-4, English translation by the author.
[33] See Rudolf [114], page 40, English translation by the author.
[34] See Bühler & Zimmerman [26].

Mean reversion was already introduced qualitatively as a process swinging around a mean interest rate value, the mean reversion level. The term *mean reversion model* was already defined quantitatively in 2.4.10. In that definition the mean reversion level was constant over time. The Ornstein-Uhlenbeck process, the process in the Vasicek model, and the process in the CIR model are examples that exhibit a constant mean reversion level. However, it is not necessary to apply a constant mean reversion level. The mean reversion level can also be dependent on time. The Ho-Lee and the Hull-White one-factor models fit in this category. Moreover, the mean reversion level can itself be given via another stochastic process. An example of this broadest definition is the Hull-White two-factor model. The Hull-White two-factor model of 1994 includes a second risk factor as a stochastically modelled mean reversion level[35] and also offers term structure consistency.

The most general term structure model is the famous Heath-Jarrow-Morton model[36], published in 1992, that extends the Ho-Lee model from 1986[37]. This model is a multi-factor model with $n \in \mathbb{N}$ independent risk factors to model the movement of the instantaneous forward rate over time. In the original article, the modelling was done for the instantaneous forward rate and not the short rate; this can easily be rewritten for the short-rate movements[38]. Although the Heath-Jarrow-Morton model is very general and does not allow for computational results without specification of the input parameters[39] (i.e., the volatility functions), all other previously presented models can be derived from it by appropriately choosing the input parameters. Extensive work was done in 1997 by Baxter who shows[40] that all commonly used term structure models are specific cases of the Heath-Jarrow-Morton model. Rudolf shows[41] how to derive the Ho-Lee model and the Hull-White one-factor model from the Heath-Jarrow-Morton model. He stresses[42] that the main difference of the Hull-White one-factor model and the Heath-Jarrow-Morton model lies in the assumption of the Markov property[43] for the short rate of the Hull-White one-factor model.

In addition to the models presented so far, there are some other models that were published in the past. Other two-factor models were proposed by Richard

[35] See Hull & White [68] or Clewlow & Strickland [31], page 225 and page 232, for a comprehensive summary.
[36] See Heath, Jarrow & Morton [53].
[37] See Clewlow & Strickland [31], page 228.
[38] See Carverhill & Pang [29].
[39] See Rudolf [114], page 63.
[40] See Baxter [5].
[41] See Rudolf [114], pages 57-63.
[42] See Rudolf [114], page 60.
[43] See, e.g., Dixit & Pindyck [40], page 63, for a further explanation of the Markov property.

in 1978[44], by Brennan and Schwartz in 1979[45], and by Schaefer and Schwartz in 1984[46]. In 1990, Black, Derman, and Toy developed another one-factor, term structure consistent model[47]. This model is also available as a two-factor model. The Longstaff-Schwartz model[48], a two-factor model published in 1992, and the Fong-Vasicek model[49], a two-factor model published in 1992 as well, also deserve to be mentioned[50]. This list does not claim to be complete.

A comprehensive overview of short-rate models can be found in Leippold & Heinzl [79]. The authors categorize various short-rate models according to their respective approaches[51]. A detailed economic approach to term structure models (including one- and two-factor models as well as discrete and continuous-time models) can be found in Babbel & Merrill [3] and a systematical comparison of multi-factor term structure models is given in Duffie & Kan [41].

Figure 3.3 provides an overview of the model categorization according to Clewlow & Strickland [31]. Again, this overview does not claim to be complete.

There is a particular categorization of term structure models which needs a final explanation: Rudolf categorizes the Vasicek, the Dothan, and the Brennan-Schwartz models as *no-arbitrage models with an implicit equilibrium condition in its calibration process*[52]. The calibration process is defined as the process to determine the model parameters for the current term structure to get a good fit. The parameters for the Vasicek and the CIR models can only be determined using current market prices, which are no-arbitrage prices.

On the other hand, Rudolf calls the Ho-Lee, the Hull-White, and the Heath-Jarrow-Morton models *pure arbitrage models* since they use the term structure as input parameter. This naming is consistent with Leippold & Heinzl [79], page 143. In contrast, Rebonato does not make this additional distinction in the naming. He calls the Vasicek and the CIR models *equilibrium models* and the Ho-Lee, the Hull-White, and the Heath-Jarrow-Morton models *no-arbitrage models.* This is the preferred categorization for this thesis.

---

[44] See Richard [111].
[45] See Brennan & Schwartz [18].
[46] See Schaefer & Schwartz [120].
[47] See Black, Derman & Toy [8].
[48] See Longstaff & Schwartz [83] or the summary in Clewlow & Strickland [31], pages 203-205.
[49] See Fong & Vasicek [43] or the summary in Clewlow & Strickland [31], pages 201-203.
[50] For a computational implementation of the Fong-Vasicek model see Selby [119].
[51] See Leippold & Heinzl [79], page 143.
[52] See Rudolf [114], page 66, English translation by the author.

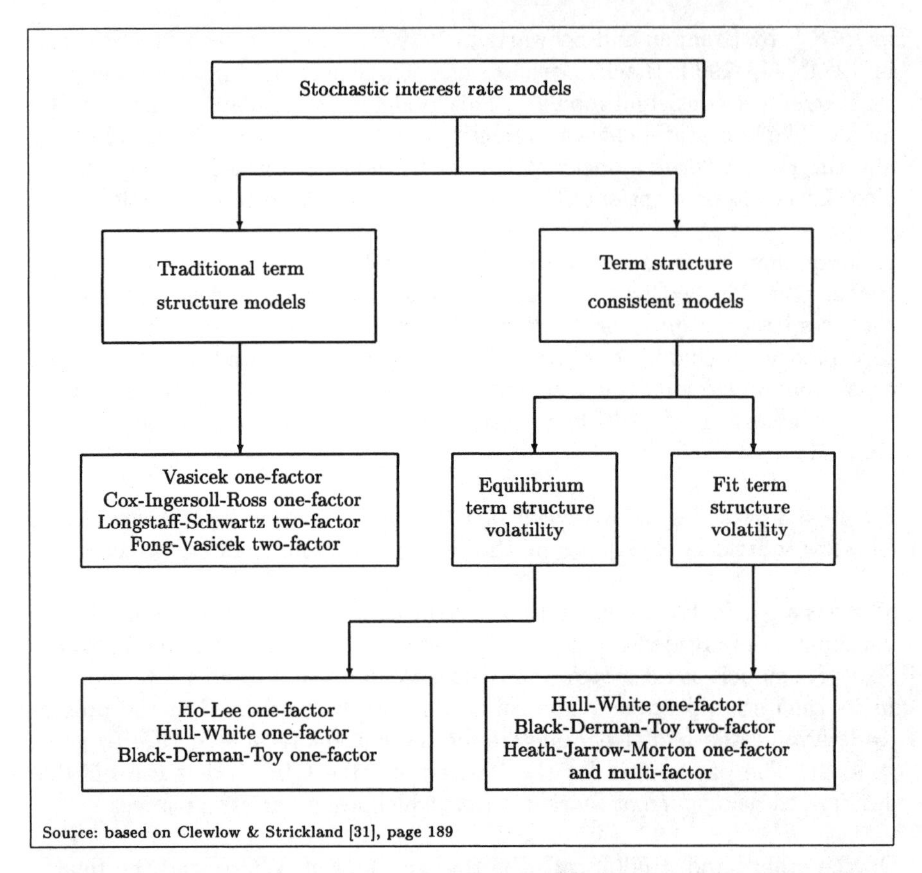

Fig. 3.3. Categorization of commonly used stochastic interest rate models.

Another categorization can be done according to the SDE for the short-rate process $(r_t)_{t \geq 0}$ of the model, which is shown in Table 3.1 for some term structure models. If the absolute changes of the interest rate are normally distributed, the model is called a *normal model*. If the absolute changes of the interest rate have a lognormal distribution, the model is called a *lognormal model*[53].

Up until now, the approach to term structure models has been dominated by mathematics and not by economics[54]. Numerical approaches to work with term structure models were scarce in the 1990s with the notable exception of two publications in 1994 by Hull and White: Hull & White [67] (*Numerical Procedures for Implementing Term Structure Models I: Single-Factor Models*) for one-factor models and Hull & White [68] (*Numerical Procedures for Imple-*

[53] For an overview of lognormal term structure models see Musiela & Rutkowski [98].

[54] See Rudolf [114], page 3.

Table 3.1. Classification of term structure models according to the short-rate SDE approach.

Model	Year published	Number of risk factors	Model type
Vasicek	1977	1	normal
Brennan-Schwartz	1979	2	lognormal
Cox-Ingersoll-Ross	1985	1	lognormal
Ho-Lee	1986	1	normal
Black-Derman-Toy	1990	1	lognormal
Heath-Jarrow-Morton two-factor	1992	2	lognormal
Hull-White one-factor	1993	1	normal
Hull-White two-factor	1994	2	normal

Source: based on Ho [56], page 17, and Rudolf [114], page 64

menting Term Structure Models II: Two-Factor Models) for two-factor models. Two very important books devoted to term structure models were published in 1996 by Rebonato (Rebonato [109], *Interest Rate Option Models*) and Jarrow (Jarrow [70], *Modelling Fixed-Income Securities and Interest Rate Options*). Rebonato explains the theory that underlies the models in great detail and offers empirical analyses and comparisons of all common models in a detail that is not to be seen anywhere else. However, the implementational aspect and numerical methods how to put the theory into practice are not a strength of this publication whereas several publications of Heath, Jarrow, and Morton (Heath, Jarrow & Morton [51], [52], and [53]) mainly focus on the implementational aspect of the different Heath-Jarrow-Morton approaches but do not deal with their interdependences.

The book *Implementing Derivatives Models* by Clewlow and Strickland[55], published in 1998, incorporates numerical procedures for term structure models but focuses more on the derivative pricing aspect and does not provide any theoretical background. The first book to combine theory and practice is Markus Rudolf's *Zinsstrukturmodelle*[56] (Interest Rate Models) published in 2000. All books require a sound understanding of modern financial engineering (i.e., of the Ito calculus) with Rebonato's book being the academically most challenging one and Clewlow & Strickland's the least demanding one. Rudolf's book deals in great detail with the Hull-White one- and two-factor models as well as the Heath-Jarrow-Morton one- and two-factor models in theory and discusses how to numerically implement them in practice.

To judge a term structure model as "good" is not simple. Ho provides a nice summary of how the choice of a term structure model depends on its desired

[55] See Clewlow & Strickland [31].
[56] See Rudolf [114].

use[57]: *What is a good model? The answer to this question has to depend on the desired use of the model. Traders often prefer a simple bond model that offers them speed and intuitive understanding of the formulation. They also need to know the values of their trades relative to the benchmarks that they continually follow. In these cases, often relative simple models are required. In managing a portfolio, or performing asset/liability management, where we need to value and manage the risk exposure, it is important that there be a consistent framework for valuing a large cross-section of bonds. In general, an arbitrage-free model is preferred because the model is calibrated with benchmark bonds that portfolio managers can observe. The choice between a one-factor and a two-factor model depends on the nature of the portfolio.*

Therefore, even though two-factor models can model a broader variety of term structure movements over time, they are not better for practical use than one-factor models. Consequently, in this thesis both one- and two-factor models are presented and used for simulations and empirical analysis.

Rebonato uses a systematical approach on how to construct a term structure model and then how to judge it based on those criteria, see Rebonato [109], pages 233-234. His criteria and approach are explained in the following:

The main features that one would like to observe in a short-rate behaviour predicted by any model, and in the implied term structure, are:

1. *the dispersion of the short rate should be consistent with one's expectations of "likely" values over a given time horizon: in particular, rates should not be allowed to become negative or to assume "implausible" large values;*
2. *very high values of rates, in historical terms, tend to be followed by a decrease in rates more frequently than by an increase; the converse is observed to be true for "unusual" low rates; a mean reversion process has therefore been suggested to account for this behaviour;*
3. *rates of different maturities are imperfectly correlated, the degree of correlation decreasing more sharply at the short end of the maturity spectrum than towards the end: ideally, the decrease in correlation with maturity implied by a model should be more pronounced in going from, say 6 to 12 months, than in going from 20 years to 20 years and 6 months;*
4. *the volatility of rates of different maturity should be different, with shorter rates usually displaying a higher volatility;*
5. *the short-rate volatility has been observed to lack homoskedasticity: in other words, the level of volatility has been observed to vary with the absolute level of the rates themselves*[58].

[57] See Ho [56], page 17.

[58] See Chan, Karolyi, Longstaff & Sanders [30]. The authors found that the best exponent γ in the expression for the volatility of the short rate σr_t^γ should be

These, of course, are only some rather crude specifications of what a "reasonable" process should look like. In setting out these "desiderata", a liberal use has been made of terms such as "reasonable", "unlikely", "usually", etc., reflecting the fact that it is very difficult, or hardly possible, to assign precise quantitative preceptions; even the requirement that interest rates should always be positive[59] *was, albeit, for a brief period of time, disobeyed in Switzerland in the 1960s.*

Since no known one- or multi-factor model manages to capture all these features at the same time, what is really important, when it comes to the choice and the practical implementation of a particular model, is not so much going through a "checklist" of met requirements but appreciating what features are essential for a given application, and which can be dispensed with. [...] *Looking back at the conditions set out before, the degree of compromise from the "ideal" model might be very considerable indeed.*

In addition to the need for compromising between the desirable features of different models when it comes to complex implementations of option pricing systems, a further word of caution should be introduced against ranking models on the basis of how many of the above criteria are met. The ultimate test of an option pricing model is its capability to determine the statistics (such as the delta and the gamma risk measure) that enables to construct a riskless portfolio, and thereby capture the option price it implies; in other terms, the hedging test is the ultimate and totally unambiguous criterion on the basis of which the superiority of one model with respect to another can be assessed.

As a consequence, only a short evaluative summary[60] is given in the following. The Vasicek and the Ho-Lee models face the problem that they can produce negative short rates, but this should not be assessed too negatively. Instead of the negativity of the short rate the total (integrated) probability of interest rates to be below zero should be considered[61]. However, a major shortcoming of the Vasicek model is that the possible shapes of the yield curve that can be

around 1.5. Such a model is, e.g., the generalized Cox-Ingersoll-Ross model that is discussed in great detail in Schulmerich [117], pages 35-43. This model allows the volatility parameter γ to be 0.5 or larger. For 0.5 the traditional CIR model emerges as a special case of the generalized CIR model.

[59] According to Rebonato [109], page 235, there are two ways to avoid negative short rates. The first is to make the short-rate volatility time decaying, the second is to integrate a mean reversion feature into the model. The first path is followed in the Ho-Lee and Black-Derman-Toy models, the second one is followed in the Vasicek and CIR models, although the Ho-Lee and the Vasicek models cannot totally exclude negative short rates.

[60] For a detailed comparison of the different models see Rebonato [109], pages 241, 242, 244, 282, 340, 351, 389, and 390.

[61] See Rebonato [109], page 241.

obtained are rather limited. The same also holds true for the CIR model that can only produce a limited number of yield curve shapes. Therefore, it cannot model a term structure that first rises (at the short end) and then decreases (at the long end)[62]. Since the Ho-Lee and the Hull-White models are term structure consistent models, they are not subject to this limitation.

Unlike with the Vasicek and the CIR models, even very complex yield curve shapes can be obtained with the Longstaff-Schwartz model, see Rebonato [109], page 340. The Hull-White two-factor model also can yield a broad range of yield curve shapes[63]. This is partly due to the fact that the short rate can become negative. However, as was already mentioned, not the short rate but the total interest rate itself should draw attention. In general, it is not necessarily an eliminative factor if the interest rate becomes negative[64]. According to Rudolf [114], page 136, the strengths of the Hull-White two-factor model warrant a stronger consideration of this model in practice and literature. To fill this gap, it is discussed in detail in this thesis.

A final aspect that has to be considered when comparing different term structure models is their ease of implementation for financial engineers. With respect to the implementation of a term structure model, Rebonato and Jarrow[65] distinguish between

- Tree structures: bushy trees and lattices
- Finite differences
- Monte Carlo simulation

Rudolf focuses on the lattice structures to model the interest rate movements and to price financial instruments[66], a method equivalent to the finite difference method[67]. Lattice structures recombine while bushy trees are non-recombining, thereby producing a major obstacle in their implementation[68].

[62] Rebonato calls this a complex term structure (see Rebonato [109], page 244) and states that the CIR model cannot obtain such a shape. However, such a shape is not uncommon in practice, see, e.g., the U.K. gilt market in 1990-1992.

[63] See Rudolf [114], page 136.

[64] In fact in the 1960s interest rates became negative in Switzerland.

[65] See Rebonato [109], part 3, and Jarrow [70], pages 202-204.

[66] See Rudolf [114].

[67] See Hull [62], page 374.

[68] In Chapter 4, the Trigeorgis log-transformed binomial tree method is presented. In its original version it contains a constant risk-free interest rate. This model will be modified to incorporate a stochastically modelled risk-free interest rate, see Section 4.4.3. As a result, the lattice structure turns into a bushy tree structure what, as will be shown in Section 5.5.1, poses major obstacles for the computational speed of the model: It is very time-consuming to run for small time steps in the tree.

Heath, Jarrow, and Morton, as well as Rudolf call the non-recombining, bushy trees just *trees* while they call recombining trees just *lattices*[69].

When implementing term structure models to price interest-rate derivatives, the problem of exponential growth in the number of nodes of the interest rate tree[70] plays an important role. The pioneer work in the area of tree complexity was done by Nelson and Ramaswamy in 1990[71]. They distinguished between computationally simple and computationally complex structures. Complex binomial trees exhibit exponential growth in the number of nodes while simple trees only exhibit linear growth. Heath shows that for an n-factor term structure model $n + 1$ branches are necessary for each node to get a solution[72]. Formally this is proven in Jarrow [70], and an intuitive insight into this relationship is given in Rudolf [114] on page 137.

Consequently, for pricing interest rate derivatives the Heath-Jarrow-Morton model poses major obstacles with respect to computer implementation due to the huge number of nodes in the tree[73]. The Ho-Lee and both Hull-White models (one- and two-factor) are computationally feasible and, therefore, the models of choice for practitioners[74]. Nelson and Ramaswamy found general conditions which are both necessary and sufficient to arrive at computationally simple trees for a term structure model. They specify the conditions in a way to ensure that the continuous-time drift and volatility of a binomial tree converge to the parameters of the corresponding diffusion process[75]. Their main research result is that the construction of a recombining binomial tree is a necessary and sufficient condition to get computational simple (and also converging) trees.

---

[69] See Heath, Jarrow & Morton [54] and Jarrow [70]. See also Rudolf [114], page 168, and Jarrow [70], page 202.

[70] See Rudolf [114], pages 166-168.

[71] See Nelson & Ramaswamy [102].

[72] See Heath [53], page 137.

[73] Computational complexity hinders the Heath-Jarrow-Morton model from being popular in practice. Carverhill and Pang modified the Heath-Jarrow-Morton model for the valuation of bond options and reduced the computational complexity, see Carverhill & Pang [29].

[74] Therefore, those models will primarily be used in Chapter 5.

[75] See Rudolf [114], page 168.

3.3 Stochastic Interest Rate Models

3.3.1 The Mean Reversion Feature in Short Rate Models

The calculus developed in Chapter 2 can now be applied to describe stochastic term structure models. The term structure models can be classified according to different criteria. One can distinguish between one-factor and multi-factor models or between equilibrium and no-arbitrage models. Finally, models can also be classified according to whether they are mean reverting or not.

Short-rate models are the core of each model for the term structure of interest rates. While the basic meaning of a short-rate model is described in Section 3.2.1, different short-rate models are introduced here and it will be explained how the term structure can be calculated from each of these short-rate models. The term *mean reversion* is central in the following, since most of the models presented here exhibit this feature. This term was already introduced mathematically in 2.4.10 and qualitatively in 3.2.1.

The mean reversion assumption can be traced to an economic assumption[76]: In phases with high interest rates, economic activity is restricted. This in return will drive the monetary policy to bring about decreasing interest rates, which will then increase economic activity and consequently interest rates. As seen in 2.4.10, mean reversion implies a long-term mean reversion level. If the current interest rates are below that level the probability for increasing interest rates is higher than for decreasing ones (and vice versa if current interest rates are above the mean reversion level).

Rudolf cites several sources of empirical evidence of mean reversion in interest rate movements[77]: Chan, Karolyi, Longstaff & Sanders [30], Leithner [80], Brown & Schaefer [23], Bühler [25], Geyer & Pichler [46] and Tobler [127]. The phenomenon of mean reversion is very common in commodity markets (see Lieskovsky, Onkey, Schulmerich, Teng & Wee [81], page 10) and in the fixed income markets for interest rate movements (see Schulmerich [117], page 10). Since a Geometric Brownian Motion does not exhibit the mean reversion feature it will not be used for interest rates[78].

However, not all models presented in Sections 3.3.2 and 3.3.3 are mean reversion models. Only the Vasicek, the Cox-Ingersoll-Ross, and the two Hull-White models are mean reverting. The Heath-Jarrow-Morton model is a general no-arbitrage model and is presented in detail in the one-factor version and the

---

[76] See Rudolf [114], page 6, and Hull & White [65].

[77] See Rudolf [114], page 6.

[78] This was first noted by Merton in 1974 who suggested an Ornstein-Uhlenbeck process instead, see Merton [94].

two-factor version. The volatility function in the Heath-Jarrow-Morton one-factor model, for example, can then be specified to yield the Ho-Lee model (not mean reverting) or the Hull-White one-factor model (mean reverting) as special cases.

3.3.2 One-Factor Term Structure Models

Vasicek model. The Vasicek model generalizes the Ornstein-Uhlenbeck process[79] by introducing a constant drift term. The short rate in the Vasicek model is given as the solution of the following SDE:

$$\left.\begin{aligned} dr_t &= \mu(r_t)dt + \sigma(r_t)dB_t,\ t \geq 0,\ r_0 \in I = \mathbb{R}, \quad \text{where} \\ \mu(x) &= \beta - \alpha x,\ x \in I,\ \alpha \in \mathbb{R}^+,\ \beta \in \mathbb{R}, \\ \sigma(x) &\equiv \sigma \in \mathbb{R}^+\ \forall x \in I. \end{aligned}\right\} \quad (3.11)$$

In the Vasicek model the short-term interest rate is the only source of uncertainty that drives the shape and development of the term structure over time. It can be shown[80]: The state space I of a process in the Vasicek model[81] is $\mathbb{R}$. This implies in particular that the short rate can become negative. The model also assumes a constant volatility, which is not true in financial markets[82].

The Vasicek model was the first term structure model proposed in Finance. It is constructed in a way that it includes mean reversion. The SDE of the model (3.11) is a one-dimensional linear SDE in the stronger sense[83]. Its solution is given according to 2.4.9 as:

$$r_t = \Phi(t)\left(r_0 + \int_0^t \Phi^{-1}(s)\beta\,ds + \int_0^t \Phi^{-1}(s)\sigma\,dB_s\right), \quad t \geq 0,$$

with $\Phi(s) = e^{-\alpha s}$, $s \geq 0$. Therefore:

$$\begin{aligned} r_t &= e^{-\alpha t}\left(r_0 + \int_0^t e^{\alpha s}\beta\,ds + \int_0^t e^{\alpha s}\sigma\,dB_s\right) \\ &= e^{-\alpha t}\left(r_0 - \frac{\beta}{\alpha}\left(1 - e^{\alpha t}\right) + \sigma\int_0^t e^{\alpha s}dB_s\right), \quad t \geq 0. \end{aligned}$$

[79] See Section 2.4.11.
[80] See, e.g., Schulmerich [117], page 32.
[81] See Rebonato [109], page 241.
[82] See, e.g., Chan, Karolyi, Longstaff & Sanders [30]. The authors show a positive correlation between the interest rate and the interest rate volatility.
[83] See Section 2.4.9.

For the expected value the Vasicek model yields:

$$E(r_t) = r_0 e^{-\alpha t} + \frac{\beta}{\alpha} - \frac{\beta}{\alpha} e^{-\alpha t} + \sigma e^{-\alpha t} E\left(\int_0^t e^{\alpha s} dB_s\right), \quad t \geq 0. \quad (3.12)$$

The expected value on the right side of (3.12) is zero according to (c) in point 2 of section 2.4.2, i.e.:

$$E(r_t) = r_0 e^{-\alpha t} + \frac{\beta}{\alpha} - \frac{\beta}{\alpha} e^{-\alpha t} \xrightarrow[t \to \infty]{} \frac{\beta}{\alpha}, \text{ since } \alpha > 0. \quad (3.13)$$

The variance $Var(r_t)$ in the Vasicek model is according to 2.4.9 (ii):

$$\begin{aligned} Var(r_t) &= Cov(r_t, r_t) \\ &= \Phi(t)\left[Var(r_0) + \int_0^t \left(\Phi^{-1}(u)\sigma(u)\right)^2 du\right]\Phi(t) \\ &= e^{-2\alpha t}\left(\sigma^2 \int_0^t e^{2\alpha u} du\right) \quad \text{since } r_0 \text{ is a constant} \\ &= \frac{\sigma^2}{2\alpha}\left(1 - e^{-2\alpha t}\right) \xrightarrow[t \to \infty]{} \frac{\sigma^2}{2\alpha} \quad \text{since } \alpha > 0. \end{aligned}$$

$\frac{\beta}{\alpha}$ is the mean reversion level of the Vasicek model. For $t \to \infty$ the stationary distribution in the Vasicek model is $N\left(\frac{\beta}{\alpha}, \frac{\sigma^2}{2\alpha}\right)$. A proof of this property can be found in Schulmerich [117], pages 139-140. For the special case $\beta = 0$ the Vasicek model turns into an Ornstein-Uhlenbeck process. All formulas above then apply with this specific choice of parameter β.

Figure 3.4 shows the simulation of a short-rate path in the Vasicek model. The path is calculated by applying the Taylor 1.5 scheme that was presented in 2.5.4. The process was simulated for the time period $[0, T]$ with N simulated realizations of the short rate in $[0, T]$. This gives a simulation step size $\Delta_s = \frac{T}{N}$. Specifically, $T = 2$ (years) and $N = 1440$ were chosen, i. e., two simulated points per day (applying a 30/360 day count method). The simulation was done using the computer simulation program that was written for this thesis.

According to the no-arbitrage pricing theory[84] the Vasicek model yields the following pricing formula for a Zero bond that matures at $t \geq 0$:

[84] See Vasicek [133] or Clewlow & Strickland [31], pages 194-195.

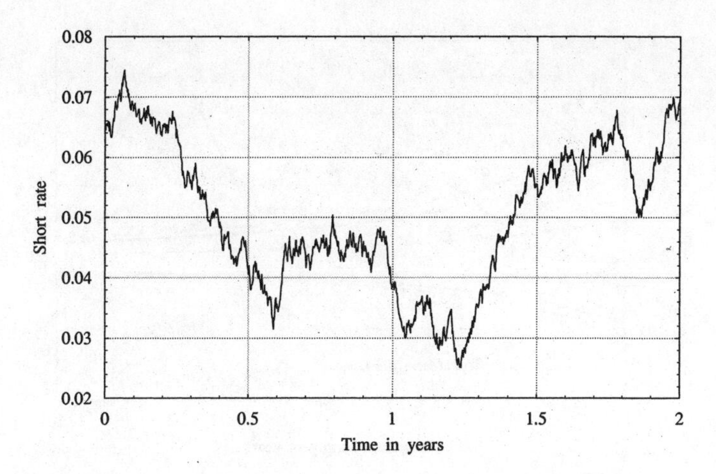

Fig. 3.4. Example of a short-rate path in the Vasicek model ($\alpha = 0.6$, $\beta = 0.03$, $\sigma = 0.02$, $r_0 = 0.065$, $T = 2$, $N = 1440$).

$$P(t,T) \quad = \quad \hat{E}_t \left[\exp\left(- \int_t^T r_\tau \, d\tau \right) \right], \quad 0 \leq t \leq T. \tag{3.14}$$

$\hat{E}_t$ is the risk-adjusted expectational value taking into account the information available at time t. The solution of (3.14) is:

$$P(t,T) = A(t,T)e^{-r_t B(t,T)} \quad \text{with}$$

$$R(t,T) = -\frac{\ln A(t,T)}{T-t} + \frac{B(t,T)}{T-t} r_t \,, \quad \text{whereby:}$$

$$A(t,T) = \exp\left\{ \frac{R_\infty}{\alpha} \left(1 - e^{-\alpha(T-t)}\right) - (T-t)R_\infty - \frac{\sigma^2}{4\alpha^3} \left(1 - e^{-\alpha(T-t)}\right)^2 \right\},$$

$$B(t,T) = \frac{1}{\alpha} \left(1 - e^{-\alpha(T-t)}\right),$$

$$R_\infty = \lim_{\tau \to \infty} R(t,\tau) \quad = \quad \frac{\beta}{\alpha} - \frac{\sigma^2}{2\alpha^2}.$$

The volatility term structure is given by:

$$\sigma_R(t,T) \quad = \quad \frac{\sigma}{\alpha(T-t)} \left(1 - e^{-\alpha(T-t)}\right). \tag{3.15}$$

Figure 3.5 shows the development of the term structure of interest rates in the Vasicek model that corresponds to the short-rate process in Figure 3.4 and Figure 3.6 the corresponding Zero bond prices for nominal 1 $.

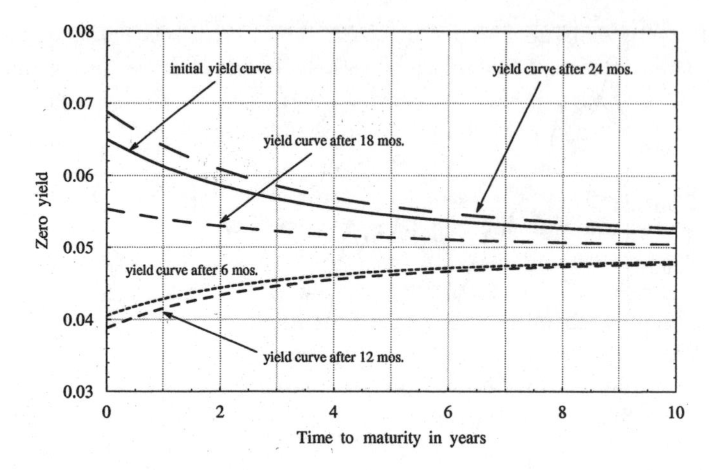

Fig. 3.5. Example of the development of the term structure of interest rates over two years in the Vasicek model, corresponding to Figure 3.4.

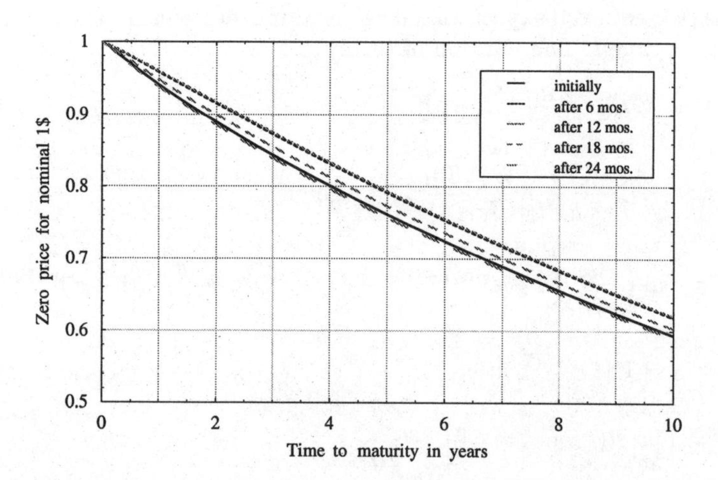

Fig. 3.6. Example of the development of the Zero prices for a nominal value of 1 \$ over two years in the Vasicek model, corresponding to Figure 3.4.

In Figure 3.7 the corresponding volatility term structure development over time is displayed. The volatility term structure is stable through time[85].

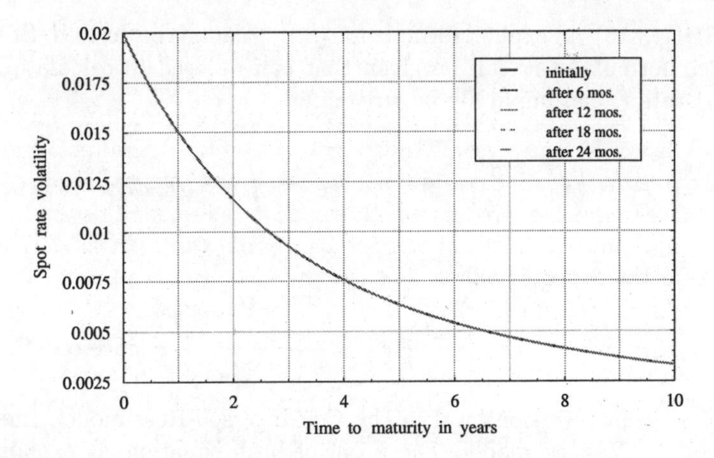

Fig. 3.7. Example of the development of the term structure of interest rate volatilities over two years in the Vasicek model, corresponding to Figure 3.4: The graphs are identical.

Cox-Ingersoll-Ross model. The short-rate process in the Cox-Ingersoll-Ross model is described via the SDE

$$\left.\begin{aligned} dr_t &= \mu(r_t)dt + \sigma(r_t)dB_t, \;\; t \geq 0, \;\; r_0 \in I = \mathbb{R}^+, \quad \text{where} \\ \mu(x) &= \beta - \alpha\, x, \;\; x \in I, \;\; \alpha \in \mathbb{R}^+, \;\; \beta \in \mathbb{R}^+, \\ \sigma(x) &= \sigma\,\sqrt{|x|} \;\; \forall x \in I, \;\; \sigma \in \mathbb{R}^+ \;\text{ with }\; 2\beta \geq \sigma^2. \end{aligned}\right\} \quad (3.16)$$

The restriction on $\sigma, \beta \in \mathbb{R}^+$ with $2\beta \geq \sigma^2$ is necessary to arrive at a stationary diffusion process[86]. As in the Vasicek model, the short-term interest rate in the CIR model is the only source of uncertainty that drives the shape and development of the term structure. It can be shown[87]: The process $(r_t)_{t\geq 0}$ has the state space[88] $I = \mathbb{R}^+$. This yields $\sigma(x) = \sigma\,\sqrt{|x|} = \sigma\,\sqrt{x}$ for the volatility.

[85] The phenomenon of a stable volatility term structure through time can also be observed in the Ho-Lee, the Hull-White one-factor, and the Hull-White two-factor models. This phenomenon is also mentioned by Clewlow and Strickland for single-factor Markovian short-rate models where the volatility structure is described by constant parameters, see Clewlow & Strickland [31], page 223.

[86] See Borkovec & Klüppelberg [12], page 10.

[87] See Schulmerich [117], page 34.

[88] See Rebonato [109], page 241.

Like the Vasicek model, the CIR model is mean reverting. For $t \to \infty$ the stationary distribution in the CIR model is the gamma distribution $\Gamma\left(\frac{2\alpha}{\sigma^2}, \frac{2\beta}{\sigma^2}\right)$. The proof can be found in Schulmerich [117], pages 140-141.

For the SDE (3.16), an explicit formula for $(r_t)_{t\geq 0}$ cannot be derived. However, an implicit formula of r_t can be given that can be used to calculate $E(r_t)$. The short rate r_t can implicitly be written as

$$r_t \;=\; \frac{\beta}{\alpha} + \left(r_0 - \frac{\beta}{\alpha}\right) e^{-\alpha t} \;+\; \sigma\, e^{-\alpha t} \int_0^t e^{\alpha s} \sqrt{r_s}\, dB_s, \quad t \geq 0,$$

and gives for the expected value:

$$E(r_t) \;=\; \frac{\beta}{\alpha} \;+\; \left(r_0 - \frac{\beta}{\alpha}\right) e^{-\alpha t} \;\xrightarrow[t \to \infty]{}\; \frac{\beta}{\alpha} \quad \text{since } \alpha > 0.$$

This is the mean reversion level in the Cox-Ingersoll-Ross model, the same level as in the Vasicek model. The proof of both equations is explained in Schulmerich [117], pages 34-40. Figure 3.8 shows a realization of a short-rate process in the CIR model.

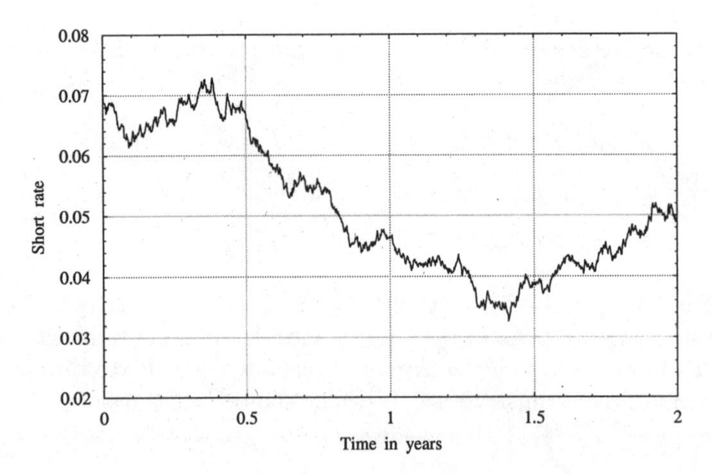

Fig. 3.8. Example of a short-rate path in the CIR model ($\alpha = 0.3$, $\beta = 0.015$, $\sigma = 0.07$, $r_0 = 0.055$, $T = 2$, $N = 1440$).

According to Clewlow & Strickland [31], page 198, the following formulas hold for the price of a Zero that matures at $T \geq 0$:

$$P(t,T) = A(t,T)e^{-r_t B(t,T)}, \quad 0 \leq t \leq T, \text{ with}$$

$$R(t,T) = -\frac{\ln A(t,T)}{T-t} + \frac{B(t,T)}{T-t} r_t, \text{ whereby:}$$

$$A(t,T) = \left(\frac{\Phi_1 \, e^{\Phi_2 (T-t)}}{\Phi_2 (e^{\Phi_1 (T-t)} - 1) + \Phi_1} \right)^{\Phi_3},$$

$$B(t,T) = \left(\frac{e^{\Phi_1 (T-t)} - 1}{\Phi_2 (e^{\Phi_1 (T-t)} - 1) + \Phi_1} \right).$$

Thereby:

$$\Phi_1 = \sqrt{\alpha^2 + 2\sigma^2}, \qquad \Phi_2 = \frac{\alpha + \Phi_1}{2}, \qquad \Phi_3 = \frac{2\beta}{\sigma^2}.$$

The volatility structure in the CIR model is given by:

$$\sigma_R(t,T) = \frac{\sigma\sqrt{r_t}}{T-t} B(t,T).$$

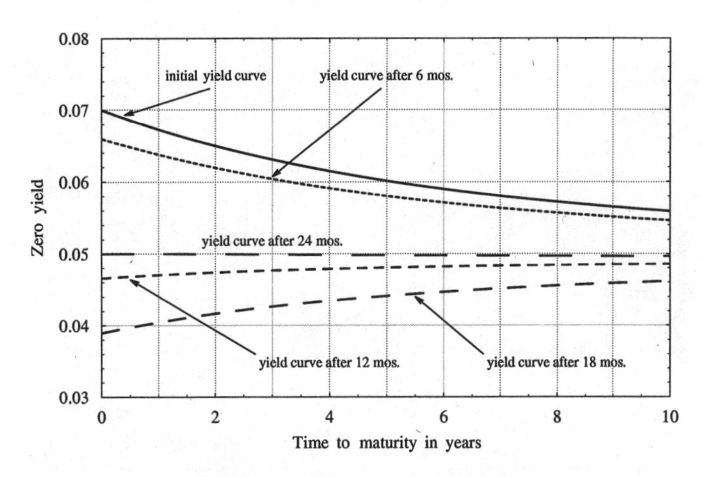

Fig. 3.9. Example of the development of the term structure of interest rates over two years in the Cox-Ingersoll-Ross model, corresponding to Figure 3.8.

The derivation of the CIR model can be found in several articles (see the original article of Cox, Ingersoll & Ross [34] or see Rebonato [109], Section 11.2, or Rudolf [114], Section 5.3). Figure 3.9 shows an example of the development of the term structure of interest rates in the CIR model that corresponds to Figure 3.8. The corresponding Zero prices for nominal 1 $ are given in Figure

3.10 and the corresponding volatility term structure[89] is displayed in Figure 3.11.

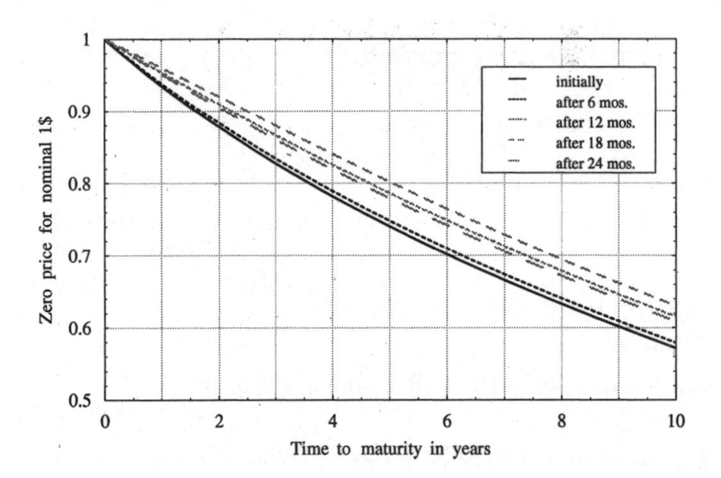

Fig. 3.10. Example of the development of the Zero prices for a nominal value of 1 $ over two years in the Cox-Ingersoll-Ross model, corresponding to Figure 3.8.

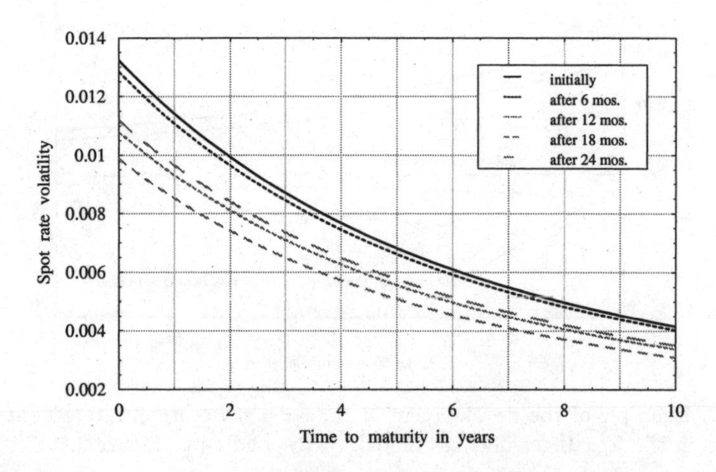

Fig. 3.11. Example of the development of the term structure of interest rate volatilities over two years in the Cox-Ingersoll-Ross model, corresponding to Figure 3.8.

[89] The volatility term structure in the Cox-Ingersoll-Ross model is not stable through time since the short-rate volatility of this model depends on the level of the short rate.

Ho-Lee model. The Ho-Lee model, published in 1986, was the first no-arbitrage model introduced in the literature[90]. Ho and Lee were the first authors who developed a model consistent with the initial term structure of interest rates. This is particularly important since practical acceptance of such a yield curve model is much higher than that of term structure models that do not reproduce the current term structure. The drawback of this model is that it is not mean reverting and interest rates can become negative[91]. The process of the Ho-Lee model is described via the SDE

$$\left.\begin{aligned} dr_t &= \theta(t)dt + \sigma dB_t, \ t \geq 0, \quad \text{where} \\ \theta(t) &= \frac{\partial f(0,t)}{\partial t} + \sigma^2 t \quad \forall t \geq 0, \ \sigma \in \mathbb{R}^+. \end{aligned}\right\} \tag{3.17}$$

$\theta(\cdot)$ represents a time-dependent drift that can be seen as an approximation[92] of the slope of the initial instantaneous forward rate curve f (as defined in Section 3.2.1). According to Clewlow & Strickland [31], page 209, the following formulas hold for the price of a Zero bond that matures at time T:

$$P(t,T) = A(t,T)e^{-r_t B(t,T)}, \quad 0 \leq t \leq T, \quad \text{with:} \tag{3.18}$$

$$A(t,T) = \exp\left\{\ln\frac{P(0,T)}{P(0,t)} - B(t,T)\frac{\partial \ln P(0,t)}{\partial t} - \frac{1}{2}\sigma^2\, t\, B(t,T)^2\right\}, \tag{3.19}$$

$$B(t,T) = T - t. \tag{3.20}$$

Given the price of the Zero bond, the yield that determines the yield curve can be calculated using equation (3.2). The volatility term structure is given by:

$$\sigma_R(t,T) = \sigma.$$

When simulating a path of the short-rate process, it has to be determined what the start value r_0 of this process ought to be. For the Ho-Lee model r_0 can be chosen such that the yield curve produced by this model for $t = 0$ coincides exactly with the observed yield curve at $t = 0$ in the capital market:

According to (3.18), it holds for the current (i.e., $t = 0$) Zero price:

$$P(0,T) = A(0,T)\, e^{-r_t B(0,T)}, \quad T \geq 0, \tag{3.21}$$

with

[90] See Clewlow & Strickland [31], page 208.
[91] See Clewlow & Strickland [31], page 209, or Rebonato [109], page 241.
[92] See Hull & White [67], page 8.

$$A(0,T) \overset{(3.19)}{=} \exp\left\{\ln \frac{P(0,T)}{P(0,0)} + B(0,T)\frac{\partial \ln P(0,t)}{\partial t}\right\}$$
$$\overset{(3.3)}{=} \exp\left\{\ln \frac{P(0,T)}{P(0,0)} + B(0,T)f(0,0)\right\}$$
$$\overset{(3.1)}{=} \exp\left\{\ln P(0,T) + B(0,T)f(0,0)\right\}$$
$$\overset{(3.1)}{=} \exp\left\{-R(0,T)\,T + B(0,T)f(0,0)\right\}, \tag{3.22}$$

$$B(0,T) \overset{(3.20)}{=} T. \tag{3.23}$$

Putting (3.22) and (3.23) into (3.21) yields for all $T \geq 0$:

$$P(0,T) = \exp\{-R(0,T)\,T + T\,(r_0 - f(0,0))\} \overset{(3.1)}{=} \exp\{-R(0,T)\,T\}.$$

This equation has to hold for all $T \geq 0$ which can only be achieved if $r_0 = f(0,0)$. Therefore, when simulating a path of the short-rate process in the Ho-Lee model with one of the simulation methods in Section 2.5, $f(0,0)$ has to be chosen as the start value r_0 to arrive at a term structure consistent model[93].

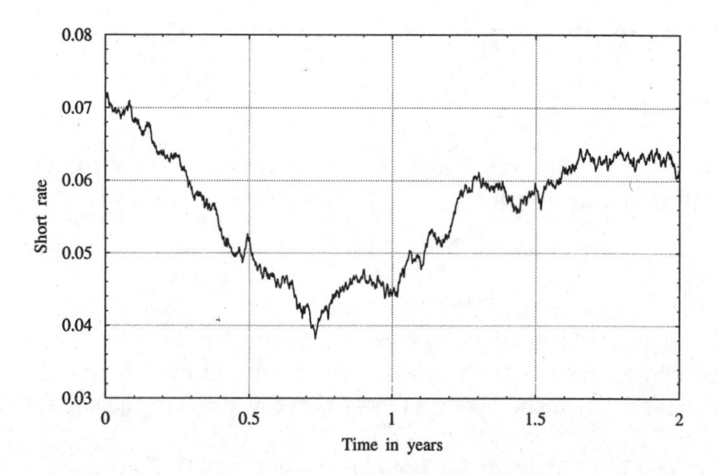

Fig. 3.12. Example of a short-rate path in the Ho-Lee model ($\sigma = 0.01$, $T = 2$, $N = 1440$, U.S. Zero yield curve from December 1, 2000 as input parameter).

[93] According to the definition of the short rate r_t and the instantaneous forward rate $f(t,T), 0 \leq t \leq T$, it follows that $r_t = f(t,t)$ holds for all $t \geq 0$. Especially, $r_0 = f(0,0)$. This relationship is not introduced in this thesis since it is not needed. Therefore, the start value r_0 had to be calculated from the formulas of the model. This will also be done for the start values in the Hull-White one-factor and two-factor models.

Figure 3.12 shows a simulated short-rate path in the Ho-Lee model. The term structure of interest rates that was used as an input parameter for the model was the U.S. Zero yield curve (spot rate) from December 1, 2000[94].

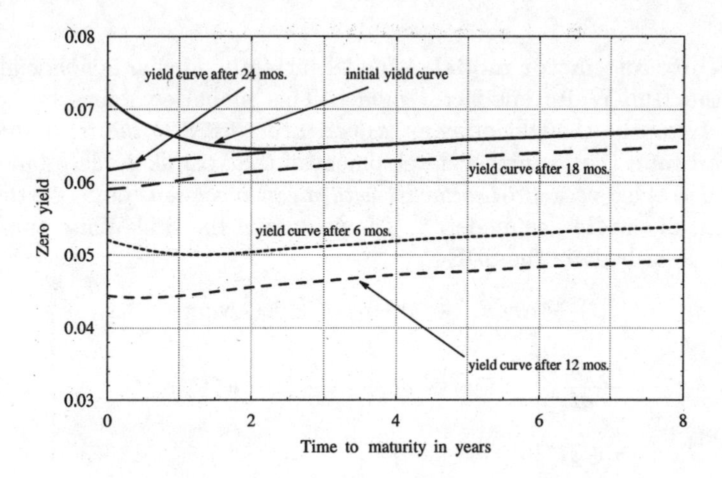

Fig. 3.13. Example of the development of the term structure of interest rates over two years in the Ho-Lee model, corresponding to Figure 3.12.

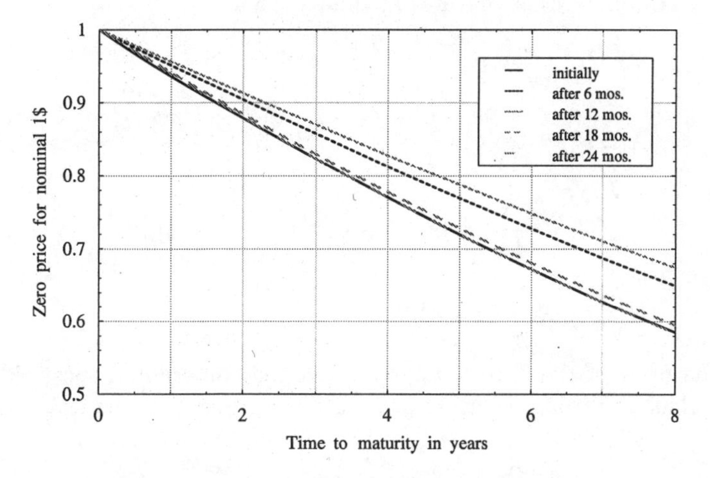

Fig. 3.14. Example of the development of the Zero prices for a nominal value of 1 $ over two years in the Ho-Lee model, corresponding to Figure 3.12.

[94] The given data for this day were the U.S. Zero yields with maturities 1 mos., 3 mos., 6 mos., 9 mos., 12 mos., 2 yrs., 3 yrs., 4 yrs., 5 yrs., 6 yrs., 7 yrs., 8 yrs., 9 yrs., and 10 yrs.

Figure 3.13 and Figure 3.14 show the corresponding development of the term structure of interest rates and the Zero prices for nominal 1 $, respectively. The corresponding spot rate volatility in the Ho-Lee model, which is the constant σ for each choice of T and t with $0 \leq t \leq T$, will not be displayed in a diagram.

Hull-White one-factor model. A model currently popular in financial practice is the Hull-White one-factor model. This model combines two earlier models. It can be thought of as a *Vasicek model fitted to the term structure of interest rates* (i.e., a further development of the Vasicek model), but it can also be perceived as a *Ho-Lee model with mean reversion* (i.e., a further development of the Ho-Lee model)[95]. The process of the Hull-White one-factor model is described via the SDE

$$\left.\begin{aligned} dr_t &= [\theta(t) - \alpha r_t]dt + \sigma dB_t, \quad t \geq 0, \quad \text{where} \\ \theta(t) &= \frac{\partial f(0,t)}{\partial t} + \alpha f(0,t) + \frac{\sigma^2}{2\alpha}\left(1 - e^{-2\alpha t}\right) \; \forall t \geq 0, \\ & \sigma \in \mathbb{R}^+, \alpha \in \mathbb{R} \setminus \{0\}. \end{aligned}\right\} \tag{3.24}$$

This model has a time-dependent mean reversion level and it can produce negative interest rates[96]. It is also important to notice that parameter α can become negative in practice, see Clewlow and Strickland [31], page 220, Figure 7.5. In the Hull-White one-factor model the following formulas hold[97] for the price of a Zero bond that matures at time T:

$$P(t,T) = A(t,T)e^{-r_t B(t,T)}, \quad 0 \leq t \leq T, \text{ with:} \tag{3.25}$$

$$A(t,T) = \exp\left\{\ln \frac{P(0,T)}{P(0,t)} - B(t,T)\frac{\partial \ln P(0,t)}{\partial t} - \frac{1}{4\alpha^3}\sigma^2 \left(e^{-\alpha T} - e^{-\alpha t}\right)^2 \left(e^{2\alpha t} - 1\right)\right\}, \tag{3.26}$$

$$B(t,T) = \frac{1}{\alpha}\left(1 - e^{-\alpha(T-t)}\right). \tag{3.27}$$

Given the price of the Zero bond, the yield that determines the yield curve can be calculated using equation (3.2). The volatility term structure is given by:

$$\sigma_R(t,T) = \frac{\sigma}{\alpha(T-t)}\left(1 - e^{-\alpha(T-t)}\right).$$

Figure 3.15 shows a simulated short-rate path of a process in the Hull-White one-factor model.

---

[95] See Hull-White [67], page 9, and Clewlow & Strickland [31], page 215.
[96] See Rebonato [109], page 282.
[97] See Clewlow & Strickland [31], page 215.

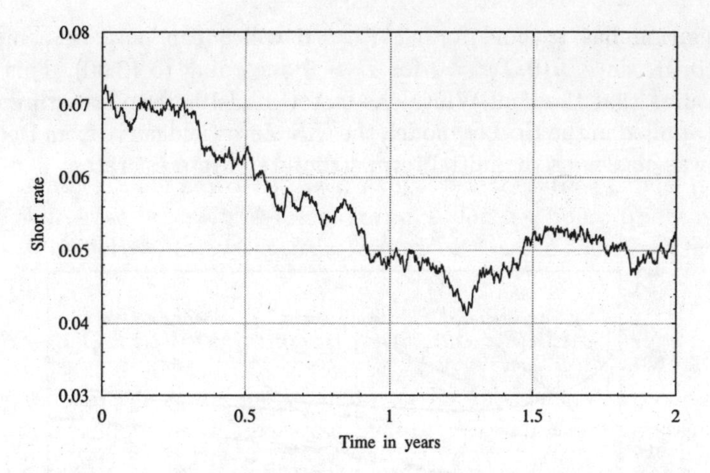

Fig. 3.15. Example of a short-rate path in the Hull-White one-factor model ($\alpha = 0.1$, $\sigma = 0.01$, $T = 2$, $N = 1440$, U.S. Zero yield curve from December 1, 2000 as input parameter).

As in the Ho-Lee model, the start value r_0 has to be chosen as $f(0,0)$, such that the term structure calculated for $t = 0$ is the initial term structure which is used as an input parameter. This derivation, which is similar to the derivation in the Ho-Lee model, will be shown in the following:

According to (3.25), it holds for the current (i.e., $t = 0$) Zero price:

$$P(0,T) \;=\; A(0,T)\, e^{-r_t B(0,T)}, \;\; T \geq 0, \tag{3.28}$$

with

$$\begin{aligned} A(0,T) &\overset{(3.26)}{=} \exp\left\{ \ln \frac{P(0,T)}{P(0,0)} + B(0,T) \frac{\partial \ln P(0,t)}{\partial t} \right\} \\ &\overset{(3.3)}{=} \exp\left\{ \ln \frac{P(0,T)}{P(0,0)} + B(0,T) f(0,0) \right\} \\ &\overset{(3.1)}{=} \exp\left\{ \ln P(0,T) + B(0,T) f(0,0) \right\} \\ &\overset{(3.1)}{=} \exp\left\{ -R(0,T)\,T + B(0,T) f(0,0) \right\}, \end{aligned} \tag{3.29}$$

$$B(0,T) \overset{(3.27)}{=} \frac{1}{\alpha} \left(1 - e^{-\alpha T}\right). \tag{3.30}$$

Putting (3.29) and (3.30) into (3.28) yields for all $T \geq 0$:

$$P(0,T) = \exp\{ -R(0,T)\,T + B(0,T)\,(r_0 - f(0,0)) \} \overset{(3.1)}{=} \exp\{-R(0,T)\,T\}.$$

This equation has to hold for all $T \geq 0$ which can only be achieved if $r_0 = f(0,0)$, since $B(0,T) \neq 0$ for $T > 0$ according to (3.30). This choice of r_0 ensures that the Hull-White one-factor model is term structure consistent. As applied in the Ho-Lee model, the U.S. Zero yield curve from December 1, 2000 was chosen as the initial term structure of interest rates.

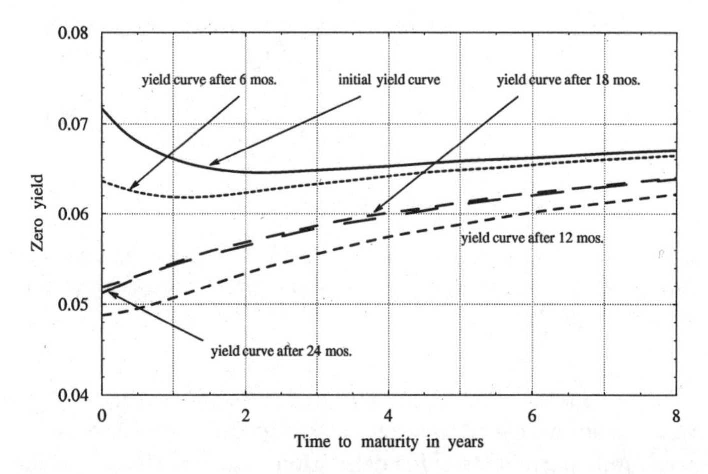

Fig. 3.16. Example of the development of the term structure of interest rates over two years in the Hull-White one-factor model, corresponding to Figure 3.15.

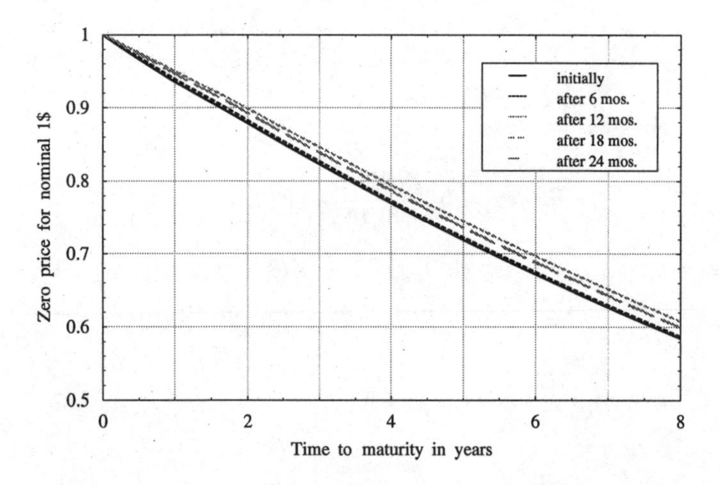

Fig. 3.17. Example of the development of the Zero prices for a nominal value of 1 $ over two years in the Hull-White one-factor model, corresponding to Figure 3.15.

Figure 3.16 shows the development of the term structure of interest rates in the Hull-White one-factor model that corresponds to the short-rate path in Figure 3.15. The corresponding Zero prices for nominal 1 \$ and the corresponding term structure of volatilities are given in Figure 3.17 and Figure 3.18, respectively.

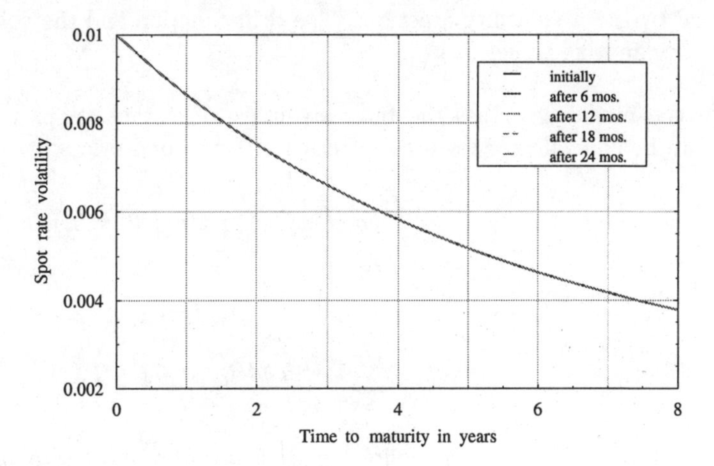

Fig. 3.18. Example of the development of the term structure of interest rate volatilities over two years in the Hull-White one-factor model, corresponding to Figure 3.15: The graphs are identical.

Heath-Jarrow-Morton one-factor model. The Heath-Jarrow-Morton model is different from all other models presented so far, in that it models the instantaneous forward rate f and not the short rate r itself. However, process $(r_t)_{t\geq 0}$ can then be derived from the instantaneous forward-rate model. The following presentation follows Clewlow and Strickland[98] as well as Rudolf[99] who all base their description on the original article of Heath, Jarrow, and Morton[100], published in 1992. The Heath-Jarrow-Morton model in its most general form is a multi-factor model for the instantaneous forward rate f. It extends the Ho-Lee model by describing the instantaneous forward rate process with $n \in \mathbb{N}$ factors via the SDE

$$\left.\begin{aligned} df(t,T) &= \alpha(t,T)dt + \sum_{i=1}^{n} \sigma_i(t,T,f(t,T))dB_t^i, \quad 0 \leq t \leq T, \quad \text{where} \\ \alpha(t,T) &= \sum_{i=1}^{n} \left\{ \sigma_i(t,T,f(t,T)) \left[\int_t^T \sigma_i(t,u,f(t,u))du \right] \right\}. \end{aligned}\right\} \qquad (3.31)$$

[98] See Clewlow & Strickland [31], pages 229-232.
[99] See Rudolf [114], pages 95-115.
[100] See Heath, Jarrow & Morton [53].

$T \geq 0$ is the maturity time point of a Zero bond and $(B_t^i)_{t\geq 0}$, $i = 1, \dots, n$, are n independent Standard Brownian Motions which are used to model n independent risk factors that influence the term structure. The volatility structure is given by n volatility functions $\sigma_i, i = 1, \dots, n$. The drift rate $\alpha(\cdot, \cdot)$ is determined by the no-arbitrage condition and depends, according to equation (3.31), on the volatility functions. Therefore, the whole process is completely determined by the n volatility functions. The drift function and the volatility functions[101] can take values in $\mathbb{R}$.

Although the model describes the term structure of instantaneous forward rates, it can be equivalently restated in terms of Zero bond prices[102]. With

$$\nu_i(t,T) \quad := \quad -\int_t^T \sigma_i(t,u)du, \;\; 0 \leq t \leq T, \tag{3.32}$$

the Zero bond price equation is:

$$\frac{dP(t,T)}{P(t,T)} \quad = \quad r_t dt + \sum_{i=1}^{n} \nu_i(t,T)dB_t^i, \;\; 0 \leq t \leq T. \tag{3.33}$$

Moreover, (3.31) implies[103] the following SDE for the short-rate process $(r_t)_{t\geq 0}$:

$$dr_t = \left[\frac{\partial f(0,t)}{\partial t} + \sum_{i=1}^{n}\left\{\int_0^t \nu_i(u,t)\frac{\partial^2 \nu_i(u,t)}{\partial t^2} + \frac{\partial \nu_i(u,t)^2}{\partial t} du\right.\right.$$

$$\left.\left. + \int_0^t \frac{\partial^2 \nu_i(u,t)}{\partial t^2} dB_u^i\right\}\right] dt + \sum_{i=1}^{n} \left.\frac{\partial \nu_i(u,t)}{\partial t}\right|_{(u=t)} dB_u^i, \;\; t \geq 0. \tag{3.34}$$

The Heath-Jarrow-Morton model is very general, allowing the volatility functions to depend on the entire forward rate curve[104]. The Ho-Lee model as well as the Hull-White one-factor model are special cases of the Heath-Jarrow-Morton model[105].

---

[101] The mathematical properties of the functions α and $\sigma_i, i = 1, \dots, n$, can exactly be specified, see Heath, Jarrow & Morton [53], pages 79-81. However, this will not be done here in order not to distract from the applicational aspect of this chapter.

[102] See Carverhill & Pang [29] or Clewlow & Strickland [31], page 231.

[103] See Clewlow & Strickland [31], page 231, equation (7.33).

[104] See Clewlow & Strickland [31], page 290.

[105] See Clewlow & Strickland [31], page 291, equations (10.5) and (10.6). For a detailed derivation see Rudolf [114], pages 57-63.

The model presented above is the most general Heath-Jarrow-Morton model available[106]. This thesis applies the one-factor and the two-factor specifiations and their numerical solution as presented in Rudolf [114], pages 95-115. In the following, n is specified as 1 which gives the Heath-Jarrow-Morton one-factor model. Since both calculations follow a similar algorithm, the introduction of this one-factor model serves as a preparation for the introduction of the Heath-Jarrow-Morton two-factor model in 3.3.3. However, this algorithm is easier to understand for a one-factor model than for a two-factor model. These two Heath-Jarrow-Morton models will not be used in Chapter 5 for empirical analysis but are implemented in the computer simulation program that was developed for this thesis.

According to Rudolf [114], Section 2.2, a binomial tree can be derived for the Heath-Jarrow-Morton one-factor model that gives the Zero bond price and volatilities by solving the tree backwards. To obtain this tree structure, time interval $[0, T]$ is divided into N equidistant time periods of lengths $\Delta := \frac{T}{N}$. Within each sub period of length Δ the difference $f(t, T) - f(t - \Delta, T)$ of the instantaneous forward rate is modelled via a binomial tree, see Figure 3.19.

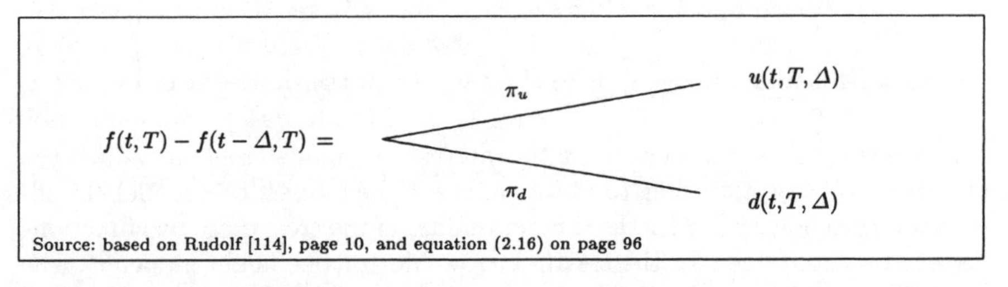

Fig. 3.19. Binomial branch in the Heath-Jarrow-Morton one-factor model.

At the end of this time period, this difference can take the value $u(t, T, \Delta)$ in the up-case with probability π_u, and the value $d(t, T, \Delta)$ in the down-case with probability π_d. π_u and π_d are the martingale probabilities in the up-case and down-case[107], respectively. The functions u and d that depend on time $t \geq 0$, maturity $T > 0$, and Δ are called *perturbation functions*. However, the functions e^u and e^d also play an important role in the derivation of the

[106] See Clewlow & Strickland [31], page 230.

[107] A detailed introduction of martingale probabilities is, e.g., given in Rudolf [114], pages 8-11, and pages 83-88. A martingale probability is the risk-adjusted probability that is used in a pricing process. Thus, the risk-adjusted probability p of the Cox-Ross-Rubinstein method in Section 2.6.2 is the martingale probability of the underlying random walk process. However, the mathematical details of a martingale probability will not be explained here since the focus is on the applicational aspect of the theory and not on the mathematical aspects thereof.

Heath-Jarrow-Morton model and are sometimes also referred to as perturbation functions.

The meaning of the perturbation functions can best be explained by comparing the stochastic case with the deterministic case[108]. In the deterministic case with a deterministic short rate r_t at time t, the following relationships hold for the price of a Zero bond:

$$P(t,T) = P(t+\Delta,T)\, e^{-r_t\,\Delta}, \tag{3.35}$$

$$P(t,t+\Delta) = e^{-r_t\,\Delta}. \tag{3.36}$$

(3.35) and (3.36) give:

$$P(t+\Delta,T) \quad = \quad \frac{P(t,T)}{P(t,t+\Delta)}. \tag{3.37}$$

If, however, the term structure of interest rates is stochastically modelled, the deterministic equation (3.37) has to be modified to account for the stochastic feature. This is done via the two following perturbation functions:

$$\left.\begin{aligned} P_{up}(t+\Delta,T) &= \frac{P(t,T)}{P(t,t+\Delta)}\, e^{u(t,T,\Delta)} && \text{in the up-state,} \\ P_{down}(t+\Delta,T) &= \frac{P(t,T)}{P(t,t+\Delta)}\, e^{d(t,T,\Delta)} && \text{in the down-state.} \end{aligned}\right\} \tag{3.38}$$

The functions e^u and e^d depend on the interest rate model used and adjust the deterministic case according to the structure of the interest rate model. To obtain the Zero bond price for the different nodes of the tree, these two functions have to be determined for the Heath-Jarrow-Morton one-factor model[109]. The perturbation functions are:

$$\left.\begin{aligned} u(t,T,\Delta) &= \ln\left(\frac{2}{1+e^{\sigma(t,T,f)2\sqrt{\Delta}}}\right) && \text{in the up-state,} \\ d(t,T,\Delta) &= \ln\left(\frac{2}{1+e^{-\sigma(t,T,f)2\sqrt{\Delta}}}\right) && \text{in the down-state.} \end{aligned}\right\} \tag{3.39}$$

Putting (3.39) and (3.36) into (3.38) yields:

$$P(t+\Delta,T) \;=\; \left\{\begin{aligned} &\frac{2e^{r_t\Delta}\, P(t,T)}{1+e^{\sigma(t,T,f)2\sqrt{\Delta}}} && \text{in the up-state,} \\ &\frac{2e^{r_t\Delta}\, P(t,T)}{1+e^{-\sigma(t,T,f)2\sqrt{\Delta}}} && \text{in the down-state.} \end{aligned}\right\} \tag{3.40}$$

[108] See Rudolf [114], pages 80-81.
[109] See Rudolf [114], pages 104-105.

Heath, Jarrow, and Morton propose two choices for the volatility function[110], whereby $\sigma \in \mathbb{R}^+$ and $\lambda \in \mathbb{R}^-$:

$$(i) \qquad \sigma(t,T,f) = \sigma(t,T) := r_t\,(T-t-\Delta)\,\sigma, \quad 0 \le t \le T,$$

$$(ii) \qquad \sigma(t,T,f) = \sigma(t,T) := \sigma e^{-\lambda(T-t-\Delta)}, \quad 0 \le t \le T.$$

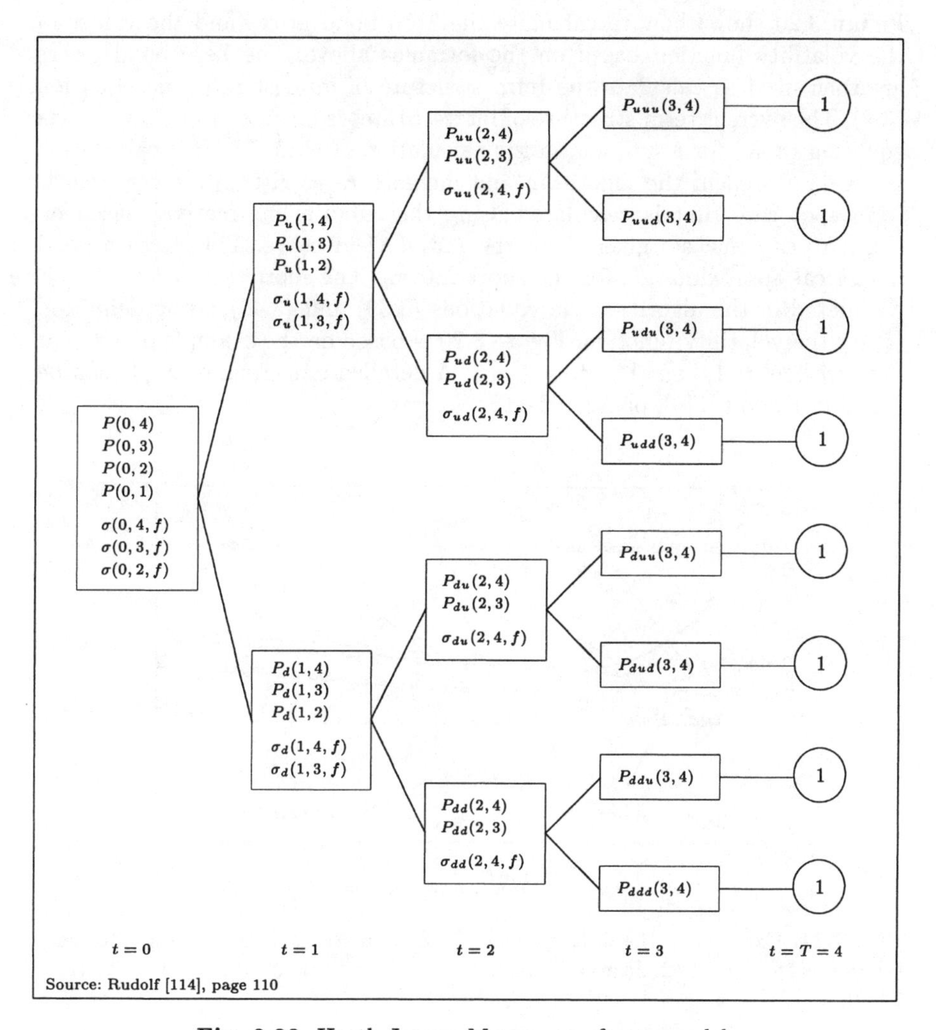

Fig. 3.20. Heath-Jarrow-Morton one-factor model.

[110] See Heath, Jarrow & Morton [52] or Rudolf [114], pages 108-109.

Since (*ii*) needs an additional parameter λ and since (*i*) is also the choice of Rudolf[111], the variation (*i*) will be used in this thesis and is implemented in the computer simulation program. To obtain the term structure of interest rates, equation (3.2) has to be applied:

$$R(t,T) \quad = \quad -\frac{1}{T-t}\ln P(t,T), \quad 0 \leq t \leq T.$$

Figure 3.20 shows how to calculate the Zero bond prices and the values of the volatility function based on the formulas above. The Zero bond prices are then used to calculate the term structure of interest rates via equation (3.2). The current term structure of interest rates is included in the model via equation (3.40) for $t = 0$, i.e., in the calculation of $P(\Delta, T)$. This calculation is the first step in the algorithm and includes r_0 as $R(0, \Delta)$, a very short-term spot rate that is calculated using the cubic spline method based on the maturity buckets given from the initial yield curve. This gives a good numerical approximation for the short rate r_0. The short rate r_t for $t > 0$ is calculated in the algorithm via equations (3.39) and (3.40), using definition (*i*) for the volatility function. Figure 3.20 is based on the example of a 4-year Zero with $N = 4$, i.e., $\Delta = \frac{T}{N} = \frac{4}{4} = 1$. A detailed numerical example can be found in Rudolf [114], pages 108-112.

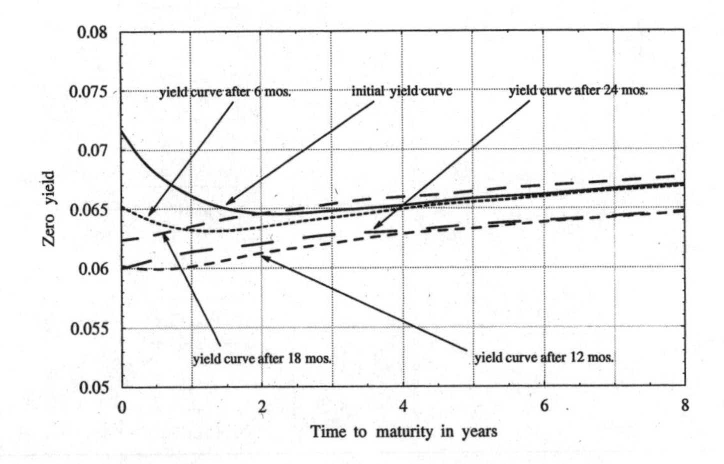

Fig. 3.21. Example of the development of the term structure of interest rates over two years in the Heath-Jarrow-Morton one-factor model ($\sigma = 0.05$, $T = 2$, $N = 1440$).

Examples of the term structure of interest rates and the corresponding Zero prices for nominal 1 $ are displayed in Figure 3.21 and Figure 3.22, respec-

[111] See Rudolf [114], pages 108-109.

tively. The U.S. Zero yields on December 1, 2000 are used as the initial values to start the algorithm.

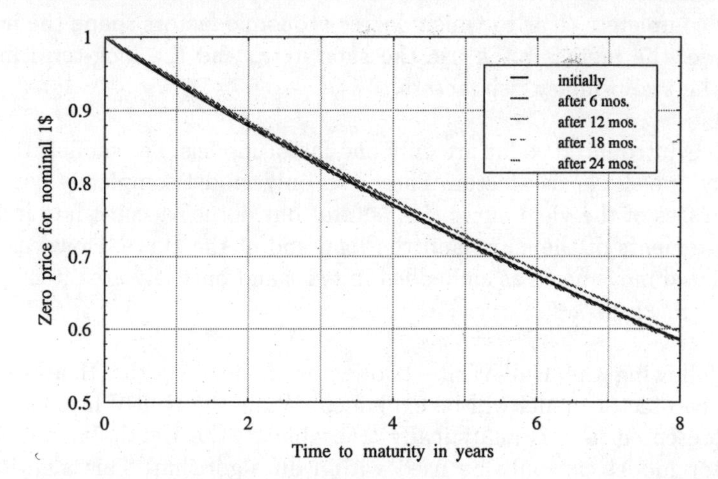

Fig. 3.22. Example of the development of the Zero prices for a nominal value of 1 $ over two years in the Heath-Jarrow-Morton one-factor model, corresponding to Figure 3.21.

3.3.3 Two-Factor Term Structure Models

This section presents selected two-factor interest rate models. These models incorporate two factors that influence the shape and movement of the term structure of interest rates over time. There is no doubt in financial practice that the term structure of interest rates is influenced by different sources of uncertainty, depending on whether the short or the long end of the curve is observed. Therefore, quantitative modelling requires more than one risk factor. It also requires risk factors which are not highly correlated. For example, the short end of the yield curve is more influenced by monetary policy than the long end of the curve. On the other hand the long end of the curve is driven more by fundamental data such as exchange rates and interest rate parity[112].

Empirical studies have shown that the movements of the short-term and the long-term interest rates over time are not completely correlated. A study undertaken by Chan, Karolyi, Longstaff, and Sanders[113] concludes that one risk factor is insufficient to describe the shape and movement of the yield curve.

[112] See Rudolf [114], page 135.

[113] See Chan, Karolyi, Longstaff & Sanders [30].

According to Litterman and Scheinkman[114] as well as Bühler and Zimmermann[115] the main factors that influence the shape of the term structure of interest rates are the level, the slope, and the curvature of the curve. Although it is not completely clear to which degree economic factors shape the interest rate curve, the models often use the short-term and the long-term interest rate as the two primary risk factors.

All models introduced so far are only one-factor models, i.e., models that employ only one single risk factor. These can only model completely correlated interest rates of the yield curve (i.e., shifts) but do not accomodate independent movements of the short and the long end of the curve. However, these uncorrelated movements as embedded in twist and butterfly are often spotted in capital markets.

In the following the Hull-White two-factor model and the Heath-Jarrow-Morton two-factor model will be explained. While the Hull-White two-factor model presented here is analytically trackable[116], the Heath-Jarrow-Morton two-factor model can only be used within an algorithm. This algorithm is based on the trinomial tree described in Rudolf [114], pages 137-148, and shows great similarity to the algorithm already presented for the Heath-Jarrow-Morton one-factor model. For the presentation of the Hull-White two-factor model, the original publication of Hull and White is used, see Hull & White [68]. In this thesis, special attention will be devoted to the Hull-White two-factor model since it has not yet gained the acceptance that it deserves for its favorable features[117].

Hull-White two-factor model. In 1994, Hull and White developed a two-factor model that extends their existing one-factor model by a variable drift term. The advantage is that this model provides an analytical solution for Zero bond prices (and, therefore, for the term structure of interest rates according to (3.2)) and for the volatility term structure. The process of the Hull-White two-factor model is described via the SDE

$$\left.\begin{aligned} dr_t &= [\theta(t) + u_t - \alpha r_t]dt + \sigma_1 dB_t^1, \ t \geq 0, \quad \text{where} \\ \theta(t) &= \frac{\partial f(0,t)}{\partial t} + \alpha f(0,t) + \frac{\sigma_1^2}{2\alpha}\left(1 - e^{-2\alpha t}\right), \ t \geq 0, \\ du_t &= -\xi\, u_t\, dt + \sigma_2 dB_t^2, \ t \geq 0, \ u_0 = 0, \\ &\sigma_1 \in \mathbb{R}^+, \alpha \in \mathbb{R} \setminus \{0\}, \sigma_2 \in \mathbb{R}^+, \xi \in \mathbb{R}^+, \alpha \neq \pm\xi. \end{aligned}\right\} \quad (3.41)$$

[114] See Litterman & Scheinkman [82].
[115] See Bühler & Zimmermann [26].
[116] See Clewlow & Strickland [31], pages 225-232.
[117] According to Rudolf [114], page 136, the strengths of the Hull-White two-factor model warrant a much stronger consideration in practice and literature.

The process $(u_t)_{t\geq 0}$ is a second diffusion process, an Ornstein-Uhlenbeck process (see 2.4.11), that reverts to mean zero with mean reversion force ξ (see 2.4.10). This process yields a stochastic mean reversion level. $(B_t^1)_{t\geq 0}$ and $(B_t^2)_{t\geq 0}$ are two Standard Brownian Motions that are assumed to be correlated with parameter $\rho \in [-1,1]$. The second stochastic process gives additional flexibility to the drift term. This allows the two-factor model to produce *a richer pattern of term structure movements and of possible volatility structures*[118]. The start value of this Ornstein-Uhlenbeck process is always $u_0 = 0$ according to Hull and White[119].

With $0 \leq t \leq T$ and T as the maturity time point of the Zero bond, the following formulas hold for the price $P(t,T)$ of this Zero[120]:

$$P(t,T) = A(t,T)\, e^{-r_t B(t,T) - u_t C(t,T)}, \tag{3.42}$$

with

$$A(t,T) = \exp\left\{ \ln \frac{P(0,T)}{P(0,t)} + B(t,T) f(0,t) - \eta \right\}, \tag{3.43}$$

$$B(t,T) = \frac{1}{\alpha}\left(1 - e^{-\alpha(T-t)}\right), \tag{3.44}$$

$$C(t,T) = \frac{1}{\alpha(\alpha-\xi)} e^{-\alpha(T-t)} - \frac{1}{\xi(\alpha-\xi)} e^{-\xi(T-t)} + \frac{1}{\alpha\xi}. \tag{3.45}$$

To calculate η the following formulas have to be used:

$$\eta = \frac{\sigma_1^2}{4\alpha}\left(1 - e^{-2\alpha t}\right) B(t,T)^2 - \rho\sigma_1\sigma_2 \left[B(0,t)C(0,t)B(t,T) + \gamma_4 - \gamma_2\right] - \frac{1}{2}\sigma_2^2 \left[C(0,t)^2 B(t,T) + \gamma_6 - \gamma_5\right], \tag{3.46}$$

$$\gamma_1 = \frac{e^{-(\alpha+\xi)T}\left[e^{(\alpha+\xi)t} - 1\right]}{(\alpha+\xi)(\alpha-\xi)} - \frac{e^{-2\alpha T}\left[e^{2\alpha t} - 1\right]}{2\alpha(\alpha-\xi)},$$

$$\gamma_2 = \frac{1}{\alpha\xi}\left[\gamma_1 + C(t,T) - C(0,T) + \frac{1}{2}B(t,T)^2 - \frac{1}{2}B(0,T)^2 + \frac{t}{\alpha} - \frac{e^{-\alpha(T-t)} - e^{-\alpha T}}{\alpha^2}\right],$$

[118] See Clewlow & Strickland [31], page 225.
[119] See Hull & White [68], page 42.
[120] See Hull & White [68], pages 45-46.

$$\gamma_3 = -\frac{e^{-(\alpha+\xi)t} - 1}{(\alpha-\xi)(\alpha+\xi)} + \frac{e^{-2\alpha t} - 1}{2\alpha(\alpha-\xi)},$$

$$\gamma_4 = \frac{1}{\alpha\xi}\left[\gamma_3 - C(0,t) - \frac{1}{2}B(0,t)^2 + \frac{t}{\alpha} + \frac{e^{-\alpha t} - 1}{\alpha^2}\right],$$

$$\gamma_5 = \frac{1}{\xi}\left[\frac{1}{2}C(t,T)^2 - \frac{1}{2}C(0,T)^2 + \gamma_2\right],$$

$$\gamma_6 = \frac{1}{\xi}\left[\gamma_4 - \frac{1}{2}C(0,t)^2\right].$$

The volatility term structure is given by:

$$\sigma_R(t,T) \quad = \quad \frac{1}{T-t}\sqrt{[B(t,T)\sigma_1]^2 + [C(t,T)\sigma_2]^2 + 2\rho\sigma_1\sigma_2 B(t,T)C(t,T)}.$$

As in the Hull-White one-factor model, this two-factor model includes the current yield curve via the instantaneous forward curve f. Again, the choice of $r_0 = f(0,0)$ as the start value for the short-rate process $(r_t)_{t\geq 0}$ ensures that for $t = 0$ the Hull-White two-factor model calculates the initial term structure of interest rates. This will be shown in the following:

According to (3.42) and since $u_0 = 0$, it holds for the current (i.e., $t = 0$) Zero price:

$$P(0,T) \quad = \quad A(0,T)\, e^{-r_0 B(0,T)}, \tag{3.47}$$

with

$$A(0,T) \overset{(3.43),(3.3),(3.1)}{=} \exp\{-R(0,T)\,T + B(0,T)f(0,0) - \eta\}, \tag{3.48}$$

$$B(0,T) \overset{(3.44)}{=} \frac{1}{\alpha}\left(1 - e^{-\alpha T}\right). \tag{3.49}$$

With $B(0,0) = 0$ according to (3.49) and with $C(0,0) = 0$ according to (3.45), the variable η simplifies to

$$\eta \overset{(3.46)}{=} -\rho\sigma_1\sigma_2(\gamma_4 - \gamma_2) - \frac{1}{2}\sigma_2^2(\gamma_6 - \gamma_5). \tag{3.50}$$

By applying $B(0,0) = 0$ and $C(0,0) = 0$, it is easy to see that for $t = 0$ the following holds: $\gamma_1 = \gamma_2 = \gamma_3 = \gamma_4 = \gamma_5 = \gamma_6 = 0$. Putting these values into (3.50) yields $\eta = 0$ for $t = 0$. Therefore, equation (3.48) simplifies to:

$$A(0,T) \quad = \quad \exp\{-R(0,T)\,T + B(0,T)f(0,0)\}. \tag{3.51}$$

Equations (3.47) and (3.51) then yield:

$$P(0,T) = \exp\{-R(0,T)\,T + B(0,T)\,(r_0 - f(0,0))\}$$

$$\stackrel{(3.1)}{=} \exp\{-R(0,T)\,T\}.$$

This equation has to hold for all $T \geq 0$ which can only be achieved if $r_0 = f(0,0)$, since $B(0,T) \neq 0$ for $T > 0$ according to (3.49). This choice of r_0 ensures that the Hull-White two-factor model is term structure consistent.

Figure 3.23 gives an example of the short-rate process for this model. The initial term structure of interest rates was the U.S. Zero yield curve from December 1, 2000. The corresponding path of the Ornstein-Uhlenbeck process is displayed in Figure 3.24.

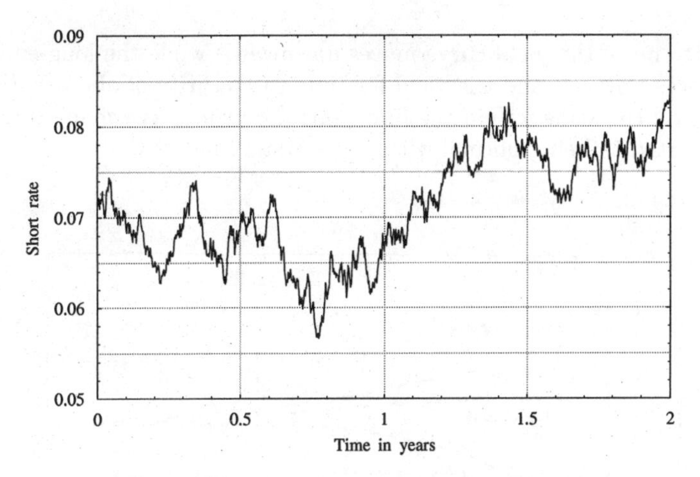

Fig. 3.23. Example of a short-rate path in the Hull-White two-factor model ($\alpha = 0.1$, $\sigma_1 = 0.015$, $\xi = 0.6$, $\sigma_2 = 0.01$, $\rho = -0.4$, $T = 2$, $N = 1440$).

Figure 3.25 shows the term structure of interest rates, Figure 3.26 the Zero prices for nominal 1 \$, and Figure 3.27 the volatility term structure over time in the Hull-White two-factor model for the short-rate process displayed in Figure 3.23. Figure 3.25 clearly shows the additional flexibility in the shape and movement of the yield curve due to the second risk factor in the model. While one-factor term structure models can only generate shift movements of the term structure, a two-factor model can also generate twist and butterfly movements. This can be seen in Figure 3.25, where the movement from the initial term structure to the term structure after 6 months and then to the one after 12 months develops a twist of the term structure within one year:

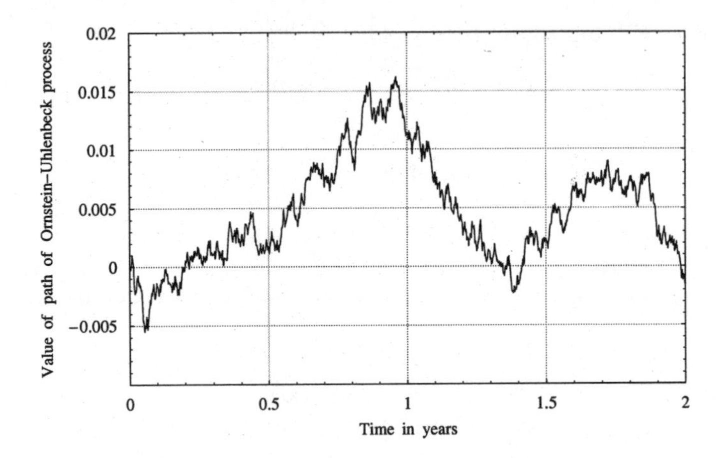

Fig. 3.24. Example of a path of the Ornstein-Uhlenbeck process $(u_t)_{t\geq 0}$ in the Hull-White two-factor model, corresponding to Figure 3.23.

The short-end of the yield curve moves downwards while the long-end of the curve moves upwards and the yield for time to maturity of about half a year almost stays the same within the first year; the yield curve shape turns from being inverted to being normal within the first 12 months.

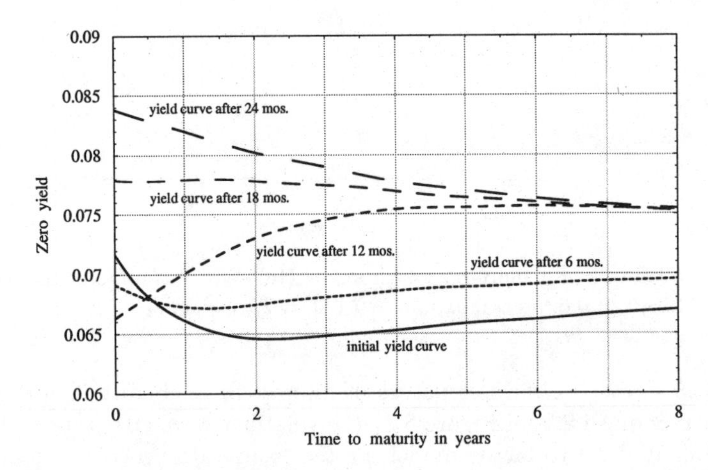

Fig. 3.25. Example of the development of the term structure of interest rates over two years in the Hull-White two-factor model, corresponding to Figure 3.23.

In this model, the volatility term structure, which is stable through time, can take various shapes depending on the choice of the model parameters. Even a humped and a monotonically increasing volatility term structure can be

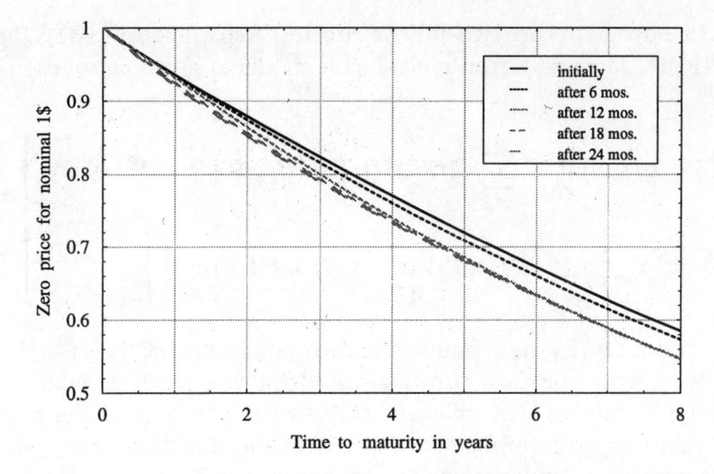

Fig. 3.26. Example of the development of the Zero prices for a nominal value of 1 $ over two years in the Hull-White two-factor model, corresponding to Figure 3.23.

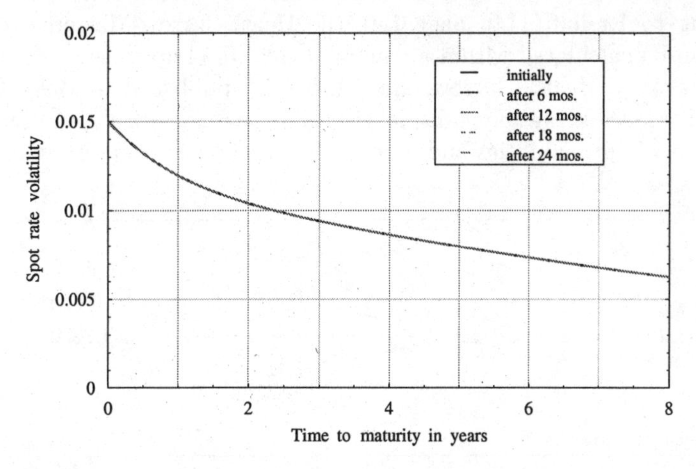

Fig. 3.27. Example of the development of the term structure of interest rate volatilities over two years in the Hull-White two-factor model, corresponding to Figure 3.23: The graphs are identical.

generated. However, as Clewlow and Strickland point out[121], it is the general feature of a volatility term structure in practice that the spot rate volatility decreases with longer times to maturity. The variability that the Hull-White two-factor model provides for the shape of the volatility term structure makes it well suited for practical use.

[121] See Clewlow & Strickland [31], page 184.

Heath-Jarrow-Morton two-factor model. According to (3.31), the most general Heath-Jarrow-Morton multi-factor model is given as solution of the SDE:

$$\left.\begin{aligned} df(t,T) &= \alpha(t,T)dt + \sum_{i=1}^{n} \sigma_i(t,T,f(t,T))dB_t^i, \;\; 0 \le t \le T, \;\text{ where} \\ \alpha(t,T) &= \sum_{i=1}^{n} \left\{ \sigma_i(t,T,f(t,T)) \left[\int_t^T \sigma_i(t,u,f(t,u))du \right] \right\}. \end{aligned}\right\} \quad (3.52)$$

$T > 0$ is the maturity time point of a Zero bond and $(B_t^i)_{t \ge 0}$, $i = 1, \ldots, n$, are n independent Standard Brownian Motions with n as the number of independent risk factors that influence the term structure. α is the drift rate, and the volatility structure is given by n volatility functions $\sigma_i, i = 1, \ldots, n$. The drift and volatility functions can take values in $\mathbb{R}$.

The specification $n = 2$ yields the Heath-Jarrow-Morton two-factor model. According to Rudolf [114], page 139, the Heath-Jarrow-Morton two-factor model can be calculated within a trinomial tree. His implementation is based on an article of Heath, Jarrow, and Morton[122] published in 1991. For the trinomial tree the situation is displayed in Figure 3.28 with π_u, π_m, and π_d as the martingale probabilities and u, m, and d as the perturbation functions.

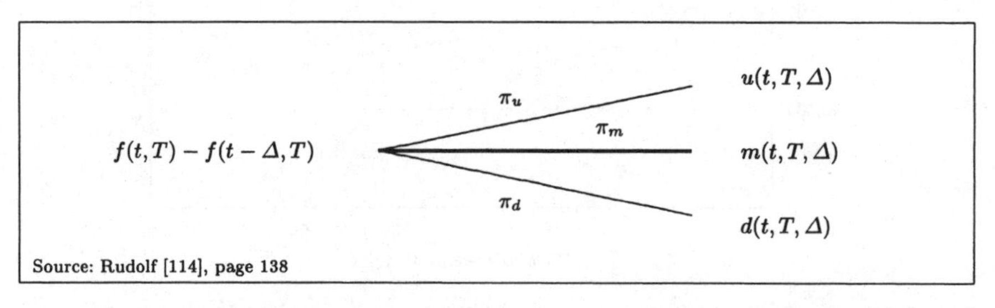

Fig. 3.28. Trinomial branch in the Heath-Jarrow-Morton two-factor model.

Under the martingale measure the following equation holds[123]:

$$E_t^{\pi}(P(t+\Delta,T)) = \frac{P(t,T)}{P(t,t+\Delta)} \left(\pi_u e^{u(t,T,\Delta)} + \pi_m e^{m(t,T,\Delta)} + \pi_d e^{d(t,T,\Delta)} \right),$$

which implies the following no-arbitrage condition[124]:

$$\pi_u \, e^{u(t,T,\Delta)} \; + \; \pi_m \, e^{m(t,T,\Delta)} \; + \; \pi_d \, e^{d(t,T,\Delta)} \quad = \quad 1. \qquad (3.53)$$

[122] See Heath, Jarrow & Morton [52].
[123] See Rudolf [114], page 139.
[124] See Rudolf [114], page 139.

This gives for the pricing of a Zero bond[125]:

$$P(t+\Delta,T) = \begin{cases} \dfrac{P(t,T)}{P(t,t+\Delta)} \cdot e^{u(t,T,\Delta)} & \text{in the up-state,} \\ \dfrac{P(t,T)}{P(t,t+\Delta)} \cdot e^{m(t,T,\Delta)} & \text{in the middle-state,} \\ \dfrac{P(t,T)}{P(t,t+\Delta)} \cdot e^{d(t,T,\Delta)} & \text{in the down-state.} \end{cases} \tag{3.54}$$

This can be solved for the three perturbation functions u, m, and d in each branch of the tree. Following the approach of Rudolf[126] and the ideas of Heath, Jarrow, and Morton[127], the values for the three states in the trinomial tree structure are:

$$\begin{aligned} u(t,T,\Delta) &= \sigma_1\sqrt{\Delta} + \mu(t,T,f), \\ m(t,T,\Delta) &= -\sigma_1\sqrt{\Delta} + \sigma_2\sqrt{2\Delta} + \mu(t,T,f), \\ d(t,T,\Delta) &= -\sigma_1\sqrt{\Delta} - \sigma_2\sqrt{2\Delta} + \mu(t,T,f). \end{aligned}$$

Function μ will be determined below using equation (3.53). According to (3.36) and (3.54), this gives for the price of a Zero bond in the respective states (up, middle, and down):

$$P(t+\Delta,T) = \begin{cases} P(t,T)e^{r_t\Delta+\sigma_1(t,T,f)\sqrt{\Delta}+\mu(t,T,f)}, \\ P(t,T)e^{r_t\Delta-\sigma_1(t,T,f)\sqrt{\Delta}+\sigma_2(t,T,f)\sqrt{2\Delta}+\mu(t,T,f)}, \\ P(t,T)e^{r_t\Delta-\sigma_1(t,T,f)\sqrt{\Delta}-\sigma_2(t,T,f)\sqrt{2\Delta}+\mu(t,T,f)}. \end{cases} \tag{3.55}$$

The application of (3.53) yields[128]:

$$\begin{aligned} \mu(t,T,f) = \ln(4) - \ln\Big(&2e^{\sigma_1(t,T,f)\sqrt{\Delta}} + e^{-\sigma_1(t,T,f)\sqrt{\Delta}+\sigma_2(t,T,f)\sqrt{2\Delta}} \\ &+e^{-\sigma_1(t,T,f)\sqrt{\Delta}-\sigma_2(t,T,f)\sqrt{2\Delta}}\Big). \end{aligned} \tag{3.56}$$

To complete the model, the volatility functions σ_1 and σ_2 need to be specified. As done in Rudolf [114], page 142, and Heath, Jarrow & Morton [52], the volatility functions are:

---

[125] See Rudolf [114], page 140, equation (3.4).
[126] See Rudolf [114], pages 140-141.
[127] See Heath, Jarrow & Morton [52], page 95.
[128] See Rudolf [114], page 139 and page 142.

$$\sigma_1(t,T,f) = \sigma_1(t,T) := -\sigma_1(T-t-\Delta), \tag{3.57}$$
$$\sigma_2(t,T,f) = \sigma_2(t,T) := \sigma_2(T-t-\Delta)\, e^{-\lambda(T-t-\Delta)}, \tag{3.58}$$

$\sigma_1, \sigma_2 \in \mathbb{R}^+$ and $\lambda \in \mathbb{R}^-$. While the σ_2-function is an exponentially decreasing volatility function similar to the one used in Heath, Jarrow & Morton [52], the choice of the σ_1-function is by definition (see Rudolf [114], page 142). The choice of this σ_1-function assures that the value of this function is zero exactly one period before the T-year Zero matures (i.e., $\sigma_1(T-\Delta,T)=1$). This implies that the bond value is 1 at maturity.

Note that for the σ_1-function the values are zero or negative. This does not imply a negative volatility for the logarithm of the Zero prices; it only describes the dynamic of the interest rate development. According to Rudolf [114], page 142, the definitions (3.57) and (3.58) ascertain that the long-term bonds move in the opposite direction in the middle- and down-states compared with the short-term bonds. This is the only way to model twists in a term structure model. Again, λ is a decay factor that needs to be specified in advance.

Figure 3.29 shows how to calculate the price and the values of the volatility functions for a 3-year Zero with $N=3$, i.e., $\Delta = \frac{T}{N} = \frac{3}{3} = 1$. A detailed numerical example can be found in Rudolf [114], pages 143-145. The current term structure of interest rates is included in the model via equation (3.55) for $t=0$, i.e., in the calculation of $P(\Delta,T)$. This calculation is the first step in the algorithm and includes r_0 as $R(0,\Delta)$, a very short-term spot rate that is calculated using the cubic spline method based on the maturity buckets given from the initial yield curve. This yields a good numerical approximation for the short rate r_0. The short rate r_t for $t>0$ is calculated in the algorithm via the equations (3.55) and (3.56), using definitions (3.57) and (3.58) for the two volatility functions.

The tree in Figure 3.29 is the basis for arriving at the term structure of interest rates via the Zero prices. Figure 3.30 shows an example of the term structure of interest rates. Figure 3.31 displays the corresponding Zero prices for nominal 1 \$. The Zero yield curve from December 1, 2000 is chosen as input parameter.

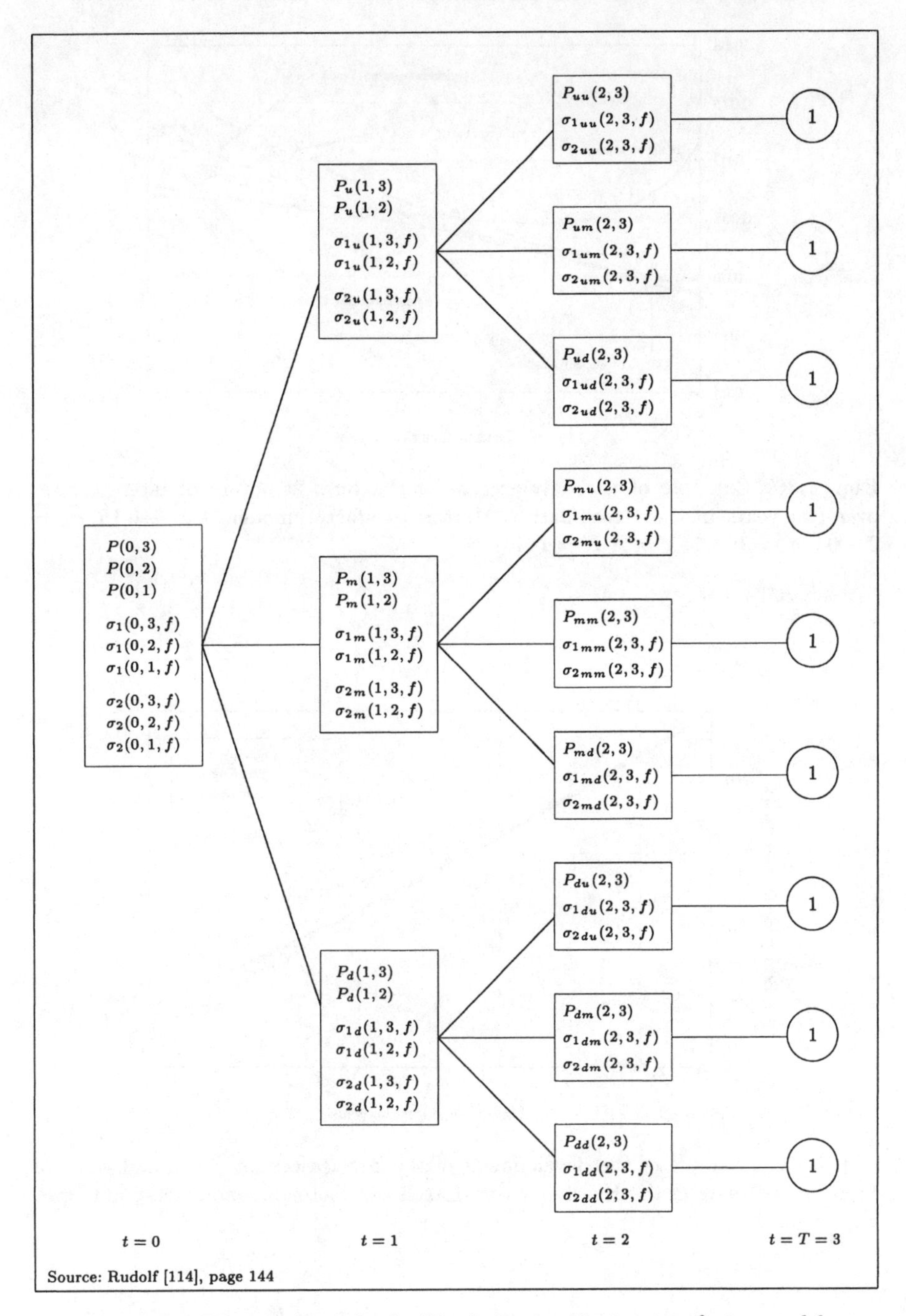

Fig. 3.29. Trinomial tree in the Heath-Jarrow-Morton two-factor model.

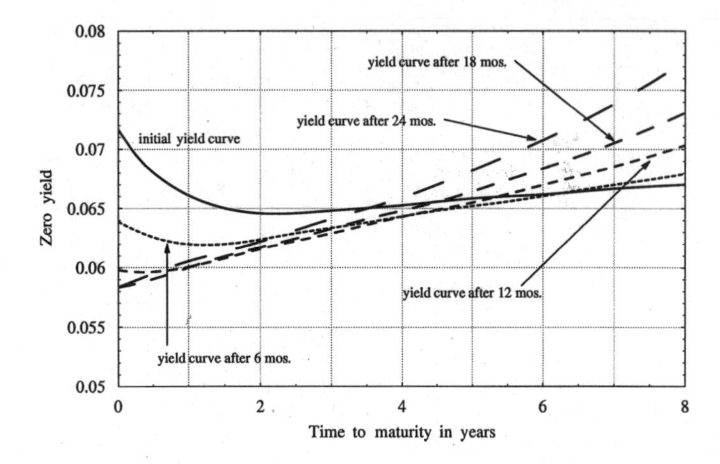

Fig. 3.30. Example of the development of the term structure of interest rates over two years in the Heath-Jarrow-Morton two-factor model ($\lambda = -0.15$, $\sigma_1 = 0.0005$, $\sigma_2 = 0.001$, $T = 2$, $N = 1440$).

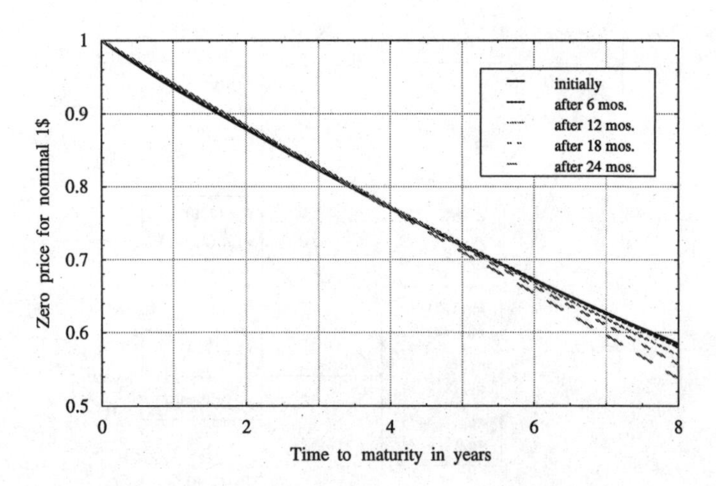

Fig. 3.31. Example of the development of the Zero prices for a nominal value of 1 $ over two years in the Heath-Jarrow-Morton two-factor model, corresponding to Figure 3.30.

3.4 Summary

This chapter presented various stochastic term structure models. In addition, an overview of the history of term structure modelling and of the differences and similarities of these models was given. Several concepts were explained: equilibrium models, no-arbitrage models, one-factor models, multi-factor models, mean reversion models, and term structure models without mean reversion. These concepts are the building blocks of this thesis. In the beginning of the following chapter, real options pricing tools are explained in detail. Some of these tools will then be modified to accommodate for the stochastic term structure models presented in Chapter 3. This will be a central chapter of the thesis since it presents various new ideas and approaches to include a non-constant interest rate in valuation tools for real options. Putting these ideas into practice is the topic of Chapter 5 where all simulation and backtesting results are stated and analyzed. All calculations are done with a computer simulation program that contains the stochastic term structure models presented in Chapter 3 and all real options valuation tools presented in Chapter 4.

The stochastic term structure models presented in this chapter are put together in the following table to provide a summary of the models, the SDEs, the parameter specifications, and the major properties.

Table 3.2. Overview of the stochastic term structure models from Chapter 3.

Describing SDE	Properties
Ornstein-Uhlenbeck Process	
$dr_t = \mu(r_t)dt + \sigma(r_t)dB_t,\ \ t \geq 0$ $r_0 \in \mathbb{R}$ constant $\mu(x) = -\alpha\, x \quad \forall x$ $\sigma(x) \equiv \sigma \quad \forall x \in \mathbb{R}$ $\alpha \in \mathbb{R}^+,\ \ \sigma \in \mathbb{R}^+$	mean reverting around zero with stationary distribution $N\left(0, \frac{\sigma^2}{2\alpha}\right)$ equilibrium model
Vasicek Model	
$dr_t = \mu(r_t)dt + \sigma(r_t)dB_t,\ \ t \geq 0$ $r_0 \in \mathbb{R}$ constant $\mu(x) = \beta - \alpha\, x \quad \forall x$ $\sigma(x) \equiv \sigma \quad \forall x \in \mathbb{R}$ $\alpha \in \mathbb{R}^+,\ \ \beta \in \mathbb{R},\ \ \sigma \in \mathbb{R}^+$	mean reverting around $\frac{\beta}{\alpha}$ with stationary distribution $N\left(\frac{\beta}{\alpha}, \frac{\sigma^2}{2\alpha}\right)$ equilibrium model
Cox-Ingersoll-Ross Model	
$dr_t = \mu(r_t)dt + \sigma(r_t)dB_t,\ \ t \geq 0$ $r_0 \in \mathbb{R}^+$ constant $\mu(x) = \beta - \alpha\, x \quad \forall x$ $\sigma(x) = \sigma\sqrt{x} \quad \forall x$ $\alpha \in \mathbb{R}^+,\ \ \beta \in \mathbb{R}^+$ $\sigma \in \mathbb{R}^+,\ \ 2\beta \geq \sigma^2$	mean reverting around $\frac{\beta}{\alpha}$ with stationary distribution $\Gamma\left(\frac{2\alpha}{\sigma^2}, \frac{2\beta}{\sigma^2}\right)$ equilibrium model
Ho-Lee Model	
$dr_t = \theta(t)dt + \sigma dB_t,\ \ t \geq 0$ $\theta(t) = \frac{\partial f(0,t)}{\partial t} + \sigma^2 t \quad \forall t \geq 0$ $\sigma \in \mathbb{R}^+,\ \ r_0 = f(0,0)$	not mean reverting term structure consistent no-arbitrage model

Describing SDE	Properties
Hull-White One-Factor Model	
$dr_t = [\theta(t) - \alpha r_t]dt + \sigma dB_t,\ \ t \geq 0$	mean reverting
$\theta(t) = \frac{\partial f(0,t)}{\partial t} + \alpha f(0,t) + \frac{\sigma^2}{2\alpha}\left(1 - e^{-2\alpha t}\right)\ \ \forall t \geq 0$	term structure consistent
$\alpha \in \mathbb{R} \setminus \{0\},\ \ \sigma \in \mathbb{R}^+,\ \ r_0 = f(0,0)$	no-arbitrage model
Hull-White Two-Factor Model	
$dr_t = [\theta(t) + u_t - \alpha r_t]dt + \sigma_1 dB_t^1,\ \ t \geq 0$	mean reverting
$du_t = -\xi\, u_t\, dt\ +\ \sigma_2 dB_t^2,\ \ t \geq 0$	term structure consistent
$\theta(t) = \frac{\partial f(0,t)}{\partial t} + \alpha f(0,t) + \frac{\sigma_1^2}{2\alpha}\left(1 - e^{-2\alpha t}\right),\ \ t \geq 0$	no-arbitrage model
$\sigma_1 \in \mathbb{R}^+,\ \ \sigma_2 \in \mathbb{R}^+,\ \ \alpha \in \mathbb{R} \setminus \{0\},\ \ \xi \in \mathbb{R}^+,$ $\alpha \neq \pm\xi,\ \ r_0 = f(0,0)$	
Heath-Jarrow-Morton Multi-Factor Model	
$df(t,T) = \alpha(t,T)dt + \sum_{i=1}^{n} \sigma_i(t,T,f(t,T))dB_t^i,\ \ 0 \leq t \leq T$	term structure consistent
$\alpha(t,T) = \sum_{i=1}^{n} \left\{ \sigma_i(t,T,f(t,T)) \left[\int_t^T \sigma_i(t,u,f(t,u))du \right] \right\}$	no-arbitrage model
$\sigma_i, i = 1, \ldots, n \in \mathbb{N}$, are n independent volatility functions with values in $\mathbb{R}$	
α function with values in $\mathbb{R}$	

4

Real Options Valuation Tools in Corporate Finance

4.1 Introduction

In Chapter 4 the most common valuation methods for real options are presented in theory. These tools are implemented in the computer simulation program and will be used in Chapter 5 to price various real options by applying a non-constant interest rate which is modelled via the term structure models introduced in the previous chapter. Chapter 4 is organized in three main sections:

- Section 4.2:
  Numerical Methods for Real Options Pricing with Constant Interest Rates
- Section 4.3:
  Schwartz-Moon Model
- Section 4.4:
  Real Options Pricing with Stochastic Interest Rates

The final section (4.5) contains, as usual, the summary of this chapter.

The very first method used to price a real option is the decision-tree analysis (DTA) which was already presented and analyzed in Section 2.6.1. But DTA exhibits two major drawbacks. First, the decision tree structure becomes complicated even in simple real world situations. Second, the DTA calculations use real probabilities and a risk-adjusted interest rate. This interest rate is constant and does not change even if there are decisions in the decision tree that change the risk structure of the project. Moreover, in usually complex investment situations, DTA turns into "decision-bush analysis" that cannot be handled appropriately.

These drawbacks are eliminated in the contingent-claims analysis (CCA) by introducing risk-adjusted probabilities, a method already presented in 2.6.2 in the context of the Cox-Ross-Rubinstein binomial tree method. This method will be used at great length and, therefore, will be reviewed briefly in 4.2.1 in the framework of a constant interest rate. Another lattice method that will be explained in detail is a log-transformed binomial lattice method introduced by Trigeorgis[1] in 1991. Both the Cox-Ross-Rubinstein method and the Trigeorgis log-transformed binomial tree method are presented in Section 4.2.1, *Numerical Methods for Real Options Pricing with Constant Interest Rates.*

Section 4.2.2, *Finite Difference Methods*, deals with various finite difference methods. Real options that can be described mathematically via a partial differential equation can be priced by using an approximation of this partial differential equation with finite (in contrast to infinitesimal) differences that replace the partial derivatives in the equation. Although finite differences can be applied directly to the partial differential equation with the price of the "twin security" or the project value as the underlying, such a finite difference method exhibits poor numerical properties[2]. However, if a log-transformation of the underlying is used and the partial differential equation is built, the resulting finite difference method exhibits advantageous numerical properties. Therefore, only log-transformed finite difference methods will be presented here. They are also implemented in the computer simulation program.

Depending on the type of approximation, distinctions can be made between various types of finite difference methods. This thesis presents the log-transformed explicit and implicit finite difference methods, two basic kinds of finite difference methods. Each method will be explained with respect to its mathematical theory, numerical properties, and practical limitations. Other methods like the Crank-Nicholson algorithm, a hybrid between the explicit and the implicit method, will not be explained here[3]. Another method that builds on the log-transformed explicit finite difference method and also includes a time-dependent risk-free rate modelled by using a short-rate process was presented in Hull & White [64]. This method will not be presented here since this thesis focuses on modelling the complete yield curve in simulations and historical backtesting and not the short-rate process only. In addition, the modification of the Cox-Ross-Rubinstein binomial tree method and the Trigeorgis log-transformed binomial tree method which is realized in this thesis by incorporating a stochastic risk-free rate is less complicated (and, therefore, more likely to be applied in practice) than the modification of the log-transformed explicit finite difference method carried out by Hull and White.

Section 4.3 presents the Schwartz-Moon model, a specific model for pricing real options in R&D of new drug development. This model uses various stochastic processes to model the underlying and the cost processes. The model has a deterministic counterpart that applies a constant interest rate and that is

[1] See Trigeorgis [132], pages 320-328.
[2] See Trigeorgis [132], page 314.
[3] For more information on the Crank-Nicholson algorithm see Trigeorgis [132], page 319.

used by Schwartz and Moon as the benchmark to their stochastic model. To introduce the idea of a non-constant interest rate, this deterministic counterpart will be used in connection with a deferred project start for systematically changing interest rates in order to evaluate the effect that such a modification has on the NPV of a project and therefore, on the manager's investment decision.

Section 4.4, *Real Options Pricing with Stochastic Interest Rates*, presents real options valuation tools that accommodate for a non-constant risk-free interest rate, and explains modifications of existing valuation methods to include a non-constant risk-free rate. In the beginning of this section, the pioneering idea of Ingersoll and Ross will be explained (4.4.1). They were the first to show the existence of a real option even in the case of deterministic future cash flows but with stochastic interest rates. Their publication in 1992 was the very first to deal with this idea by applying the Ito calculus to investigate an option to defer on the purchase of a Zero bond, the simplest situation possible. Even such a simple real option exhibits extreme mathematical complexity which makes it difficult to be applied in financial practice.

The idea of this thesis is not to mathematically analyze other types of simple real options, but to modify existing numerical tools for real options (especially complex real options) which are common to the world of Corporate Finance today, and to implement these tools in a computer simulation program. Thus, the thesis fills the gap mentioned by Trigeorgis, who stressed the lack of software tools to handle complex real options in practice (see Trigeorgis [132], page 375, Section 12.3 - Future Research Directions): *Developing generic options-based user-friendly software packages with simulation capabilities that can handle multiple real options as a practical aid to corporate planners.*

First, a modification of the Cox-Ross-Rubinstein binomial tree model is presented in 4.4.2 to incorporate stochastic interest rates into real options pricing. Although the idea itself is not new, the method applied here is different from the other approaches. While Hull includes a variable interest rate in the Cox-Ross-Rubinstein binomial tree via the currently observable forward rates that he derives from the current spot rates[4], this thesis models the whole yield curve stochastically and includes the risk-free interest rate with the appropriate maturity into the model. This means that Hull assumes the current forward rates to become the future spot rates needed for the tree in future time periods while this thesis models the complete future term structure development via stochastic term structure models. Second, a modification of the Trigeorgis log-transformed binomial tree method that accommodates for stochastic interest rates is presented in 4.4.3.

[4] See Hull [62], page 356.

4.2 Numerical Methods for Real Options Pricing with Constant Interest Rates

Numerical methods are the main focus of this thesis. Although analytical methods to price real options are plentiful (see the overview in Section 2.6.3), their application to real options practice is rather limited. To quote Trigeorgis[5]: *Real-life investments are often more complex in that they may involve more than one option simultaneously. In such cases, analytic solutions may not exist and it might not even be possible to write down the set of partial differential equations describing the underlying stochastic processes. Valuing each option separately and adding up the individual results is often inappropriate since multiple options may in fact interact.* In contrast, numerical methods can easily be applied to various complex situations encountered in Corporate Finance practices. Moreover, they can easily be modified to include a stochastically modelled interest rate, which is often not feasible for analytical models due to mathematical complexities.

The first type of numerical methods explained below are lattice methods. Lattices, also called trees, focus on the approximation of the underlying stochastic process. The algorithm needs a current start value for the underlying and calculates the option value or project value for this particular start value. The second type of numerical methods are finite difference methods that approximate the partial differential equation which describes the real options situation mathematically. The resulting grid in the finite difference methods allows for various start values of the underlying, and the method gives the option value for each of these start values in the end.

4.2.1 Lattice Methods

This section explains the Cox-Ross-Rubinstein binomial tree method, the classical pricing tool for options (financial and real), as well as the Trigeorgis log-transformed binomial tree method. While the first method can easily be applied to options with non-constant interest rates, the modification of the Trigeorgis model is not equally easy in practice because of its numerical challenges and its computational intensity.

Cox-Ross-Rubinstein binomial tree. The Cox-Ross-Rubinstein binomial tree method[6] for option valuation, published in 1979, is *the* classical tool for option pricing that can also be easily tailored to incorporate complex real options situations. It is also easy to implement in a computer simulation program. The theoretical elegance of the model is complemented by its practical value as it fulfills numerical criteria like stability and consistency for

[5] See Trigeorgis [130], page 310.
[6] See Cox, Ross & Rubinstein [36].

certain choices of the tree parameters[7]. The Cox-Ross-Rubinstein binomial tree method was already explained in the context of contingent-claims analysis (CCA) in Section 2.6.2. Therefore, the focus now is on the numerical algorithm, its implementation and numerical properties. In the following the necessary definitions will be repeated:

V := total value of the project,

S := price of the twin security that is almost perfectly correlated with V,

E := equity value of the project for the shareholder,

r_f := risk-free interest rate,

p := risk-neutral probability for up-movements of V and S per period,

u := multiplicative factor for up-movements of V and S per period,

d := multiplicative factor for down-movements of V and S per period.

S is the price of the so-called *twin security* that can be traded in the capital markets and that exhibits the same risk profile as the project with value V. The maturity of the real option is $T > 0$ and the interval $[0, T]$ is divided into $N \in \mathbb{N}$ equally spaced subintervals. This gives the time points $0 = t_0 < t_1 < \ldots < t_N = T$ with the equal subinterval length of $\Delta t = \frac{T}{N}$, i.e., $t_j = j\Delta t$ for $j = 0, 1, \ldots, N$.

If, in the beginning of a subinterval with length Δt, the underlying value is S, it is assumed that in the end of this time interval the underlying can take two values, uS and dS, with risk-neutral probabilities p and $1-p$, respectively. Note, that $u > 1$ and $d < 1$. This is exactly the same as with the method of contingent-claims analysis in Chapter 2. The continuous-time counterpart is that S is a Geometric Brownian Motion with volatility σ. In a risk-neutral world the return of the underlying is the risk-free interest rate. Assuming a continuously compounded, annualized risk-free rate r_f this means

$$Se^{r_f \Delta t} = puS + (1-p)dS,$$

which is equivalent to

$$e^{r_f \Delta t} = pu + (1-p)d. \tag{4.1}$$

According to Hull [62], equation (11.4) and page 344, the variance of the change in the underlying over this time period yields

$$e^{2r_f \Delta t + \sigma^2 \Delta t} = pu^2 + (1-p)d^2. \tag{4.2}$$

[7] See Trigeorgis [130], page 320 and pages 324-325.

Equations (4.1) and (4.2) impose two conditions on p, u, and d. A third condition is given by Cox, Ross, and Rubinstein[8] via $u = \frac{1}{d}$. According to Hull [62], page 345, it can be shown that these three conditions imply

$$u = \exp\left(\sigma\sqrt{\frac{T}{N}}\right) = \frac{1}{d} \quad \text{and} \quad p = \frac{e^{r_f \Delta t} - d}{u - d}. \tag{4.3}$$

The goal is to replicate the pay-off structure of the project after time period Δt through the purchase of n shares of the twin security and through issuing a Δt-year Zero bond with nominal value B and risk-free return r_f. The issue of the Zero is equivalent to getting a loan of value B by paying the (annualized) interest rate r_f. Therefore, a positive B indicates a bond issue. The replicating portfolio develops as shown in Figure 4.1.

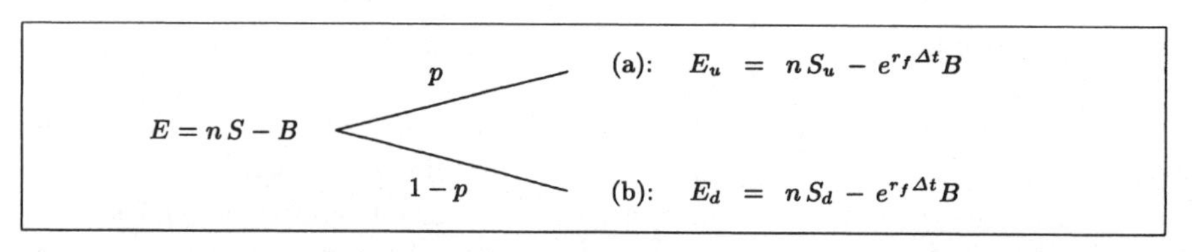

Fig. 4.1. The concept of a replicating portfolio in the Cox-Ross-Rubinstein binomial tree method.

When comparing Figure 4.1 to Figure 2.6 in which the replicating portfolio was introduced in the context of CCA, there seems to be a small difference with respect to the risk-free interest rate r_f. In CCA the interest rate factor used was $1 + r_f$ for the length of 1 year with r_f as the effective risk-free interest rate for this period. Here, $e^{r_f \Delta t}$ is used as the interest rate factor for the period of length Δt with r_f as the continuously compounded risk-free interest rate, the approach used by Hull[9]. In fact this is not a difference since the r_f values are defined differently in Chapter 2 and here. The notation does not cause confusion, especially since in the following only the approach of Figure 4.1 will be used with r_f as the continuously compounded, annualized risk-free interest rate[10].

By choosing the replicating portfolio approach $E = nS - B$ (i.e., E is replicated via n shares of stocks and the issue of a Δt-year Zero worth nominal B),

[8] See Cox, Ross & Rubinstein [36].
[9] See Hull [62], Section 15.1.
[10] The interest rate delivered by the continuous-time stochastic term structure models which were explained in Chapter 3 is also continuously compounded. Therefore, the results from Chapter 3 and the Cox-Ross-Rubinstein binomial tree method (as well as the Trigeorgis log-transformed binomial tree method introduced later) fit nicely together.

the equations (a) and (b) of Figure 4.1 have to be solved simultaneously to determine n and B. The solution to this problem is given in Figure 4.2.

$$\mathrm{n} = \frac{E_u - E_d}{S_u - S_d} \qquad \mathrm{E} = \frac{pE_u + (1-p)E_d}{e^{r_f \Delta t}}$$

$$\mathrm{B} = \frac{E_u S_d - E_d S_u}{(S_u - S_d)\, e^{r_f \Delta t}} \qquad \mathrm{p} = \frac{e^{r_f \Delta t} - d}{u - d}$$

Fig. 4.2. The solution of a replicating portfolio in the Cox-Ross-Rubinstein binomial tree method.

Figure 4.3 displays the complete tree for real options pricing with a constant risk-free interest rate r_f and a start value S_0 for the underlying. It shows the complete tree for the underlying and the variables used for this method. The formulas from Figure 4.2 have to be used for every single tree starting from the last time subinterval back to the first one.

Figure 4.4 restates the choice of indices for the state and time axis. i is the state index and j is the time index. By using this simple and symmetrical notation, it is easy to derive closed-form solutions for simple option cases[11]. In the case of complex real options it cannot be expected to arrive at a closed-form solution. However, the Cox-Ross-Rubinstein binomial tree method can be modified to accommodate for complex real options as shown below.

As mentioned, the Cox-Ross-Rubinstein binomial tree method can easily accommodate for various types of complex real options. The distinguishing points are the specifications of E at each node in the tree. Since this method involves working backwards through the tree, the terminal conditions at the nodes with time index $j = N$ (time T) have to be specified first. Using the formulas in Figure 4.2, one gets the values for E at each of the nodes with time index $j = N - 1$. Depending on the real option(s) included in the investment project, this value E has to be modified from E to E' whereby this E' will then be used when stepping downwards to time index $j = N - 2$, etc.

In this thesis, three real options cases will be analyzed (see Section 1.1):

1. **Case 1:** Option to abandon the project at any time during the construction period for a salvage value X. Since such a real option is an American put option, the salvage value is the strike price of this option.

[11] See Cox & Rubinstein [37], Section 5.

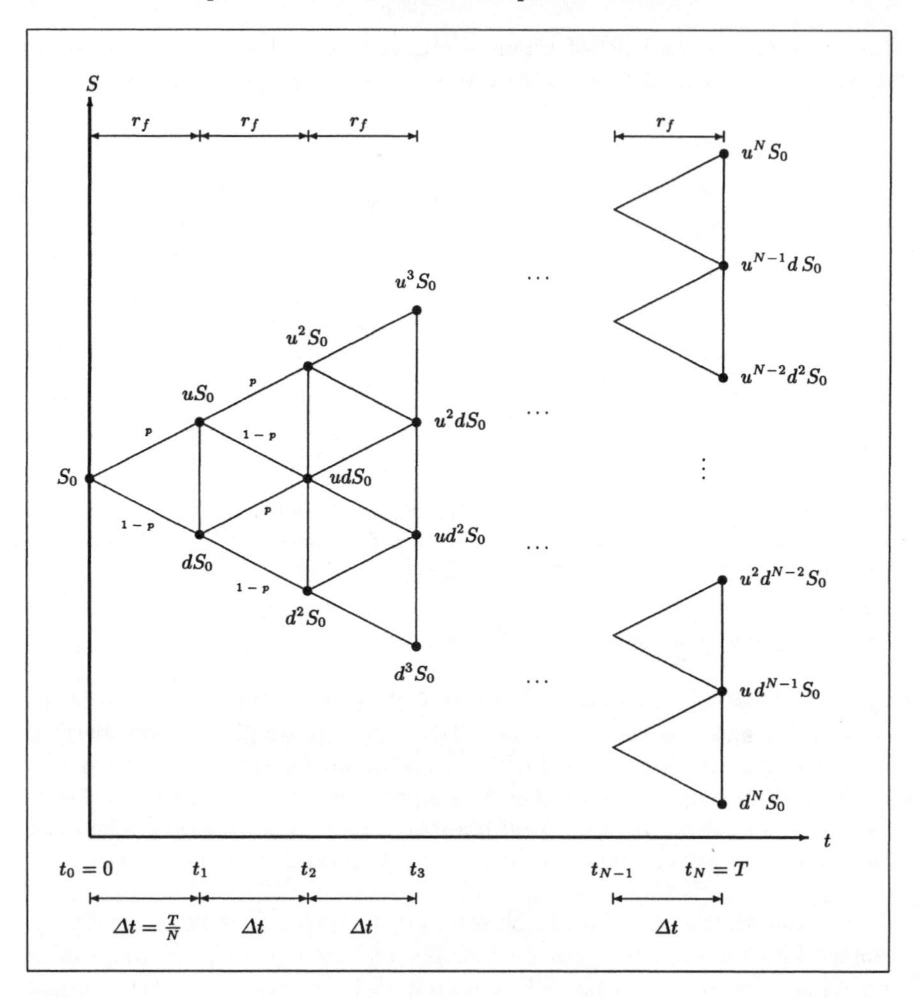

Fig. 4.3. Basic variables of the Cox-Ross-Rubinstein binomial tree method.

2. **Case 2:** Option to abandon the project at any time during the construction period for a salvage value X (case 1) and option to expand the project once by an expand factor e (e.g., expand project by $e = 20\%$) for an expand investment at the end of the construction period[12] denoted with I_e.

[12] The expand investment is assumed to be a portion of the initial investment cost I_0. In Chapter 5 it will always be assumed that the initial investment cost is $I_0 = 110$. More explanations about this are given in Section 5.5.

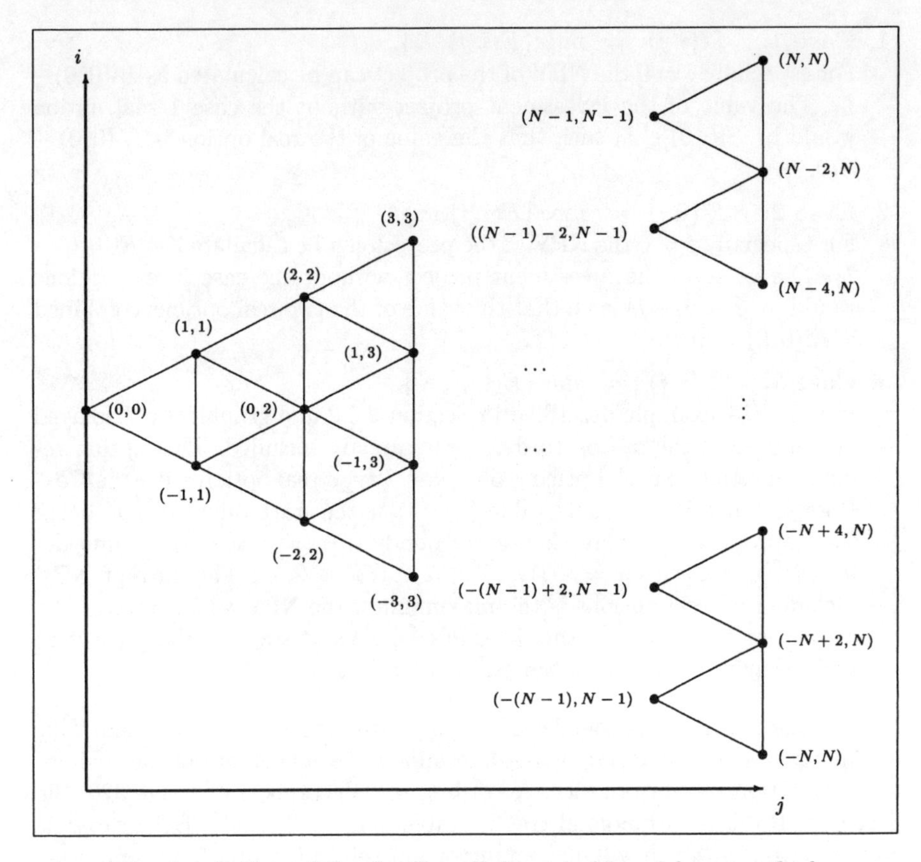

Fig. 4.4. Indices of the Cox-Ross-Rubinstein binomial tree method.

3. **Case 3:** Complex real option in case 2 combined with an option to defer the project start by exactly one year. Thus, the project can start today or in exactly one year from today if the investment in one year has a positive NPV. If, in one year from now, the NPV will be negative, the project will not be started at all.

Using the notation $E(i, j)$ for the value of E at node (i, j) and $S(i, j)$ for the value of S at node (i, j) (see Figure 4.4 for the indices), the terminal conditions (i.e., $j = N$) of these three cases are for each i:

1. **Case 1:** $E(i, N) = max\{S(i, j), X\}$.
2. **Case 2:** $E(i, N) = max\{S(i, j), S(i, j)(1 + e) - I_e, X\}$.
3. **Case 3:** $E(i, N) = max\{S(i, j), S(i, j)(1 + e) - I_e, X\}$.

At each node (i, j) with $j \in \{0, 1, \ldots, N - 1\}$, the following modifications to E have to be made for each i:

1. **Case 1:** $E'(i,j) = max\{E(i,j), X\}$.
   For especially $j = 0$ the NPV of the project can be calculated as $E'(0,0) - I_0$. The value of the investment project without the case 1 real option would be $S(0,0) - I_0$ such that the value of the real option is $E'(0,0) - S(0,0)$.
2. **Case 2:** $E'(i,j) = max\{E(i,j), X\}$.
   For especially $j = 0$ the NPV of the project can be calculated as $E'(0,0) - I_0$. The value of the investment project without the case 2 real options would be $S(0,0) - I_0$ such that the value of the two real options combined is $E'(0,0) - S(0,0)$.
3. **Case 3:** $E'(i,j) = max\{E(i,j), X\}$.
   In the CCA example described in Section 2.6.2 and graphically displayed in Figure 2.8, the option to defer was already included. This option requires multiple parallel pricing of a case 2 type real option with the Cox-Ross-Rubinstein tree using value $S(i,j^*)$ as the start value with j^* being the appropriate j index that corresponds to $t = 1$ year (the time deferred[13]), and with $i = -j^*, -j^* + 2, \ldots, j^* - 2, j^*$. The current NPV including all real options is the maximum of the NPV when starting the project at $t = 0$ and of the discounted NPVs of starting the project at $t = 1$ year as described in the example of Section 2.6.2.

Finally, the numerical properties of this algorithm need to be analyzed. This is important since these properties determine the applicability of the real options valuation tool in practice. According to Trigeorgis [130], page 320, the Cox-Ross-Rubinstein binomial tree is stable and consistent[14]. Efficiency[15] is another criterion which will play an important role in Chapter 5, the empirical simulation part.

As shown by Trigeorgis[16] the Cox-Ross-Rubinstein binomial tree method is stable for

$$\Delta t \leq \frac{\sigma^2}{\left| r_f - \frac{1}{2}\sigma^2 \right|^2}. \tag{4.4}$$

It is also consistent within this limit but with a variance that is downward biased by $1/N$ times the square of the mean[17]. As will be shown in Chapter 5,

[13] See Trigeorgis [130], page 158.

[14] *Consistency*, also called *accuracy*, refers to the idea that the discrete-time process used for calculation has the same mean and variance for every time-step size as the underlying continuous process. *Stability*, or *numerical stability*, means that the approximation error in the computations will be dampened rather than amplified.

[15] *Efficiency* refers to the number of operations, i.e., the computational time, needed for a given approximation accuracy. It is crucial for an algorithm to be efficient in order to be of any value for financial practice.

[16] See Trigeorgis [130], pages 324-325.

[17] See Trigeorgis [132], page 335, Table 10.5.

the efficiency of the algorithm is very high even for complex real options that contain defer options. Therefore, its usefulness for Corporate Finance practice is very high.

Trigeorgis log-transformed binomial tree. The log-transformed binomial tree method was developed by Trigeorgis, a pioneer in the field of real options. This section explains the theory behind the model used in this thesis, which is based on Trigeorgis [130] (the original publication from 1991) but tailored to the cases analyzed here. However, the notation used in this thesis is slightly different from the notation in the original article in order to clarify the algorithm. Trigeorgis' goal was *to approximate the underlying continuous diffusion process when there is a series of exercise prices, non-proportional dividends, and interactions among a variety of options embedded requiring a series of capital outlays to generate expected cash flows*[18].

Trigeorgis developed his model for a constant risk-free interest rate r_f. In Section 4.4.3 this original model will be modified to accommodate for a stochastic risk-free interest rate. Regardless of whether the interest rate is constant or stochastic, the model can easily accommodate for complex real options. In Chapter 5, both models will be numerically analyzed for various cases of real options practice.

Let $V = (V_t)_{t \geq 0}$ be the (gross) present value of the expected future cash flows from immediately undertaking the investment project[19]. V is the *gross or naked project value, not including any required investment cost outlays or any embedded real options. To the firm, V represents the market value of a claim on the future cash flows from installing the project now*[20]. This project is assumed to follow a diffusion process (a Geometric Brownian Motion) given as the solution of the following SDE:

$$dV_t \;=\; \alpha V_t \, dt \;+\; \sigma V_t \, dB_t, \;\; t \geq 0. \tag{4.5}$$

$\alpha \in \mathbb{R}$ is the instantaneous expected return on the project, $\sigma \in \mathbb{R}^+$ is the instantaneous standard deviation, and $(B_t)_{t \geq 0}$ is the Standard Brownian Motion. Let $T > 0$ be the length of the investment project. Time period $[0, T]$ will now be divided into $N > 1$ equally spaced time intervals $[t_{j-1}, t_j]$ whereby time point t_j is defined as $t_j := j\Delta t$, $\Delta t := \frac{T}{N}$ for $j = 1, \ldots, N$.

With $Y_t := \ln(V_t), t \geq 0$, a new process $Y = (Y_t)_{t \geq 0}$ is defined. This is the so-called *log-transformation* that gives the model its name and its advantageous properties as shown later. In any infinitesimal time interval dt the process Y

[18] See Trigeorgis [130], page 310.

[19] This means that V is the same V as in the case of CCA and the Cox-Ross-Rubinstein tree.

[20] See Trigeorgis [130], page 311.

follows an Arithmetic Brownian Motion. Under risk-neutrality (i.e., $\alpha = r_f$) this transformation yields:

$$dY = \ln\left(\frac{V_{t+dt}}{V_t}\right) = \left(r_f - \frac{1}{2}\sigma^2\right) dt + \sigma dB_t, \quad t \geq 0,$$

for the infinitesimal time period dt. It can be shown:

$$dY \overset{i.i.d.}{=} N\left(\left[r_f - \frac{1}{2}\sigma^2\right] dt, \sigma^2 dt\right).$$

Applying the transformation $K := \sigma^2 dt$ yields

$$dY \overset{i.i.d.}{=} N(\mu K, K) \quad \text{with} \quad \mu := \frac{r_f}{\sigma^2} - \frac{1}{2}. \tag{4.6}$$

This procedure is called *to transform time as to be expressed in units of variance*[21]. The log-transformed process $(Y_t)_{t\geq 0}$ can be approximated with a discrete-time version. Within each subinterval $t_j - t_{j-1}, j = 1, \ldots, N$, this discrete-time version is a Markov random walk that goes up by ΔY with probability P and that goes down by $-\Delta Y$ with probability $1 - P$.

For the remainder the definition $H := \Delta Y$ will be used as in the original article. According to Trigeorgis [130], page 311, the following holds:

$$E(\Delta Y) = 2PH - H, \tag{4.7}$$
$$Var(\Delta Y) = H^2 - (2PH - H)^2. \tag{4.8}$$

For the discrete-time process to be consistent with the continuous-time process (4.5), the following two consistency criteria have to hold:

(*i*) mean of the continuous process = mean of the discrete process,

(*ii*) variance of the continuous process = variance of the discrete process.

According to (4.6), (4.7), and (4.8) this yields $2PH - H = \mu K$, and $H^2 - (\mu K)^2 = K$. These two equations can be used to derive P and H:

$$2PH - H = \mu K \implies P = \frac{1}{2}\left(1 + \frac{\mu K}{H}\right), \tag{4.9}$$

$$H^2 - (\mu K)^2 = K \implies H = \sqrt{K + (\mu K)^2}. \tag{4.10}$$

As mentioned, the definition $Y_t = \ln(V_t), t \geq 0$, endowes this method with its advantageous properties as shown in the following:

[21] See Trigeorgis [132], page 321.

$$
\begin{aligned}
P &\overset{(4.9)}{=} \frac{1}{2}\left(1+\frac{\mu K}{H}\right) \overset{(4.10)}{=} \frac{1}{2}\left(1+\frac{\mu K}{\sqrt{K+(\mu K)^2}}\right) \\
&\leq \frac{1}{2}\left(1+\frac{|\mu K|}{\sqrt{K+(\mu K)^2}}\right) \\
&\leq \frac{1}{2}\left(1+\frac{|\mu K|}{|\mu K|}\right), \quad \text{since} \quad K = Var(\Delta Y) \geq 0 \\
&= 1. \qquad (4.11)
\end{aligned}
$$

Moreover:

$$
\begin{aligned}
H^2 - (\mu K)^2 &\overset{(4.8),(4.9)}{=} Var(\Delta Y) \geq 0 \Longrightarrow H^2 \geq (\mu K)^2 \\
\Longrightarrow |H| \geq |\mu K| &\overset{H>0}{\Longrightarrow} H \geq |\mu K| \\
\Longrightarrow -H \leq -|\mu K| \leq |\mu K| \leq H &\overset{H>0}{\Longrightarrow} -1 \leq \tfrac{\mu K}{H} \leq 1.
\end{aligned}
$$

This gives:

$$
P \overset{(4.9)}{=} \frac{1}{2}\left(1+\frac{\mu K}{H}\right) \geq \frac{1}{2}(1-1) = 0. \qquad (4.12)
$$

(4.11) and (4.12) yield $0 \leq P \leq 1$. Since $P + (1 - P) = 1$ like probabilities, the conditions that ensure unconditional stability in this weighted average numeric scheme are fulfilled with no external constraints on K and H[22]. Moreover, this method is consistent due to (*i*) and (*ii*). To quote Trigeorgis[23]: *The* [...] *transformation of the state and time variables were designed so as to guarantee stability as well as consistency of the discrete-time approximation to the continuous process. As noted, for any weighted average numeric scheme - such as the binomial model or the explicit finite difference method* [...] *- to be stable (i.e., for the error not to blow up) the weights must be constrained between* 0 *and* 1 *and add up to* 1 *(like probabilities)* [...].

Thereby, Trigeorgis has reached the goal he defined for the algorithm[24]: *The goal is to design a consistent, stable, and efficient binomial tree method for valuing complex investments with potentially interacting options. The method is applicable both to the valuation of complex financial options* [...] *and to the valuing of capital budgeting projects with multiple real options.*

22 See Trigeorgis [130], page 312. The same can be seen for the finite difference methods. A log-transformation of the underlying provides the log-transformed finite difference methods with very good numerical properties in opposite to finite difference methods for the underlying itself without any up-front transformation.

23 See Trigeorgis [132], pages 321-322.

24 See Trigeorgis [132], page 320.

Figure 4.5 displays the important notations and definitions for the Trigeorgis log-transformed binomial tree. The discretization of the continuous process $(Y_t)_{t\geq 0}$ is denoted with $Y(i)$ whereby i is the state index of the underlying. Especially, $Y(i)$ is independent from the time state index j.

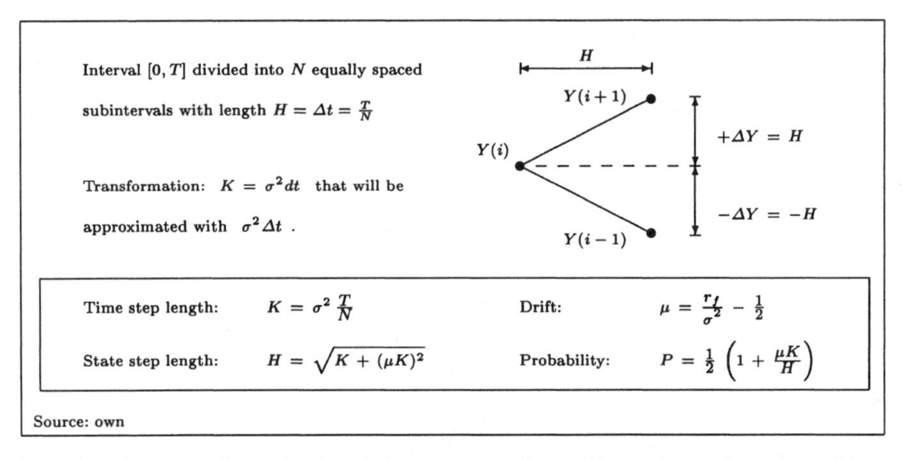

Fig. 4.5. Log-transformed binomial tree approach by Trigeorgis: basic relationships.

Figure 4.6 shows graphically the Trigeorgis log-transformed binomial tree method for the whole binomial tree and provides an overview of the notation used for the most general form of the model. Several new variables are also introduced in this figure. The variables C_l denote the external cash flows that occur during the life of the project. At any node where a cash inflow (from the perspective of the initiator of the project) takes place, the value V is decreased by this cash amount. This means for the algorithm that at these time points the *path triangle followed by* $Y = ln(V)$ *is extended downwards by an amount determined by the cash flow drop*[25]. The cash flow also determines the lower limit $M_1(j)$ and the upper limit $M_2(j)$ for the state variable i at time state j, i.e., $i \in \{M_1(j), M_1(j)+1, \ldots, M_2(j)-1, M_2(j)\}$. The variable $R = R(i,j)$ is the total investment opportunity value for node (i,j), i.e., the combined value for the project and its embedded real options.

The choice of the time index j and the state index i are explained separately in Figure 4.7[26].

[25] See Trigeorgis [130], page 321.

[26] This choice of indices i and j is different from the choice of indices that will be made in the finite difference methods as will be displayed in Figure 4.11 later in this section.

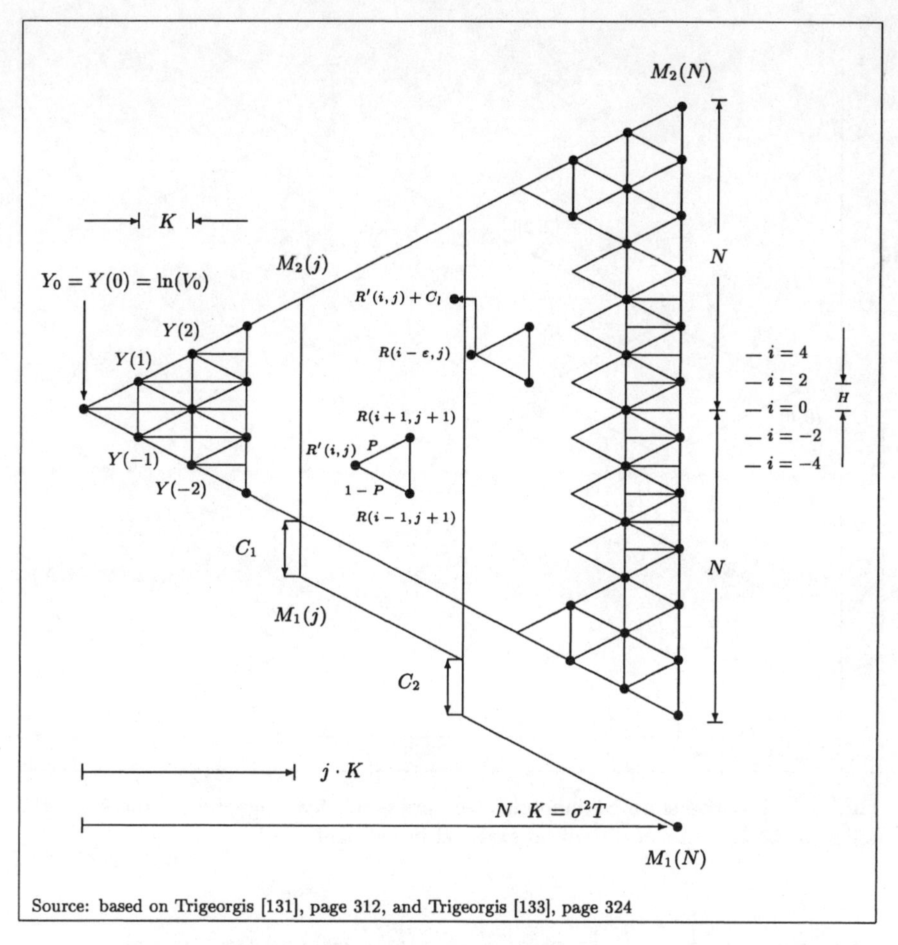

Fig. 4.6. Log-transformed binomial tree approach by Trigeorgis.

A further explanation is best provided by a quotation of Trigeorgis[27]: *For each time step j (j = N-1, ..., 1) and every second state i [the algorithm] calculates the total investment opportunity value* [...] *using information from step* $j+1$ *or earlier. That is, between any two consecutive periods, the value of the opportunity in the earlier period* (j) *at state i,* $R'(i)$ *- where the prime indicates a new or revised value - is determined iteratively from its expected end-of-period values in the up- and down-states calculated in the previous time step* $(j+1)$*, discounted back one period of length* [...] $\frac{K}{\sigma^2}$ *at the risk-free interest rate* r_f*,*

[27] See Trigeorgis [130], pages 312-314.

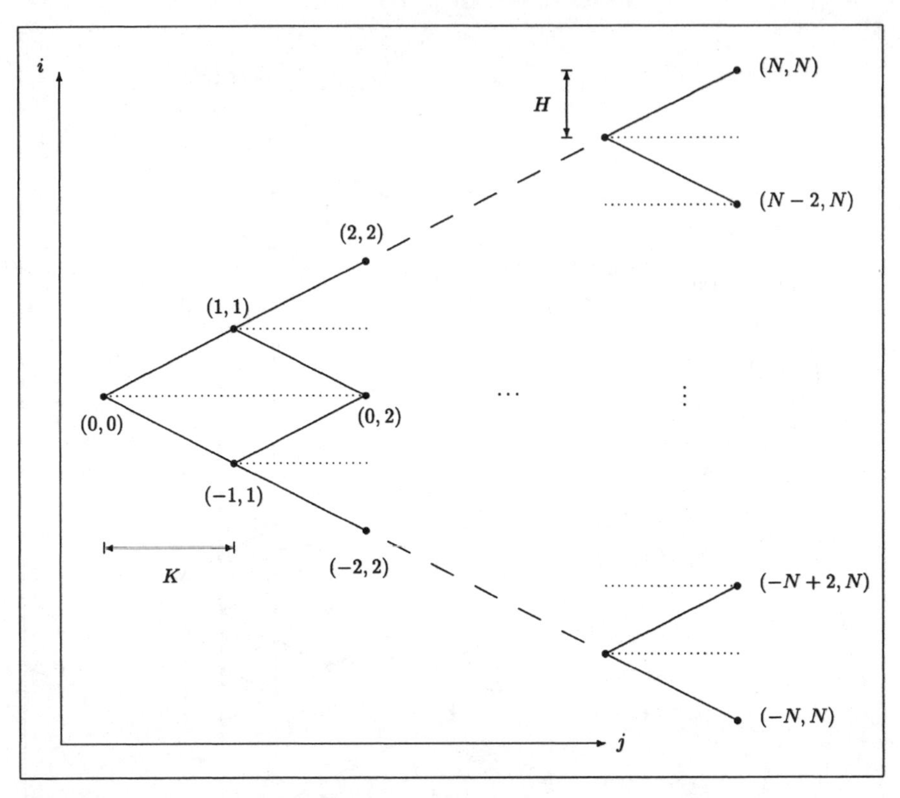

Fig. 4.7. Log-transformed binomial tree approach by Trigeorgis: indices for the nodes of the tree in the case of no external investments.

$$R'(i,j) = \exp\left\{-r_f \frac{K}{\sigma^2}\right\} [P \cdot R(i+1,j+1) + (1-P) \cdot R(i-1,j+1)]. \quad (4.13)$$

[...] *Adjustments for project cash flows* [...], *for the various real options embedded in the project, and for exogenous competitive arrivals (jumps) must be made at appropriate time within the backward iterative process* [...].

The model needs two types of adjustments:

(*i*) Adjustments due to external cash flows

(*ii*) Adjustments due to embedded real options

Adjustments (*i*) due to external cash flows are denoted with C_l in Figure 4.6 (the index l in this figure indicates that several external cash flows can occur). They can occur because of dividend payments or investment outlays. While dividend payments are not included in this thesis, investment outlays due to expanding the project are part of case 2 and case 3 in the analysis of Chapter 5.

With V the project value (excluding the real options) was denoted. At a node with an external cash flow, i.e., a node where an adjustment is necessary, the value before and after the cash flow adjustment has to be calculated. The superscript $^+$ denotes the time immediately *after* a cash flow occured (also known as *ex-dividend time*). The superscript $^-$ denotes the time immediately *before* a cash flow occured (also known as *cum-dividend time*). A cash inflow $C > 0$ (from the perspective of the initiator of the project who receives this money which is seen as a dividend the project pays) changes the value V^- of the underlying that prevails immediately before the cash flow occurs such that the new, *ex-dividend time* value V^+ is

$$V^+ = V^- - C.$$

The value of the option component is unchanged[28]. Therefore, the total opportunity value R' immediately after the cash flow is

$$R'(V^+) = R(V^- + C) - C.$$

As introduced above, R' stands for a new or revised value of R. For implementation purposes, the binomial lattice has to be adjusted in the following way according to Trigeorgis[29]: [...] *at each time a cash inflow is received, the "path triangle" followed by $Y(:= \ln(V))$ is extended downwards [see Figure 4.6] by an amount determined by the cash flow drop. The revised ("ex-dividend") total value of the investment opportunity (asset plus option) at state index i, $R'(i)$, is then obtained from the opportunity value corresponding to some index ε nodes lower, $-\varepsilon$ (with ε depending on the cash flow amount C), that is*

$$R'(i,j) = R(i - \varepsilon, j) - C.$$

In general, with discrete non-proportional cash flows, each node in the "tree" is shifted by a different amount (ε differs for each i). Only in the special case when cash flow is a constant proportion of asset (project) value is the whole tree shifted by a constant amount. Note that the shift is done in a way that the original ("predividend") node structure is always maintained (i.e., nodes recombine).

Adjustments (*ii*) due to the embedded real option(s) depend on the real option(s) inherent in the investment project. In the presence of real options (simple and complex), the general equation (4.13) has to be modified. These adjustments are summarized in Table 4.1. $V(i) = e^{Y(i)}$ is the project value at node (i, j) that is independent from j, and X is the salvage value.

[28] See Trigeorgis [132], page 325.

[29] See Trigeorgis [132], pages 325-326.

Table 4.1. Adjustment of the total opportunity value according to the type of the real option.

Type of real option	**Adjustment**
Switch use (or abandon for salvage X)	$R'(i,j) = \max\{R(i,j), X\}$
Expand by e through investing an additional amount I_e	$R'(i,j) = R(i,j) + \max\{eV(i) - I_e, 0\}$
Contract by c, saving I_c	$R'(i,j) = R(i,j) + \max\{I_c - cV(i), 0\}$
Abandon by defaulting on an investment I	$R'(i,j) = \max\{R(i,j) - I, 0\}$
Defer until next period	$R'(i,j) = \max\{e^{-r_f \frac{\sigma^2}{K}} E(R(i,j+1)), R(i,j)\}$

Source: based on Trigeorgis [130], page 323

The implementation of the Trigeorgis log-transformed binomial lattice scheme is done in four steps. Figure 4.8 shows these four steps in an overview for the most general form of the Trigeorgis log-transformed model. This model will then be specified for the three cases that will be analyzed in Chapter 5 of this thesis.

When step 4 is completed, i.e., $j = 0$ is reached, $R'(0,0)$ gives the investment value including all embedded real options for the start value $V(0) = V_0$. It is important to notice that the Trigeorgis model also allows to adjust for exogenous jumps, thereby modelling competitive entry. Since this is not part of the simulation in the following chapter, the theory of these jumps will not be explained here[30].

After having presented the Trigeorgis log-transformed binomial lattice method in its most general form, it will now be specified for the three cases that will be analyzed in Chapter 5:

[30] For more details on this see Trigeorgis [132], pages 328-329.

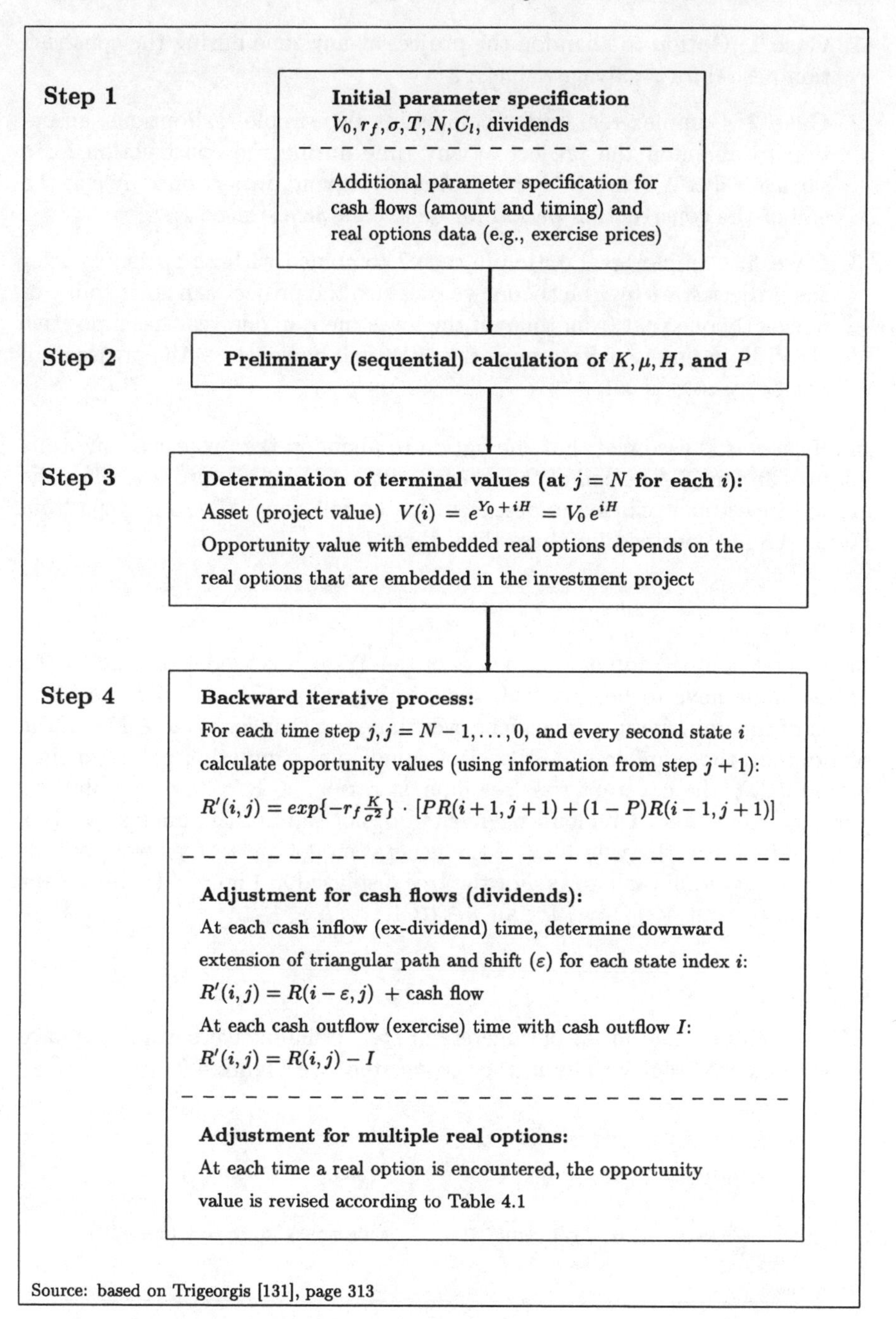

Fig. 4.8. Flow chart of the Trigeorgis log-transformed binomial lattice algorithm in its most general form.

1. **Case 1:** Option to abandon the project at any time during the construction period for a salvage value X.
2. **Case 2:** Complex real option comprising two simple real options: an option to abandon the project at any time during the construction for a salvage value X (case 1) and an option to expand project once by e at the end of the construction period for an expand investment I_e.
3. **Case 3:** Complex real option in case 2 combined with an option to defer the project start by exactly one year. Thus, the project can start today or in exactly one year from today if the investment in one year has a positive NPV. If, in one year from now, the NPV will be negative, the project will not be started at all.

In all cases it is assumed that the option to abandon the project at any time matures at time T, the end of the construction period of the project, when the expand investment can be undertaken. In the following, the four steps from Figure 4.8 are specified for these three cases.

Step 1:

In the first step the input parameters of the Trigeorgis model that affect the option value have to be specified: the starting value $V_0 \in \mathbb{R}^+$ for the simulation of the underlying's movement over time, the time length $T \in \mathbb{R}^+$ of the real option, the number $N \in \mathbb{N}\backslash\{1\}$ of equidistant simulation points in time period $[0, T]$, the constant risk-free interest rate $r_f \in \mathbb{R}^+$, and the volatility parameter $\sigma \in \mathbb{R}^+$. Dividend payments are not part of the analysis in this thesis. Moreover, the values C_l of the general model are zero as well, not allowing for external cash flows. For the tree displayed in Figure 4.6, this means $M_1(j) = -j$ and $M_2(j) = j$ for all $j \in \{0, 1, \ldots, N\}$.

Step 2:

After specifying the initial parameters in step 1 the lattice parameters have to be calculated sequentially in the second step, see Figure 4.9.

1. Time step length:	$K = \sigma^2 \frac{T}{N}$	2. Drift:	$\mu = \frac{r_f}{\sigma^2} - \frac{1}{2}$
3. State step length:	$H = \sqrt{K + (\mu K)^2}$	4. Probability:	$P = \frac{1}{2}\left(1 + \frac{\mu K}{H}\right)$

Source: own

Fig. 4.9. Basic relationships within the Trigeorgis log-transformed binomial tree approach.

Note that the time steps are independent from the chosen risk-free rate r_f. However, both the drift and the state step length depend on r_f. The relevance of this observation will become evident in Section 4.4.3 when this model is modified to accommodate for a non-constant risk-free interest rate.

Step 3:
In step 3 the underlying asset value for each node is calculated. As can be seen in Figure 4.7, the index j, $0 \leq j \leq N$, denotes the time step within the lattice. The index i, $M_1(j) \leq i \leq M_2(j)$, is the total number of ups minus downs at a specific time step j. Since $Y = \ln(V)$ and $Y(i) = Y_0 + iH$ according to the construction of the lattice, the asset value is given as

$$V(i) = e^{Y_0+iH} \quad \text{for } i \in \mathbb{N} \text{ with } -j = M_1(j) \leq i \leq M_2(j) = j$$

for each $0 \leq j \leq N$. Moreover, the terminal boundary conditions $R(i, N)$ for the last time period have to be specified for all i. This specification depends on the investment project and its embedded real option(s). For the three cases analyzed in this thesis, this means for $-j = M_1(j) \leq i \leq M_2(j) = j$:

1. **Case 1:** $R'(i, N) = max\{V(i), X\}$
2. **Case 2:** $R'(i, N) = max\{V(i), V(i)(1+e) - I_e, X\}$
3. **Case 3:** $R'(i, N) = max\{V(i), V(i)(1+e) - I_e, X\}$

Step 4:
The backward iterative process is the core in the final step 4. It starts at $j = N - 1$ and goes down to $j = 0$, which will give the NPV of the project. At each $j < N - 1$, the value $R(i, j)$ has to be adjusted for the real options embedded in the investment project.

1. **Case 1:** $R'(i, j) = max\{V(i), X\}$.
2. **Case 2:** $R'(i, j) = max\{R(i, j) + max\{eV(i) - I_e, 0\}, X\}$
3. **Case 3:** $R'(i, j) = max\{R(i, j) + max\{eV(i) - I_e, 0\}, X\}$.

   The option-to-defer feature is realized in a way best expressed by Trigeorgis himself: The start of the investment is deferred by one more period *if the discounted expected opportunity value from initiating the project next period $(j+1)$ exceeds the value from exercising or investing in the current period (j). [This] requires multiple parallel runs of the rest of the program already adjusted for further options and pairwise comparisons conditional on project initiation in successive periods when moving backwards starting from the end of the deferrable period [...] to the present* [31].

[31] See Trigeorgis [130], page 323.

Finally, the total investment value is given with $R'(0,0)$, which includes all embedded real options for the start value V_0. To determine the current NPV of the investment project, the initial investment cost I_0 has to be substracted, i.e., $\text{NPV} = R'(0,0) - I_0$.

4.2.2 Finite Difference Methods

In the following, the basic principles of the finite difference methods will be explained by deriving an algorithm for the pricing of a real option to abandon a project at any time for a salvage value. The finite difference methods presented in this thesis (and implemented in the computer simulation program) are based on the approximation of the partial differential equation that describes the development of the real options value with respect to the log-transformed underlying instead of being based on the plain underlying itself. Therefore, the finite difference methods are called log-transformed finite difference methods.

Compared with finite difference methods without a log-transformation of the underlying, the log-transformed finite difference methods exhibit advantageous properties: *A logarithmic transformation of the state variable can improve certain computational attributes of finite-difference (and other) methods, such as stability and efficiency*[32].

In particular, the log-transformed implicit and explicit finite difference methods will be derived here with the description being based on the original article of Brennan and Schwartz[33]. Other finite difference methods like the Crank-Nicholson algorithm are not topics of this thesis[34].

The log-transformed finite difference method is described for an American put option. In the language of real options, this is an option to abandon a project at any time for a salvage value[35]. The salvage value is the strike price of the put option and will be denoted with X.

---

[32] See Trigeorgis [130], page 315. A good comparison of the log-transformed finite difference methods with the non log-transformed finite difference methods is given in Trigeorgis [132], Section 10.2.

[33] See Brennan & Schwartz [17].

[34] Various authors deal with finite difference methods which are different from the explicit and implicit ones. For example: Clewlow & Strickland [31], pages 56-76; Deutsch [39], Chapter 10; Hull [62], Section 15.8; Hull & White [64]; Trigeorgis [132], Section 10.2; Wilmott [135], Part 6. It is also worth mentioning that the notation varies between the different authors.

[35] This is case 1 of the three real options cases that will be analyzed in Chapter 5.

Let $S = S_t$, $t \geq 0$, be the price of the underlying at time t. The price is assumed[36] to follow the stochastic differential equation

$$dS = r_f S dt + \sigma S dB_t, \ t \geq 0. \tag{4.14}$$

r_f is the constant risk-free interest rate (the drift of the price process), and σ is the constant volatility of the price process. The starting point of the finite difference theory is the partial differential equation that describes the development of the option value over time depending on the underlying value. This partial differential equation will result when considering a portfolio that comprises a short position in the option (value $F = F(S,t), t \geq 0$) and a long position of $\frac{\partial F}{\partial S}$ units of the underlying. Let P denote the value of this portfolio. Then

$$P = -F + \frac{\partial F}{\partial S} S. \tag{4.15}$$

Using the Ito formula yields after some calculations[37]:

$$\frac{1}{2}\sigma^2 \frac{\partial^2 F}{\partial S^2} + r_f S \frac{\partial F}{\partial S} + \frac{\partial F}{\partial t} - r_f F = 0, \quad t \geq 0. \tag{4.16}$$

The boundary conditions of this partial differential equation determine the type of the option[38]. The American put option with strike price X described here has the following boundary conditions:

$$F(S,T) = \max(X - S, 0), \ F(0,T) = X, \text{ and } \lim_{S \longrightarrow \infty} \frac{F(S,T)}{S} = 0. \tag{4.17}$$

[36] Under the assumption of risk-neutrality, all assets earn the risk-free interest rate, see Clewlow & Strickland [31], page 6.

[37] See Clewlow & Strickland [31], Chapter 1.3. Equation (4.16) is not the most general form of the partial differential equation since it is also possible to include dividend payments of the underlying and the option in the equation. In this case the partial differential equation would be

$$\frac{1}{2}\sigma^2 \frac{\partial^2 F}{\partial S^2} + (r_f - \delta) S \frac{\partial F}{\partial S} + \frac{\partial F}{\partial t} - r_f F + d = 0, \quad t \geq 0,$$

with d as the constant dividend yield of the option (as a security) and δ as the constant dividend yield of the underlying. But since this thesis does not consider dividend payments, d and δ are both set to zero. In the case of an American call option, this simplification means that the call price is the same for European and American call options since it is never optimal to exercise the call early according to Merton [94]. However, in the case of an American put option early exercise might be optimal even if the underlying pays no dividends, as explained in 2.6.3. Therefore, the price of an American put option and a European put option are generally different.

[38] See Trigeorgis [132], page 312, and Hull [62], Chapter 11.5.

Although equation (4.16) could now be approximated, it will not be applied here since such an approach yields a finite difference scheme with poor numerical properties. For example, generating finite difference methods based on (4.16) yields coefficients that depend on the state of the underlying variable, which results in an unstable algorithm[39]. If, however, the approach is modified by a log-transformed underlying process instead of being based on the underlying process itself, the resulting log-transformed finite difference methods exhibit advantageous properties. Especially, the coefficients of the resulting algorithm are independent from the state of the underlying.

To determine the log-transformed partial differential equation, the underlying process is transformed via $Y := \ln(S)$. With the definition $W(Y,t) := F(S,t)$, this yields the relationships

$$\frac{\partial F}{\partial S} = \frac{\partial W}{\partial Y} e^{-Y}, \quad \frac{\partial^2 F}{\partial S^2} = \left(\frac{\partial^2 W}{\partial Y^2} - \frac{\partial W}{\partial Y}\right) e^{-2Y}, \quad \text{and} \quad \frac{\partial F}{\partial t} = \frac{\partial W}{\partial t}. \tag{4.18}$$

This gives the following partial differential equation for the log-transformed underlying process $Y = \ln(S)$:

$$\frac{1}{2}\sigma^2 \frac{\partial^2 W}{\partial Y^2} + \left(r_f - \frac{1}{2}\sigma^2\right) \frac{\partial W}{\partial Y} + \frac{\partial W}{\partial t} - r_f W = 0, \; t \geq 0. \tag{4.19}$$

To quote Brennan and Schwartz[40]: [...] *(4.19) unlike (4.16) is a partial differential equation with constant coefficients. This simplifies the numerical analysis, and* [...] *makes it possible to employ an explicit finite difference approximation to (4.19), whereas the explicit finite difference approximation to (4.16) is in general unstable.*

The explicit and implicit finite difference methods will now be applied on partial differential equation (4.19). To do this, the observation period has to be specified first and is set from today (time point $t = 0$) to time point T. This period will be divided into N equally spaced time intervals with length $k = \Delta t = \frac{T}{N}$. The price of the underlying (either the project value or the value of a twin security with the same risk profile) is assumed to stay in a range $[S_{min}, S_{max}]$ with $0 < S_{min} < S_{max}$. The log-transformed range $[\ln(S_{min}), \ln(S_{max})]$ is divided into $M \in \mathbb{N}$ equally spaced state intervals with length

$$h := \frac{\ln(S_{max}) - \ln(S_{min})}{M}.$$

[39] See Schulmerich [118], Section 4.3.1.

[40] See Brennan & Schwartz [17], pages 462-463.

Consequently, the underlying price itself is no longer equally spaced but simply the log-transformed price. Figure 4.10 displays the situation for the underlying state and the time state in the log-transformed case. It will be compared with the case of S itself being the underlying.

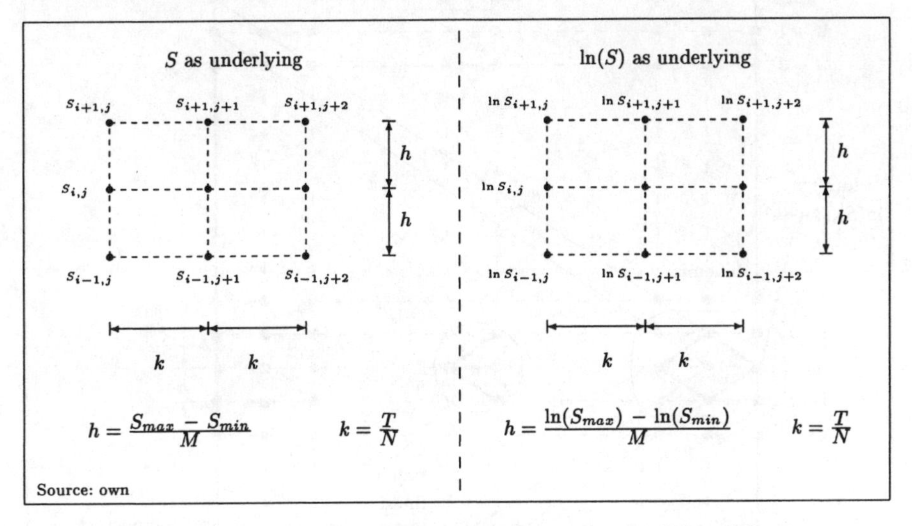

Fig. 4.10. Development of the underlying over time for the finite difference versus the log-transformed finite difference methods.

Several methods exist regarding how to discretize equation (4.19). Each method determines another algorithm that can be used to value the option. Common to all these methods is the grid structure to price the option and the choice of indices to name stock value and option value at each node of the tree.

Figure 4.10 shows graphically: Each node is identified by two indices (i, j) where i denotes the underlying state and j the time state in the grid. In this thesis the notation is $S_{i,j}$ for finite difference methods and $S(i, j)$ for the Cox-Ross-Rubinstein and the Trigeorgis log-transformed binomial tree methods. The same notation is used for the real option values and the total project values.

Figure 4.11 gives the general overview of the log-transformed finite difference methods presented in this thesis.

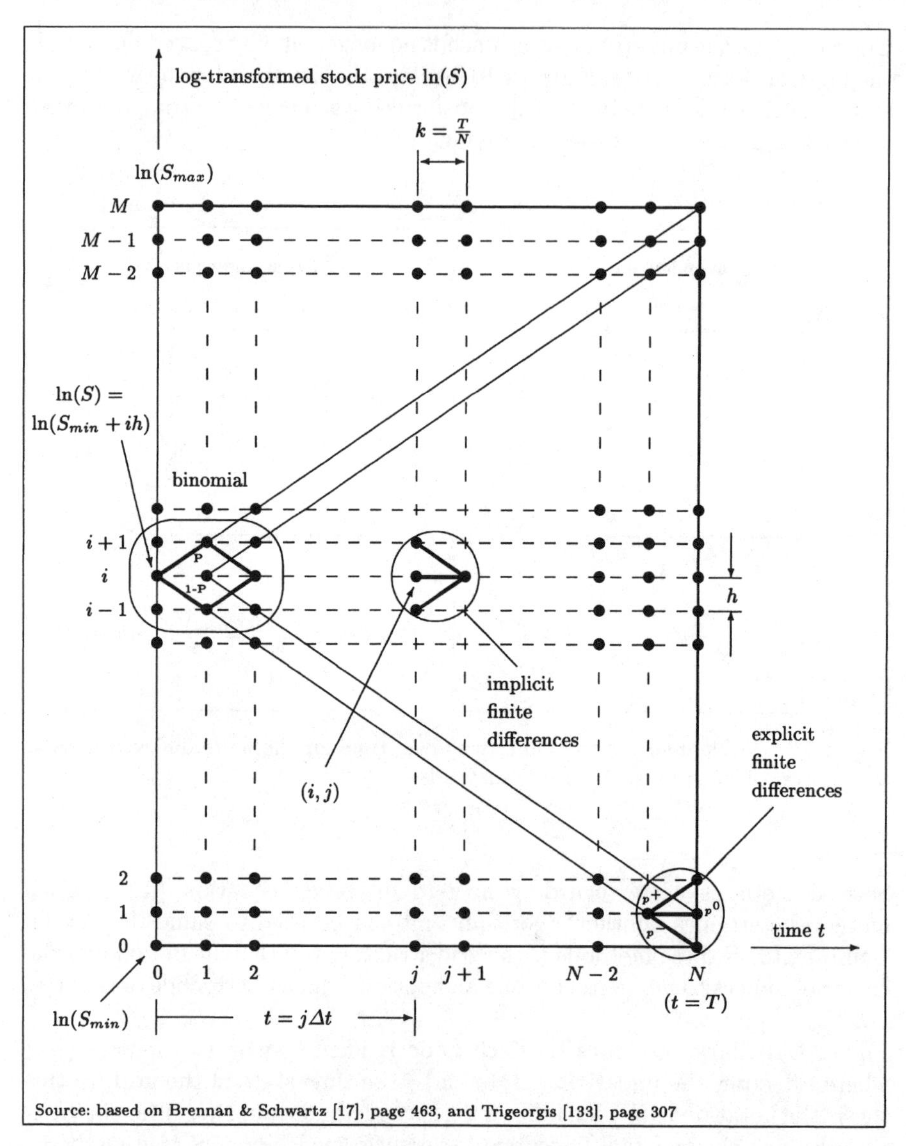

Fig. 4.11. Grid for the log-transformed finite difference methods.

Log-transformed explicit finite difference method. The following discretization at time point $j+1$ is used for the log-transformed explicit finite difference method whereby $j \in \{0, 1, \ldots, N-1\}$ and $i \in \{1, 2, \ldots, M-1\}$. This discretization is called a *discretization at time point* $j+1$ since the first and second derivatives of F with respect to Y contain the time index $j+1$:

$$\left.\begin{aligned} W &\approx W_{i,j}, \\ \frac{\partial W}{\partial Y} &\approx \frac{W_{i+1,j+1} - W_{i-1,j+1}}{2h}, \\ \frac{\partial^2 W}{\partial Y^2} &\approx \frac{W_{i+1,j+1} - 2W_{i,j+1} + W_{i-1,j+1}}{h^2}, \\ \frac{\partial W}{\partial t} &\approx \frac{W_{i,j+1} - W_{i,j}}{k}. \end{aligned}\right\} \tag{4.20}$$

Putting (4.20) into equation (4.19) yields for each $j \in \{0, 1, \ldots, N-1\}$ and for each $i \in \{1, 2, \ldots, M-1\}$:

$$p_i^- W_{i-1,j+1} + p_i^0 W_{i,j+1} + p_i^+ W_{i+1,j+1} = W_{i,j}(1 + r_f k). \tag{4.21}$$

Thereby:

$$\left.\begin{aligned} p^- &:= \tfrac{1}{2}\left[\frac{\sigma^2 k}{h^2} - \frac{k}{h}\left(r_f - \frac{\sigma^2}{2}\right)\right], \\ p^0 &:= 1 - \frac{\sigma^2 k}{h^2}, \\ p^+ &:= \tfrac{1}{2}\left[\frac{\sigma^2 k}{h^2} + \frac{k}{h}\left(r_f - \frac{\sigma^2}{2}\right)\right]. \end{aligned}\right\} \tag{4.22}$$

It is important to notice that the coefficients p^-, p^0, and p^+ in (4.22) are independent from index i, the underlying state index, and independent from index j, the time index[41]. However, using (4.21) and (4.22) does not yet allow to solve the valuation problem completely since for time $j \in \{0, 1, \ldots, N-1\}$ the $M-1$ equations only give the values for $M-1$ unknown variables but not the boundary values $W_{0,j}$ and $W_{M,j}$. But since the algorithm runs from right to left in the grid, these boundary values are needed. The boundary conditions have already been specified in (4.17) for the continuous-time situation. They can also be written in the following form:

$$\left.\begin{aligned} \frac{\partial F}{\partial S} &= \lambda_u \quad \text{for the upper boundary,} \\ \frac{\partial F}{\partial S} &= \lambda_l \quad \text{for the lower boundary.} \end{aligned}\right\} \tag{4.23}$$

[41] If no log-transformation took place before approximating the partial differential equation, i.e., if partial differential equation (4.16) was approximated, the coefficients p^-, p^0, and p^+ are not independent from index i any more but still independent from j, see Schulmerich [118], Section 4.3.2.

In the case of an option to abandon a project at any time for a salvage value, the specifications $\lambda_u = 0$ and $\lambda_l = -1$ are necessary[42]. In terms of the discretization of the explicit finite difference method it holds for each $j \in \{0, 1, \ldots, N-1\}$:

$$\left.\begin{aligned} \frac{F_{M,j} - F_{M-1,j}}{S_{M,j} - S_{M-1,j}} &= \lambda_u, \\ \frac{F_{1,j} - F_{0,j}}{S_{1,j} - S_{0,j}} &= \lambda_l. \end{aligned}\right\} \tag{4.24}$$

According to the specifications of λ_u and λ_l this gives

$$F_{M,j} = F_{M-1,j} \quad \text{and} \quad F_{1,j} - F_{0,j} = -(S_{1,j} - S_{0,j}) \tag{4.25}$$

for all $j \in \{0, 1, \ldots, N-1\}$. Since $W(Y,t) = F(S,t)$, and according to the definition of h, the following relationships hold for each $i \in \{0, 1, \ldots, M\}$ and $j \in \{0, 1, \ldots, N\}$:

$$W_{i,j}(Y_{i,j}, t) = F_{i,j}(S_{i,j}, t) \quad \text{and} \quad S_{i,j} = S_{min}\, e^{ih}. \tag{4.26}$$

(4.25) and (4.26) yield as the boundary conditions in the discrete situation:

$$W_{M,j} = W_{M-1,j} \quad \text{and} \quad W_{1,j} - W_{0,j} = S_{min}\,(1 - e^{h}). \tag{4.27}$$

Bundling the two equations from (4.27) and the $M-1$ equations from (4.21) gives $M+1$ equations that determine the $M+1$ unknown variables $W_{i,j}$ for each $j \in \{0, 1, \ldots, N-1\}$. On each node (i, j) the option condition needs to be specified. For the option to abandon the project at any time for a salvage value X, $X - S_{i,j}$ has to be compared with $F_{i,j}$. The $F_{i,j}$ used for the next calculation step at time $j-1$ has to be the maximum of these two values. Finally, this provides the values $W_{i,0}$ for $i = 0, 1, \ldots, M$, and also yields a distribution of the current real options value depending on the underlying project value in the range $[S_{min}, S_{max}]$. Therefore, the log-transformed explicit finite difference method works just like a lattice approach[43].

Finally, the numerical properties of this algorithm have to be analyzed. The log-transformed explicit finite difference scheme only exhibits stability if the coefficients p^-, p^0, and p^+ do not become negative[44]. According to Trigeorgis [132], page 316, the stability is satisfied if

[42] For this to hold true, the values for S_{min} and S_{max} have to be chosen such that the put option is deeply in the money and deeply out of the money, respectively.

[43] See Trigeorgis [132], page 315.

[44] According to Brennan & Schwartz [17], page 464: *For the stability of the explicit solution, it is necessary that the coefficients of (4.19) are nonnegative (McCracken*

$$k \leq \frac{h^2}{\sigma^2} \quad \text{and} \quad h \leq \frac{\sigma^2}{\left|r_f - \frac{1}{2}\sigma^2\right|}. \tag{4.28}$$

According to Hull [62], page 376, the numerically most efficient choice of the input parameters is $h = \sigma\sqrt{3k}$.

However, there are also some disadvantages to this algorithm[45]: *The variance of the [...] process [dY] is a downward-biased estimate of the variance of the continuous diffusion process, $\sigma^2 k$, by the square of the mean jump $(r_f - \frac{1}{2}\sigma^2)k$.*

If the conditions in (4.28) do not hold, the results of this algorithm can oscillate extremely. This will be shown later in Figure 4.12 in comparison with the log-transformed implicit finite difference method (Figure 4.13).

Log-transformed implicit finite difference method. As opposed to the log-transformed explicit finite difference method, the discretization at time point j is used for the log-transformed implicit finite difference method whereby $j \in \{0, 1, \ldots, N-1\}$ and $i \in \{1, 2, \ldots, M-1\}$:

$$\left.\begin{aligned} W &\approx W_{i,j+1}, \\ \frac{\partial W}{\partial Y} &\approx \frac{W_{i+1,j} - W_{i-1,j}}{2h}, \\ \frac{\partial^2 W}{\partial Y^2} &\approx \frac{W_{i+1,j} - 2W_{i,j} + W_{i-1,j}}{h^2}, \\ \frac{\partial W}{\partial t} &\approx \frac{W_{i,j+1} - W_{i,j}}{k}. \end{aligned}\right\} \tag{4.29}$$

This discretization complicates the algorithm but results in better numerical properties. Putting (4.29) into equation (4.19) yields for each $j \in \{0, 1, \ldots, N-1\}$ and for each $i \in \{1, 2, \ldots, M-1\}$:

$$p^- W_{i-1,j} + p^0 W_{i,j} + p^+ W_{i+1,j} = (1 - r_f k)\, W_{i,j+1}. \tag{4.30}$$

Thereby:

& Dorn [90]). While appropriate choice of h and k may guarantee this for (4.19), the corresponding coefficients of the explicit approximation to (4.16) depend on i, and will be negative for sufficiently large values of i, so that this explicit finite difference approximation may not be applied to the untransformed equation (4.16).

[45] See Trigeorgis [130], pages 316-317.

$$\left.\begin{array}{lcl} p^- & := & -\frac{1}{2}\left[\frac{\sigma^2 k}{h^2} + \frac{k}{h}\left(r_f - \frac{\sigma^2}{2}\right)\right], \\ p^0 & := & 1 + \frac{\sigma^2 k}{h^2}, \\ p^+ & := & -\frac{1}{2}\left[\frac{\sigma^2 k}{h^2} - \frac{k}{h}\left(r_f - \frac{\sigma^2}{2}\right)\right]. \end{array}\right\} \quad (4.31)$$

Again, the coefficients p^-, p^0, and p^+ in (4.31) are independent from i and j. The linear equation system (4.30) comprises $M-1$ equations but $M+1$ unknown variables. Therefore, to solve the linear equation system (4.30), the two boundary conditions are needed. Of course, they are the same as for the log-transformed explicit finite difference scheme, i.e., for an option to abandon a project at any time for a salvage value, the two boundary conditions are[46]

$$W_{M,j} = W_{M-1,j} \quad \text{and} \quad W_{1,j} - W_{0,j} = S_{min}\,(1 - e^h). \quad (4.32)$$

Bundling (4.30) and (4.32) yields a linear equation system with $M+1$ equations and $M+1$ unknown variables. To solve this system, it is advisable to introduce a more suitable notation. Therefore, the matrix $D \in \mathbb{R}^{M+1,M+1}$ and the vectors $F_{j+1} \in \mathbb{R}^{M+1}$ and $F_j^{r_f} \in \mathbb{R}^{M+1}$, $j = 0, 1, \ldots, N-1$, will be defined as follows:

$$D := \begin{pmatrix} -1 & 1 & & & & \\ p^- & p^0 & p^+ & & & \\ & p^- & p^0 & p^+ & & \\ & & \ddots & \ddots & \ddots & \\ & & & p^- & p^0 & p^+ \\ & & & & -1 & 1 \end{pmatrix}$$

$$F_j := (F_{0,j}, F_{1,j}, \ldots, F_{M,j})^T,$$

$$F_{j+1}^{r_f} := (1 - r_f k) \cdot \left(S_{min}\left[1 - e^h\right], F_{1,j+1}, \ldots, F_{M-1,j+1}, 0\right)^T.$$

Using these definitions, the linear equation system (4.30) with the coefficients (4.31) and the boundary conditions (4.32) can be written as

$$D\,F_j = F_{j+1}^{r_f}, \quad j = 0, 1, \ldots, N-1.$$

[46] For this to hold true, the values for S_{min} and S_{max} have to be chosen such that the option is deeply in the money and deeply out of the money, respectively.

In the case of $det(D) \neq 0$, the inverted matrix D^{-1} can be built[47] such that the solution of the linear equation system is given as

$$F_j \quad = \quad D^{-1} \, F_{j+1}^{r_f} \quad \text{for } j = 0, 1, \ldots, N-1.$$

On each node (i, j) the option condition needs to be specified. For the option to abandon the project at any time for a salvage value X, $X - S_{i,j}$ has to be compared with $F_{i,j}$. The $F_{i,j}$ used for the next calculation step at time $j - 1$ has to be the maximum of these two values. This, finally, gives all values $F_{i,j}$ for the grid. In particular, it gives the values $F_{i,0}$ as the current value of the real option for the underlying state $i \in \{0, 1, \ldots, M\}$.

The log-transformed implicit finite difference method needs only the last condition of (4.28) as a stability criterion[48]:

$$h \;\leq\; \frac{\sigma^2}{\left| r_f \;-\; \frac{1}{2}\sigma^2 \right|}. \tag{4.33}$$

Therefore, the log-transformed implicit finite difference method is better suited in practice since it leads to results in cases where the log-transformed explicit finite difference method no longer produces stable results. An example for the parameters $S_{min} = 42.5$, $S_{max} = 125$, $M = 40$, $\sigma = 0.1$, $T = 2$, $N = 24$, $r_f = 0.03$, $X = 75$ yields

$$k = \frac{T}{N} = 0.08333 \qquad \text{and} \qquad h = \frac{\ln(S_{max}) - \ln(S_{min})}{M} = 0.02697.$$

These parameter choices only fulfill the last stability condition in (4.28) for the log-transformed explicit finite difference method but not the first one, since k is larger than $\frac{h^2}{\sigma^2} = 0.0727$. The resulting total project value including the option to abandon the project at any time for a salvage value can be seen in Figures 4.12 and 4.13 for the log-transformed explicit and implicit finite difference methods, respectively.

However, even the log-transformed implicit finite difference method can give variable results if the stability condition (4.33) is not fulfilled[49].

[47] A simple Gauss algorithm can be used here, see, e.g., Stoer [125], page 149.

[48] See Brennan & Schwartz [17], page 470, equation (48).

[49] This is, e.g., the case if $S_{min} = 1$, $S_{max} = 1000$, $M = 20$, $\sigma = 0.1$, $T = 5$, $N = 1000$, $r_f = 0.08$, and $X = 75$ are chosen as input parameters.

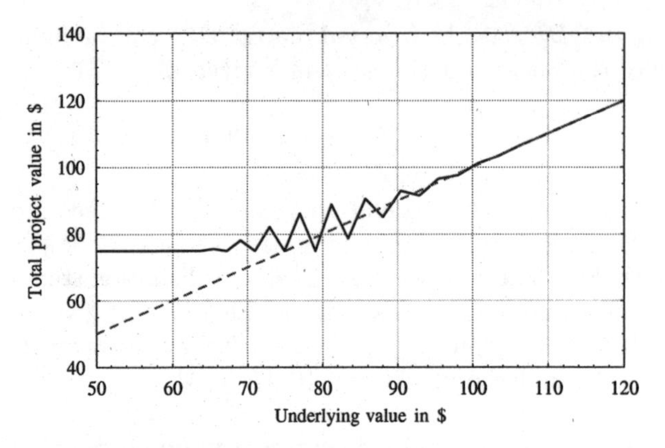

Fig. 4.12. Total value of the investment project including the option to abandon the project at any time for a salvage value X calculated with the log-transformed explicit finite difference method when the stability conditions are not fulfilled ($S_{min} = 42.5$, $S_{max} = 125$, $M = 40$, $\sigma = 0.1$, $T = 2$, $N = 24$, $r_f = 0.03$, $X = 75$).

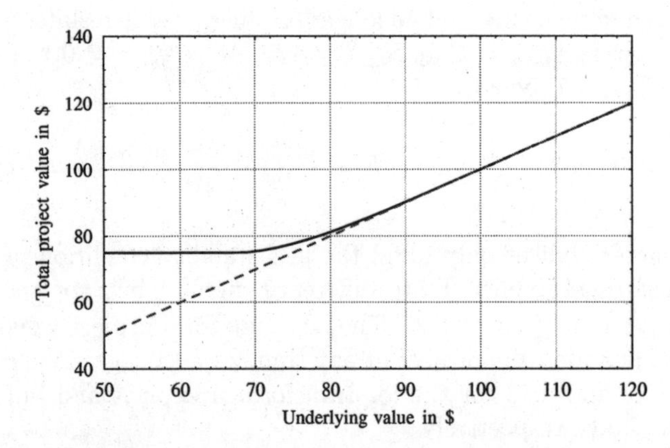

Fig. 4.13. Total value of the investment project including the option to abandon the project at any time for a salvage value X calculated with the log-transformed implicit finite difference method when the stability condition is fulfilled ($S_{min} = 42.5$, $S_{max} = 125$, $M = 40$, $\sigma = 0.1$, $T = 2$, $N = 24$, $r_f = 0.03$, $X = 75$).

4.3 Schwartz-Moon Model

This section introduces the Schwartz-Moon model[50] for evaluating research and development projects in new drug development processes. The model is presented here as an introduction to real options pricing with non-constant interest rates.

Investment projects in R&D are typical areas of real options application for several reasons. The most important one is that these projects are long-term investments[51] that exhibit many uncertainties. For example, uncertainties about the investment cost, future payoffs from the investment, and unforeseen events like competitor entry are crucial parts of such projects. Because of these uncertainties, it is very difficult to quantitatively analyze investments in R&D.

The Schwartz-Moon model is a stochastic model that describes the value of an investment project (the *asset*) depending on the cost process to complete the investment project. This model does not offer a closed-form solution but involves solving a number of numerical procedures. It includes an NPV solution of the model which in the case of a deterministic instead of a stochastic world can be used as a starting point of the numerical solution. Moreover, this so-called *static solution* can also be used as a benchmark for the stochastic model itself. The deterministic NPV solution especially is also a solution of the continuous-time differential equation in certain circumstances.

While the Schwartz-Moon model applies a constant risk-free interest rate for the stochastic and the static solutions, it is presented here with the purpose to include a non-constant risk-free interest rate within an option to defer (deferred project start). This will and can only be done for the static solution.

The Schwartz-Moon model approaches investments in R&D by modelling the investment project via three stochastic processes:

1. the investment cost process,
2. the process that describes the future payoffs from the investment, and
3. the process that models catastrophic events which might end the investment project.

The idea of the Schwartz-Moon model is based on a model developed by Pindyck[52]. Its basic approach is to distinguish two stages. The first stage is the stage to complete the investment/asset. The second stage is the income

[50] See Schwartz & Moon [121].

[51] In the U.S. the development process for a successful drug takes on average 10-12 years, see Brennan & Schwartz [121], page 87.

[52] See Pindyck [106].

generating stage which starts after completion of the investment. Changes in the expected cost of completion of the project are assumed to follow a diffusion process with variance proportional to the level of investment. This means that learning about the expected total cost to completion of the project can only occur with investment[53]. In the following, the three sources of uncertainty within the model are described mathematically:

1. **First source of uncertainty: the investment cost process** $(K_t)_{t\geq 0}$

   Let $(\tilde{K}_t)_{t\geq 0}$ be the cost to completion of the project. Then, $K_t := E(\tilde{K}_t), t \geq 0$, gives the expected cost to completion, a variable that depends on $\tilde{K}_t$ which itself is a stochastic variable. Therefore, $\tilde{K}_t$ can be seen as a stochastic variable for each $t \geq 0$, i.e., $(K_t)_{t\geq 0}$ is a stochastic process. This (expected) investment cost process $(K_t)_{t\geq 0}$ is assumed to follow the SDE

$$dK_t \quad = \quad -I\,dt \; + \; \beta\sqrt{I\,K_t}\,dB_t^1, \quad t \geq 0, \quad K_0 \in \mathbb{R}^+ \text{ constant.}$$

   I is a non-negative constant that gives the rate of investment[54] and $\beta \in \mathbb{R}$ is a parameter that calibrates the volatility term in the stochastic differential equation to real world data via the equation[55] $Var(\tilde{K}_t) = \frac{\beta^2}{2-\beta^2}[E(\tilde{K}_t)]^2, t \geq 0$. $(B_t^1)_{t\geq 0}$ is the Standard Brownian Motion that is assumed to be uncorrelated with the market portfolio and with the aggregated wealth. $I = 0$ yields $dK_t = 0$, which means that the cost to completion no longer changes. The total cost of the project can be obtained by integrating all infinitesimal investment costs along the path. Notice that in this case the total cost of the project is not the present value of the cost at $t = 0$ since no discounting takes place for these cost.

2. **Second source of uncertainty: the asset value process** $(V_t)_{t\geq 0}$

   The second source of uncertainty in the model is the present value of the project after its successful completion. The completion of the project gives an asset (i.e., the completed project) with the present value V_t in case the project is completed at time t. V_t is the expected future net cash flow from the project after its completion at time t. This present value process $(V_t)_{t\geq 0}$ (present value with respect to time t) is assumed to be given via the SDE

$$dV_t \quad = \quad \mu\,V_t\,dt \; + \; \sigma\,V_t\,dB_t^2, \quad t \geq 0, \quad V_0 \in \mathbb{R} \text{ constant.}$$

   $\sigma \in \mathbb{R}^+$ is the instantaneous standard deviation of the proportional changes in the value received at completion, and $(B_t^2)_{t\geq 0}$ is a Standard Brownian Motion that is assumed to be uncorrelated with $(B_t^1)_{t\geq 0}$.

[53] See Schwartz & Moon [121], page 86.
[54] The value of I will later be determined in an optimization procedure.
[55] See Schwartz & Moon [121], page 90, equation (6.2).

This implies that the two processes $(K_t)_{t \geq 0}$ and $(V_t)_{t \geq 0}$ are uncorrelated. $(B_t^2)_{t \geq 0}$ could, however, be correlated with the market portfolio or aggregated wealth. The instantaneous drift $\mu \in \mathbb{R}$ characterizes the particular investment project. It can be positive or negative[56].

3. **Third source of uncertainty: catastrophic events**

   The third and last source of uncertainty relates to the possibility of a catastrophic event that might end the project completely. E.g., a competitor might introduce a similar drug while the company is still in its research stage. Such a source of uncertainty can be modelled with a Poisson process which is assumed to be independent of the two other processes $(K_t)_{t \geq 0}$ and $(I_t)_{t \geq 0}$. In this Poisson process, the Poisson parameter $\lambda \in (0, 1)$ denotes the probability per unit time that the project's value will jump to zero.

The value of the investment project is modelled as a contingent claim with its underlying being the value of the asset obtained at the completion of the project and the expected cost to completion. Although the basic model can be extended to multi-stage investment projects with different characteristics, only the basic model is introduced here for the purpose of this thesis[57], following the explanations of Schwartz and Moon[58].

Let $F = F(V, K)$ be the function that represents the value of the investment opportunity depending on two variables V and K. V is the estimated value of an asset to be obtained at some time in the future, and K is the expected value of the random cost to completion of that asset. $F(V, K)$ must satisfy the following partial differential equation[59]:

$$\max_I \left[\frac{1}{2}\sigma^2 V^2 \frac{\partial^2 F}{\partial V^2} + \frac{1}{2}\beta^2 IK \frac{\partial^2 F}{\partial K^2} + \right.$$

$$\left. (\mu - \eta)V \frac{\partial F}{\partial V} - I \frac{\partial F}{\partial K} - (r_f + \lambda)F - I \right] = 0. \qquad (4.34)$$

$(\mu - \eta)$ is the risk-adjusted drift for the asset value, and r_f is the constant risk-free interest rate. According to equation (6.5) on page 92 of Schwartz & Moon [121], an investment takes place at the maximum rate I_m or not at all:

---

[56] See Schwartz & Moon [121], page 90.

[57] For a discussion of a multi-stage valuation extension see Schwartz & Moon [121], Chapter 6.5.

[58] See Schwartz & Moon [121], Chapter 6.3.4.

[59] According to Schwartz & Moon [121], page 91, the arguments for (4.34) are standard in literature and used by several authors, see, e.g., Merton [95], Brennan & Schwartz [20], or Dixit & Pindyck [40].

$$I = \begin{cases} I_m & \text{for } \frac{1}{2}\beta^2 K \frac{\partial^2 F}{\partial K^2} - \frac{\partial F}{\partial K} - 1 \geq 0, \\ 0 & \text{otherwise.} \end{cases}$$

Equation (4.34) is an elliptic partial differential equation[60] with a free boundary along the line $V^*(K)$ such that $I = I_m$ when $V > V^*(K)$, else $I = 0$. The set of critical asset values, $V^*(K)$, must be found as part of the solution for $F(V, K)$ and satisfy

$$\frac{1}{2}\beta^2 K \frac{\partial^2 F}{\partial K^2}(V^*, K) - \frac{\partial F}{\partial K}(V^*, K) - 1 = 0.$$

For $V > V^*(K)$, the value of the investment opportunity must satisfy

$$\frac{1}{2}\sigma^2 V^2 \frac{\partial^2 F}{\partial V^2} + \frac{1}{2}\beta^2 I_m K \frac{\partial^2 F}{\partial K^2} +$$

$$(\mu - \eta) V \frac{\partial F}{\partial V} - I_m \frac{\partial F}{\partial K} - (r_f + \lambda) F - I_m = 0. \tag{4.35}$$

Otherwise, it must satisfy

$$\frac{1}{2}\sigma^2 V^2 \frac{\partial^2 F}{\partial V^2} + (\mu - \eta) V \frac{\partial F}{\partial V} - (r_f + \lambda) F = 0. \tag{4.36}$$

Thereby, the boundary conditions are:

$$F(V, 0) = V, \quad F(0, K) = 0, \quad \text{and} \quad \lim_{K \to \infty} F(V, K) = 0.$$

It is not the goal of this thesis to solve this continuous-time model numerically. Instead, the attention will now be turned to the traditional NPV approach which is embedded in the ideas introduced so far[61]. Moreover, the Schwartz and Moon NPV model will be extended by introducing an option to defer the investment start and by introducing a non-constant risk-free interest rate instead of a constant one.

In the deterministic world, the time to completion of the project is $T = \frac{K}{I_m}$. K is the deterministic, undiscounted cost of completion, and I_m is the constant investment rate. The appropriate discount rate[62] for the asset value is $r_f + \lambda - \mu + \eta$, and the appropriate discount rate for the expected cost to completion is $r_f + \lambda$. Then the static present value at time $t = 0$ of the cost to completion is

[60] See Schwartz & Moon [121], page 92.
[61] See Schwartz & Moon [121], Chapter 6.3.5.
[62] See Schwartz & Moon [121], page 93.

$$PV_0(K) \;=\; \int_0^T \frac{K}{T} e^{-(r_f+\lambda)t}\, dt \;=\; \frac{K}{T(r_f+\lambda)}\left[1 - e^{-(r_f+\lambda)T}\right]. \quad (4.37)$$

According to Schwartz & Moon [121], page 93, ***the probability of a catastrophic event can be interpreted as an annual "tax rate" λ on the value of the project since on average a fraction λ of the project would be lost every year.***

With V_T denoting the present value of the asset after completion at time T, the present value of V_T at time $t = 0$ is:

$$PV_0(V_T) \;=\; V_T\, e^{-(r_f+\lambda-\mu+\eta)T}. \quad (4.38)$$

According to (4.37) and (4.38) and since $T = \frac{K}{I_m}$, it holds for the NPV of the project at time $t = 0$:

$$\left.\begin{aligned} NPV_0 &= PV_0(V_T) - PV_0(K) \\ &= V_T\, e^{-(r_f+\lambda-\mu+\eta)T} \;-\; \frac{I_m}{r_f+\lambda}\left[1 - e^{-(r_f+\lambda)T}\right]. \end{aligned}\right\} \quad (4.39)$$

One can verify that this solution in the NPV world satisfies the partial differential equation (4.35) in the case of $\sigma = 0 = \beta$. According to Schwartz and Moon, ***the NPV solution [...] not only provides a benchmark with which to compare the value of the project and the optimal investment strategy under uncertainty, but also serves as a starting point for the numerical procedure used to solve the system of elliptic partial differential equations (4.35) and (4.36) subject to the appropriate boundary conditions***[63].

As stated earlier, it is not the goal of this thesis to follow this path but to modify the deterministic NPV solution by applying a non-constant risk-free interest rate in the case of an option to defer. The purpose is to show how an option can be created by including a non-constant risk-free interest rate in the model.

In the following, the risk-free interest rate r_f is not constant but moves over time. The constant risk-free interest rate in the Schwartz-Moon model poses, more precisely, a flat term structure that is constant over time. In Chapter 5 three scenarios are analyzed: First, the basic case of a constant and flat term structure without a defer option. Second, an initially flat yield curve shifts monotonically upwards by 50 bps per quarter and shifts monotonically downwards by -50 bps per quarter, respectively. Then, the project's NPV is calculated depending on when the project starts. T^d denotes the time deferred before the building stage of the R&D investment starts. It is assumed that

[63] See Schwartz & Moon [121], page 93.

the project can be started at time $t = T^d$ instead of $t = 0$ without changing the parameters of the various processes of the model. In the third scenario an initially flat yield curve shifts monotonically upwards by 100 bps per quarter and shifts monotonically downwards by -100 bps per quarter, respectively.

This modified model now has two different interest rates. The first (annualized) interest rate $r_f^{(1)}$ is the risk-free rate of equation (4.39) that prevails at time $t = T^d$. The second (annualized) interest rate $r_f^{(2)}$ is the risk-free rate that prevails today with a time to maturity of T^d. While $r_f^{(2)}$ can be observed in the current market, the risk-free rate $r_f^{(1)}$ is a future interest rate that is unknown today. Figure 4.14 shows the new situation graphically.

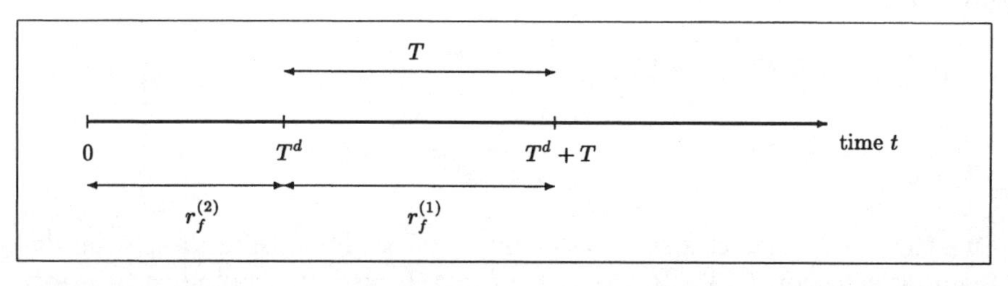

Fig. 4.14. Option to defer in the Schwartz-Moon model.

The current NPV, including the option to defer, can be calculated as:

$$\left.\begin{aligned} NPV_{T^d}^{defer} &= V_{T^d+T}\ e^{-(r_f^{(1)}+\lambda-\mu+\eta)T} \\ &\quad - \frac{I_m}{r_f^{(1)}+\lambda}\left[1 - e^{-(r_f^{(1)}+\lambda)T}\right] \\ &\quad \text{(defer period is } [0, T^d], \\ &\quad \text{and investment period ends at } T^d + T), \\ NPV_0^{defer} &= e^{-r_f^{(2)} \cdot T_d}\ NPV_{T^d}^{defer}. \end{aligned}\right\} \qquad (4.40)$$

These formulas will be used in Chapter 5 for the analysis of an option to defer applying non-constant interest rates in the context of the Schwartz-Moon model.

4.4 Real Options Pricing with Stochastic Interest Rates

So far various real options pricing models and a concrete model for R&D development, the Schwartz-Moon model, have been presented. For all these models, the risk-free rate used for discounting is constant over the life of the option. Since this life time can be very long (in R&D the drug development process takes up to 10-12 years[64]), the assumption of a constant risk-free interest rate is questionable[65]. Therefore, a non-constant risk-free interest rate was introduced in the deterministic NPV solution of the Schwartz-Moon model in a very basic way to value a deferred project start within the Schwartz-Moon framework.

Figure 4.15 displays the down-swing of the U.S. Zero yield curve with 1 month to 10 years of maturity using monthly data between December 2000 and November 2001.

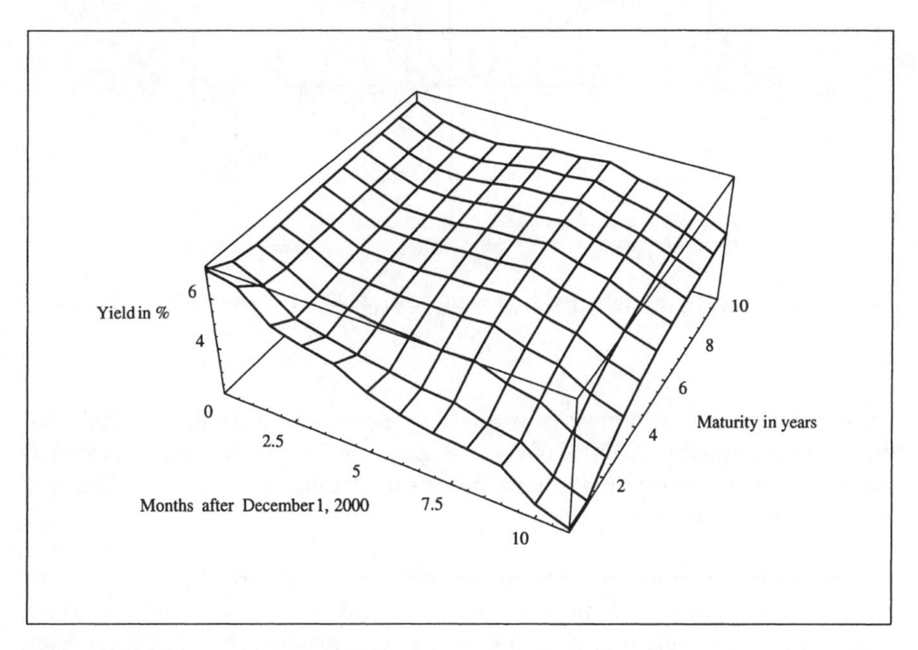

Fig. 4.15. Term structure of U.S. Zero yields, December 2000 - November 2001, monthly data.

During this time, the short end of the yield curve shifted downwards by over 1% per quarter while the long end shifted downwards only slightly. The result

[64] See Schwartz & Moon [121], page 87.

[65] See Section 2.6.4 or Alvarez & Koskela [1], page 1: *If the exercise of such [irreversible] investment opportunities takes a long time, the assumed constancy of the interest rate is questionable.*

was a very steep, upward sloping yield curve at the end of 2001. This shape did not change much until the beginning of 2004 when this thesis was completed.

Figure 4.16 displays the movement of the 1-mos. U.S. Zero yield between December 1, 2000 and November 30, 2001 using daily data. Such an extreme movement with a very steep yield curve at the end of 2001 raises the question whether a constant risk-free rate can still be applied in real options pricing tools.

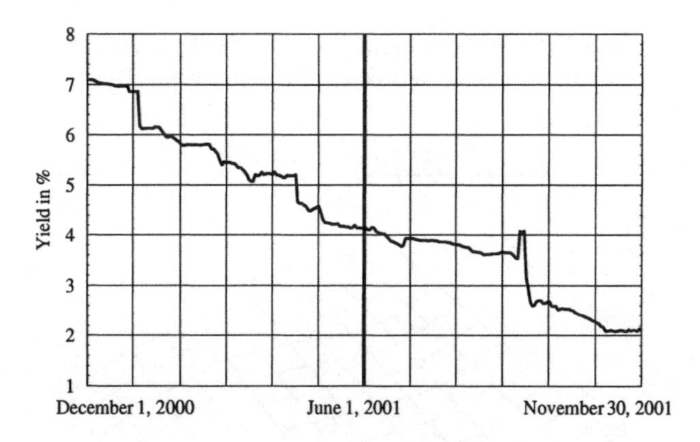

Fig. 4.16. 1-mos. U.S. Zero yield, December 1, 2000 - November 30, 2001, daily data.

During this time, the interest rate volatility increased substantially. Table 4.2 displays the annual volatilities of the U.S. Zero bonds over the last six years for various times to maturity. It shows how dramatically the volatility changed, especially for short-term yields.

For irreversible investments, Alvarez and Koskela mentioned that *interest rate variability in general can be important from the point of view of exercising real investment opportunities*[66]. However, when introducing real options pricing methods in Chapter 4, a constant risk-free interest rate has always been assumed so far. This was mandatory to derive the basic finite difference methods[67]. Without a mandatory reason, Trigeorgis log-transformed binomial tree model also assumed a constant risk-free interest rate over the whole life of the investment project. The traditional Cox-Ingersoll-Ross model that converges

[66] See Section 2.6.4 or Alvarez & Koskela [1], page 2.

[67] Hull and White developed a modified explicit finite difference method that allows for a time-dependent interest rate, see Hull & White [64]. As already explained, this method will not be presented in this thesis.

Table 4.2. Volatilities of U.S. Zero yields between April 1, 1997 and March 31, 2003 (in %).

	Time period					
Time to maturity	04/01/97 - 03/31/98	04/01/98 - 03/31/99	04/01/99 - 03/31/00	04/01/00 - 03/31/01	04/01/01 - 03/31/02	04/01/02 - 03/31/03
1 mos.	0.100	0.316	0.498	0.560	1.092	0.242
2 mos.	0.086	0.302	0.482	0.614	1.047	0.244
3 mos.	0.088	0.299	0.492	0.666	1.020	0.255
4 mos.	0.090	0.302	0.481	0.706	1.000	0.273
5 mos.	0.098	0.308	0.492	0.755	0.970	0.291
6 mos.	0.111	0.313	0.513	0.780	0.957	0.316
7 mos.	0.124	0.316	0.523	0.803	0.941	0.341
8 mos.	0.138	0.320	0.534	0.827	0.928	0.366
9 mos.	0.153	0.326	0.546	0.852	0.917	0.394
10 mos.	0.167	0.326	0.552	0.860	0.899	0.422
11 mos.	0.180	0.328	0.557	0.870	0.884	0.448
12 mos.	0.194	0.331	0.564	0.879	0.869	0.476
2 yrs.	0.306	0.384	0.567	0.861	0.643	0.689
3 yrs.	0.354	0.406	0.553	0.806	0.531	0.723
4 yrs.	0.351	0.385	0.511	0.741	0.464	0.715
5 yrs.	0.367	0.364	0.500	0.706	0.420	0.692
6 yrs.	0.380	0.347	0.488	0.678	0.392	0.665
7 yrs.	0.390	0.329	0.477	0.651	0.370	0.638
8 yrs.	0.398	0.313	0.467	0.633	0.355	0.611
9 yrs.	0.404	0.297	0.458	0.616	0.346	0.589
10 yrs.	0.410	0.282	0.449	0.599	0.338	0.564

Source: own calculations based on capital markets data

to the Black-Scholes formula in limit[68], assumes a constant risk-free rate as well. For Hull, this is a simplification that is not warranted in the case of a steeply upward or downward sloping yield curve which, according to the pure expectations theory, can be interpreted as a future up- or down-movement of the yield curve: *The usual assumption when American options are being valued is that interest rates are constant. When the term structure is steeply upward or downward sloping, this may not be a satisfactory assumption. It is more appropriate to assume that the interest rate for a period of length Δt in the future equals the current forward rate for that period* [69].

While this quotation expresses the necessity to include a non-constant interest rate in the real options pricing model, its approach to model the interest rate is different from the one followed in this thesis. The approach in this thesis is

[68] See Cox & Rubinstein [37], Section 5.6.
[69] See Hull [62], page 356.

to model the complete term structure of interest rates (and not just the short rate) via stochastic term structure models, in order to determine the correct maturity bucket of the generated future yield curves, which is to be included into the real options pricing methods (depending on the chosen time step size of the model). Hull's idea is based on the pure expectations theory, assuming that the current forward interest rate is a valid predictor of the future interest rate.

The main goal here is to present an in-depth numerical analysis of complex real options situations common in Corporate Finance practice and to evaluate the corresponding real options pricing methods through historical backtesting. The various situations are analyzed by applying non-constant (especially, stochastically modelled) risk-free rates to simulation scenarios. The two primarily used pricing tools are modifications of the Cox-Ross-Rubinstein binomial tree method and the Trigeorgis log-transformed binomial tree method. However, before explaining these two modified models for real options valuation, the first article that showed the importance of stochastic interest rates in real options valuation will be introduced. In 1992, Ingersoll and Ross published their article *Waiting to Invest: Investment and Uncertainty* as the first article to stress the notion that a real option can also be created solely because of a stochastic yield curve and not because of uncertain future cash flows. They explained how a simple real option[70] can be created by assuming a constant future cash flow structure together with stochastic interest rates.

As Ingersoll and Ross pointed out[71], they *are not the first to recognize that delaying a project can be desirable, but* [...] *are the first to observe that this need nothing to do with changes in the cash flows of the project itself or with the effects of certain changes in interest rates.* Both authors provided a nice overview of the historic development of research on investment delay. The goal is not to restate this here but to just refer to Ingersoll & Ross [69], page 3. However, none of the approaches they mentioned does include a stochastic interest rate, which can have a huge impact on project valuation[72]: *By contrast with this literature, the central theme of our work is that, even for the simplest projects with deterministic cash flows, interest rate uncertainty has a significant effect on investment.*

Another model which includes a stochastic yield curve in a pricing model was presented by Sandmann[73] in 1993. This pricing model for European options is a discrete-time model based on a binomial tree for both the underlying and

[70] Ingersoll and Ross used a T-year Zero that pays 1 $ at maturity to show how this can be viewed as a way to mimic an investment project.

[71] See Ingersoll & Ross [69], page 3.

[72] See Ingersoll & Ross [69], page 3.

[73] See Sandmann [115].

the risk-free rate. The term structure movement was described via a model introduced by Sandmann and Sondermann[74] in 1991. In consequence, the complete pricing model allows four distinct states per time step, which is the reason why this model is called a *four-state model.*

It is worth mentioning that in Sandmann's model the upward and downward movements u and d of the underlying stock price are not independent from the movements of the interest rate[75]: [...] *the consequences of the assumption of a stochastic term structure is that the measure p* [...] *under which the discounted stock price process is a martingale, becomes at this point state- and time-dependent. The next observation is that, if the possible returns u and d of the stock over one period are assumed to be constant, then in fact, either the stock or the spot rate will become a dominant security with positive probability. To avoid this, as in the simple case, it is necessary to require [that it holds for the risk-free interest rate $r_{i,j}$ with time step i and state step j]*

$$u = u(r_{i,j}) > 1 + r_{i,j} > d(r_{i,j}) = d.$$

[...] *This means that u and d are at least functions of the spot rate.*

In this thesis the risk-free interest rates and the stock price movements are independent of each other. This is not a contradiction of the ideas of Sandmann but inherent in the Cox-Ross-Rubinstein binomial tree as Hull already pointed out[76]: *[The introduction of a non-constant risk-free interest rate] does not change the geometry of the tree [for the underlying] since u and d do not depend on [the risk-free interest rate].*

4.4.1 Ingersoll-Ross Model

In 1992, Ingersoll and Ross proposed their pioneering idea which is that a real option cannot only be created when the future cash flows are uncertain but also if the future cash flows are deterministic and the interest rate is stochastic. The authors investigated how the value of a real option and the optimal investment time behave for an option to defer under a stochastically modelled yield curve of a special type. In the following, the basic ideas of this article will be explained, closely following the presentation of Ingersoll and Ross[77].

The general idea of how interest rate uncertainty can create value within an option to defer was already presented in Section 1.1. Now, a simple model of this effect will be introduced that Ingersoll and Ross describe in their article.

[74] See Sandmann & Sondermann [116].
[75] See Sandmann [115], page 206.
[76] See Hull [62], pages 356-357.
[77] See Ingersoll & Ross [69] and Trigeorgis [132], pages 197-199.

Assume that at time $t \geq 0$ someone will invest the amount I in a project that generates the one-time cash inflow of 1 $ at time $t + T$ with $T > 0$. Let $r_s, 0 \leq t \leq s \leq T$, denote the infinitesimal interest rate, the short rate, for time s. This interest rate r_s gives the interest generated during the infinitesimal small time interval $[s, s + ds]$. Therefore, the net present value $NPV(t)$ of this project at time t is

$$NPV(t) \quad = \quad P(t) - I \quad = \quad \hat{E}\left(\exp\left\{-\int_t^{t+T} r_s \, ds\right\}\right) \quad - \quad I,$$

where $\hat{E}$ is the expected value with respect to the risk-adjusted stochastic process[78]. Such a project, started at time t, has a current net present value $NPV_0(t)$ of

$$NPV_0(t) \quad = \hat{E}\left([P(t) - I] \exp\left\{-\int_0^t r_s \, ds\right\}\right).$$

The main idea is now to calculate $NPV_0(t)$ *for all* t in order to find the optimal time t^* for which NPV_0 is the largest[79]. To determine this time point analytically, a concrete term structure model needs to be specified.

Ingersoll and Ross applied a specific Cox-Ingersoll-Ross model[80] with drift zero to model the short-rate process $(r_t)_{t \geq 0}$ of the term structure via the following stochastic differential equation:

$$dr_t \quad = \quad \sigma \sqrt{r_t} \, dB_t \, , \quad t \geq 0. \tag{4.41}$$

$\sigma \in \mathbb{R}^+$ is the constant infinitesimal volatility of the short-rate process. A comparison of equation (4.41) with the specification of the Cox-Ingersoll-Ross model in (3.16) of Chapter 3 shows that in (3.16) the two parameters α and β of the drift term are supposed to be positive while in (4.41) these parameters are both assumed to be zero. The reason is that in (3.16) of Chapter 3, the Cox-Ingersoll-Ross model was specified in a way to arrive at a mean reverting model that exhibits stationarity[81]. According to Borkovec & Klüppelberg [12], page 10, and Schulmerich [117], page 35, the Cox-Ingersoll-Ross diffusion process is only stationary for $\beta, \sigma \in \mathbb{R}^+$ with $2\beta \geq \sigma^2$. The stationary distribution of a Cox-Ingersoll-Ross model is derived in Schulmerich [117], pages 140-141. In (3.16) of Chapter 3, the parameter α also needed to be positive because otherwise a mean reversion feature would be impossible.

---

[78] This is the so-called *equivalent martingale measure*.

[79] The term *for all* is only a qualitative description in order not to introduce the mathematical term of a *stopping time*. More information on this subject can be found in Karatzas & Shreve [72] and Rogers & Williams [113], which are both very theoretical and mathematically thorough.

[80] See Cox, Ingersoll & Ross [34] or Section 3.3.2 of this thesis.

[81] The term *stationarity* is explained in Section 2.3.4.

Through the choice of the drift as zero the analysis can concentrate on the effects of a random interest rate change on the investment decision[82]. This would be difficult in case of a drift not equal to zero, i.e., in case of an anticipated direction of the short-rate movements.

The crucial point in this real options model is stated by Ingersoll and Ross as follows[83]: *While traditional capital budgeting theory might erroneously suggest that this investment should be taken as soon as its NPV is positive, a misreading of option-pricing theory suggests the opposite extreme - the paradox that this option to invest should never be exercised. The underlying asset (a point-input, point-output investment) pays no dividends or other disbursements while the option is alive, and we know that it is never optimal to exercise a call option on a stock with no cash dividends until it expires (see Merton [93]). Such reasoning is faulty in this case because the underlying asset of the investment option is not like other assets. Compare this asset - that is, the value of the project at the time the commitment is made - with a zero-coupon bond with the same original maturity. As time passes, the price of the bond changes with changes in the interest rate* and *increases as its maturity shortens. Only the first effect is present in the potential investment. As long as the commitment is not made, the project's payoff comes no closer, so its present value does not tend to rise. As a consequence, the present value of the investment will tend to lag behind the value of the zero-coupon bond. This lag in value is similar to the drain in price created by a continuous dividend stream and provides an incentive for "early exercise".*

If investment I is made, the project can be thought of as a Zero bond that is issued at time t for the price I and that pays 1 $ at maturity time point $t+T$ to its holder, the initiator of the project. According to Cox, Ingersoll, and Ross[84] the price of the Zero at time t is analytically given as:

$$P(t, t+T) \quad = \quad e^{-r_t B(t,T)}, \quad 0 \leq t < T,$$

where

$$B(t,T) \;=\; B(T) \;:=\; \frac{2(e^{\gamma T}-1)}{(\gamma-\lambda)(e^{\gamma T}-1)+2\gamma}$$

with

[82] See Ingersoll & Ross [69], page 5: *This process restricts the one in Cox, Ingersoll & Ross [34] to a zero-expected change rather than their more general mean-reverting drift. This special case was chosen both to simplify the analysis and to focus on the effects of interest rate uncertainty. Clearly, there will be a tendency to delay (accelerate) investment when the interest rate is expected to fall (rise), ceteris paribus. Setting the drift to zero allows to concentrate on the effects of uncertainty and not expected rate movements on investment decisions. The qualitative properties of this example would also hold true with their more general drift term.*

[83] See Ingersoll & Ross [69], pages 5-6.

[84] See Cox, Ingersoll & Ross [34].

$$\gamma := \sqrt{\lambda^2 + 2\sigma^2} > |\lambda| \geq \lambda.$$

Parameter λ is a measure for the price of interest-rate risk[85]. Note that the formula for $P(t,T)$ is slightly different from that in Section 3.3.2 since in the problem at hand the term λ is included. It is important to notice that $B(T)$ is always positive since $\gamma > 0$ and $T > 0$ (i.e., $e^{\gamma T} > 1$) as well as $\gamma > \lambda$ (i.e., $\gamma - \lambda > 0$).

In the work of Ingersoll and Ross the term *acceptance rate* $r^* = r^*(T, I)$ is very important. If the interest rate of the market is larger than r^*, no investment is undertaken; if the interest rate is smaller than or equal to r^*, the investment will be carried out. Therefore, the acceptance rate is the interest rate at which the commitment to the project is made. The term *acceptance rate* has to be distinguished from the term *break-even rate* which is the short rate $r^{(0)} = r^{(0)}(T, I)$ at which NPV_0 of the project is 0.

For the acceptance rate the solution is[86]:

$$r^* = \frac{1}{B(T)} \ln\left(\frac{\nu - B(T)}{\nu I}\right) \quad \text{with} \quad \nu := \frac{\lambda + \gamma}{\sigma^2}. \tag{4.42}$$

For the break-even rate the solution is[87]:

$$r^{(0)} = -\frac{1}{B(T)} \ln(I). \tag{4.43}$$

Acceptance rate and break-even rate are linked according to (4.42) and (4.43) via:

$$r^* = r^{(0)} + \frac{1}{B(T)} \ln\left(\frac{\nu - B(T)}{\nu}\right) < r^0. \tag{4.44}$$

This relationship between the acceptance rate and the break-even rate results since:

$$B(T) \text{ is always positive,}$$

$$\frac{\nu - B(T)}{\nu} < 1, \text{ i.e., } \ln\left(\frac{\nu - B(T)}{\nu}\right) < 0.$$

Economically, equation (4.44) states that the project is only undertaken when it is some distance "in the money" and not at its break-even rate[88]. The value R of the real option to be able to freely choose the start of the project, i.e., of an *option to defer*, can be calculated as[89]

---

[85] For detailed information on parameter λ see Ingersoll & Ross [69], page 5.
[86] See Ingersoll & Ross [69], page 7, equation (7).
[87] See Ingersoll & Ross [69], page 8, equation (8).
[88] See Ingersoll & Ross [69], page 8.
[89] See Ingersoll & Ross [69], page 8, equation (10).

$$R(r_t) \quad = \quad e^{-\nu(r_t - r^*)} \left[e^{-B(T)r^*} - I \right].$$

The model presented so far is still fairly easy. But since the derivation of the formulas draws heavily on stochastic calculus methods, even this level of difficulty is likely to pose a barrier to broad acceptance in Corporate Finance practice. Moreover, Ingersoll and Ross stress that *the project acceptance rule will not be this simple for more general stochastic processes*[90]. Therefore, this thesis is devoted to numerical methods to incorporate a stochastic risk-free interest rate into real options pricing tools as opposed to deriving analytical solutions (if these can be derived at all).

4.4.2 A Modification of the Cox-Ross-Rubinstein Binomial Tree

All common real options pricing tools introduced in Section 4.2 use a constant risk-free interest rate. To be more specific, the risk-free interest rate was assumed to be the same for all time periods and constant over time, i.e., a flat and constant term structure of interest rates over time was assumed.

Including a stochastic risk-free interest rate is not a new idea. However, it has never been thoroughly analyzed in the context of complex real options situations by using numerical simulations nor has it ever been evaluated based on historical backtesting. Hull mentions a specific approach to use a non-constant risk-free rate in the Cox-Ross-Rubinstein binomial tree method in his book *Options, Futures, and Other Derivatives*[91] and enumerates circumstances when such an approach is recommended (as already partially cited in this thesis):

> *The usual assumption when American options are being valued is that interest rates are constant. When the term structure is steeply upward or downward sloping, this may not be a satisfactory assumption. It is more appropriate to assume that the interest rate for a period of length Δt in the future equals the current forward interest rate for that period. The process for a non-dividend-paying stock is then*
>
> $$dS_t \quad = \quad r_f(t)\, S_t\, dt \; + \; \sigma\, S_t\, dB_t, \quad t \geq 0. \qquad (4.45)$$
>
> *We can construct a binomial tree as before with* **a** *being a function of time:*
>
> $$a(t) \quad = \quad e^{r_f(t)\Delta t}, \quad t \geq 0. \qquad (4.46)$$
>
> *This does not change the geometry of the tree since u and d do not depend on a. The probabilities at the nodes are*[92]

[90] See Ingersoll & Ross [69], page 7.

[91] See Hull [62], pages 356-357.

[92] According to Hull [62], page 357, these probabilities are always positive for a sufficiently large number of time steps.

$$p = \frac{a(t) - d}{u - d} \quad and \quad 1 - p = \frac{u - a(t)}{u - d}.$$

It is important to notice that r_f is now a function that gives the risk-free interest rate as a function of time $t \geq 0$ and is no longer a constant. This function is not the short-rate process $(r_t)_{t \geq 0}$ but just a time-dependent function. The specification of function r_f will be described below.

The situation of a steeply upward or downward sloping term structure has been the case several times in the past years as can be seen in Figure 4.17.

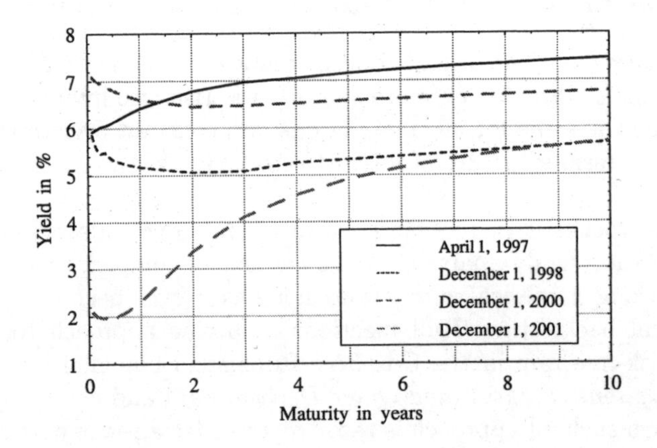

Fig. 4.17. Term structures of interest rates derived from U.S. Zero bonds for various dates.

Moreover, capital markets saw tremendous movements in interest rates in the last 6 years as calculated in Table 4.2. To illustrate this increase in volatility, Figure 4.18 displays the volatility for several times to maturity (TTM).

According to Ingersoll and Ross these interest rate movements heavily affect the value of investments[93]: [...] *even for the simplest projects with deterministic cash flows, interest rate uncertainty has a significant effect on investment. While uncertain changes in cash flows and learning can cause some projects to be delayed, the effect of interest rate uncertainty is ubiquitous and critical to understanding investment at the macroeconomic level.* However, the effects of interest rate changes have not yet been comprehensively investigated for various complex real options situations and different stochastic term structure models by using simulations and historical backtesting. Furthermore, the

[93] See Ingersoll & Ross [69], page 3.

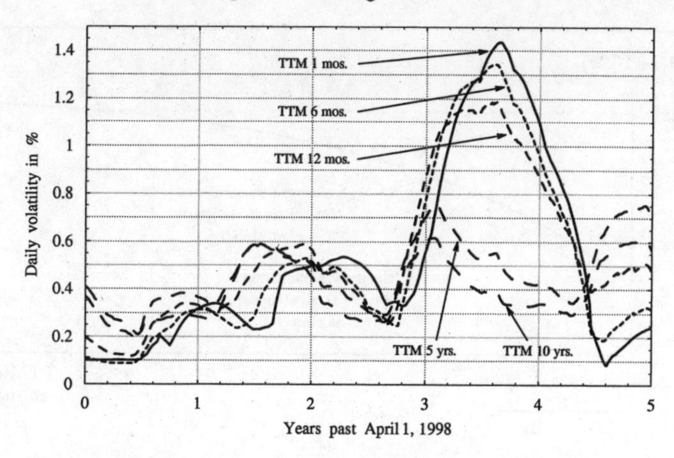

Fig. 4.18. Daily volatility of risk-free U.S. Zero yields based on daily data of the preceeding 12 months.

different term structure models for commonly used complex real options have not yet been systematically compared with each other.

The classic Cox-Ross-Rubinstein binomial tree method for options valuation was already presented in Section 2.6.2 in the framework of CCA and in Section 4.2.1. This model will now be modified by introducing a variable risk-free rate. The development of the term structure of interest rates will be modelled stochastically over time, and then the appropriate risk-free interest rate will be applied in the real options valuation tool. This idea is different from Hull's approach since Hull wants to use the current forward rate for a future time period of length Δt as the future spot rate prevailing at the beginning of that time period.

The approach followed in this thesis is to exactly model the complete future term structure according to a stochastic term structure model. Figure 4.19 explains graphically how the appropriate risk-free interest rate is extracted from a generated future term structure. As a reminder, $R(t, s)$, $0 \leq t \leq s$, gives the continuously compounded yield at time t for a Zero that matures at time s (see Section 3.2.1). This means that at time point t the time to maturity of the Zero is $s - t$. $\Delta\tau$ is the time step size of the discrete future stochastic term structure of interest rates which is calculated according to the theoretical continuous-time models by using the approximation methods from Section 2.5.

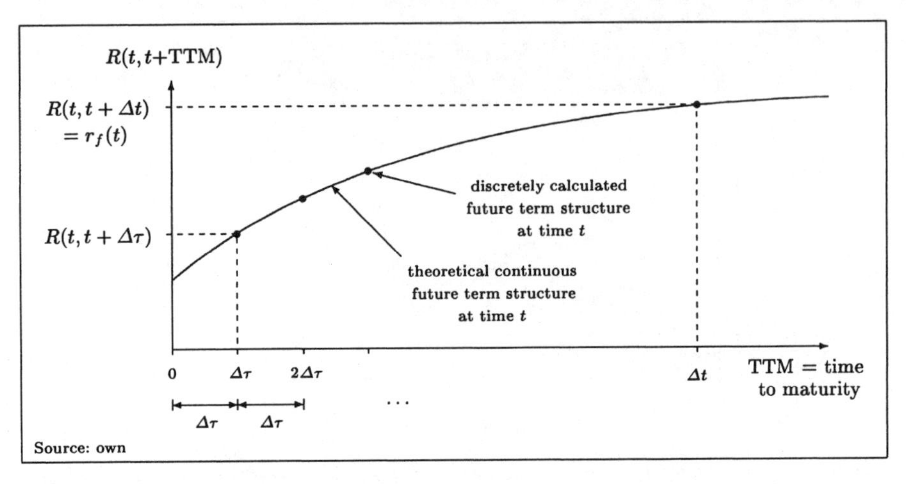

Fig. 4.19. Extracting the appropriate risk-free interest rate from the future term structure at time t for a real options valuation tool.

After having calculated the appropriate future risk-free rate, solving the tree backwards is straight forward as in the basic model with a constant risk-free interest rate. The only adjustments that have to be made at the nodes are the adjustments depending on the real options embedded in the investment project. However, these adjustments are the same as in the ordinary Cox-Ross-Rubinstein binomial tree. The formulas and indices of the Cox-Ross-Rubinstein lattice including a non-constant risk-free rate are listed below:

Let $T > 0$ be the length of the investment project. To construct a binomial lattice, $N > 1$ time points $0 = t_0 < t_1 < \ldots < t_N = T$ have to be specified with these time points assumed to be equidistant, i.e., $t_j - t_{j-1} = \Delta t := \frac{T}{N}$ for $j = 1, \ldots, N$. This gives N subintervals of equal length. The time interval $[t_{j-1}, t_j]$ is named *subinterval/interval* j or *subperiod/period* j, $1 \leq j \leq N$. Δt has to be a multiple of Δ_T in order to allow for the stochastic risk-free rate to be numerically integrated into the Cox-Ross-Rubinstein model appropriately.

The following definitions will be used[94]:

V := total value of the project

S := price of the twin security that is almost perfectly correlated with V

$r_f^{(j)}$:= risk-free interest rate in time period j

[94] See also Section 2.6.2.

$p^{(j)}$:= risk-neutral probability for up-movements of V and S in time period j

u := multiplicative factor for up-movements of V and S per period

d := multiplicative factor for down-movements of V and S per period

Note that compared with the original Cox-Ross-Rubinstein binomial tree model introduced in Section 4.2.1, there is only one difference in this tree model: r_f and p are now dependent on the time period j. The continuously compounded risk-free interest rate is chosen according to the stochastic term structure models introduced in the previous chapter. This means that the risk-free rate $r_f^{(j)} := r_f((j-1)\Delta t)$ for time period j is the interest rate $R((j-1)\Delta t, j\Delta t)$ with the notation introduced in Section 3.2.1 and applied in Figure 4.19.

With this approach the risk-free rate (and, therefore, the risk-adjusted probability) will be reset/updated at each subinterval. However, modifications of this idea are possible. The risk-free rate (and, therefore, the risk-adjusted probability) may only be updated at the beginning of each new quarter and then remain constant for all subintervals of this quarter before being updated again at the beginning of the following quarter. This approach offers computational advantages which make it the method of choice for the computer simulation program. However, the description here is done for interest rate updates at each subinterval.

The price S of the underlying and the project value V develop as in the traditional Cox-Ross-Rubinstein model[95], i.e., if in the beginning of a time interval with length Δt the underlying value is S, it is assumed that in the end of this time interval the underlying can take two values, uS and dS, whereby $u > 1$ and $d < 1$. In the case of a non-constant risk-free rate, the risk-neutral probabilities p for the up-case and $1-p$ for the down-case depend on the prevailing risk-free rate in the corresponding time interval. Therefore, the superscript $p^{(j)}$ is necessary.

The continuous-time counterpart is that S is a Geometric Brownian Motion with volatility σ. In a risk-neutral world, the return of the underlying is the risk-free interest rate. With the continuously compounded, annualized risk-free rate $r_f^{(j)}$ for subinterval j this means

$$Se^{r_f^{(j)}\Delta t} = p^{(j)}uS + (1-p^{(j)})dS,$$

which is equivalent to

[95] See Hull [62], page 357, and the beginning of this section.

$$e^{r_f^{(j)} \Delta t} = p^{(j)} u + (1 - p^{(j)}) d. \tag{4.47}$$

According to Hull [62], equation (11.4), page 344 and pages 356-357, the variance of the change in the underlying within subinterval j is

$$e^{2 r_f^{(j)} \Delta t + \sigma^2 \Delta t} = p^{(j)} u^2 + (1 - p^{(j)}) d^2. \tag{4.48}$$

Equations (4.47) and (4.48) impose two conditions on $p^{(j)}$, u, and d. A third condition is given (on each of the N subintervals separately) by Cox, Ross, and Rubinstein[96] via $u = \frac{1}{d}$. According to Hull [62], page 345 and pages 356-357, it can be shown that these three conditions imply

$$u = \exp\left(\sigma \sqrt{\frac{T}{N}}\right) = \frac{1}{d} \quad \text{and} \quad p^{(j)} = \frac{e^{r_f^{(j)} \Delta t} - d}{u - d}. \tag{4.49}$$

The replicating portfolio now has to incorporate the stochastic risk-free interest rate $r_f^{(j)}$. This is pictured in Figure 4.20.

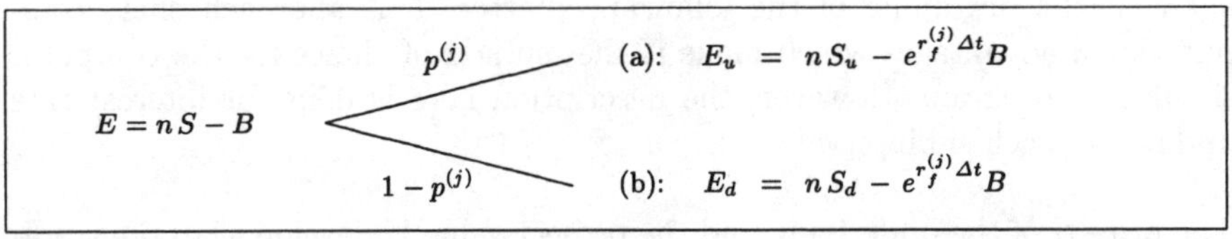

$$E = nS - B \quad \begin{cases} p^{(j)}: & \text{(a):} \quad E_u = n S_u - e^{r_f^{(j)} \Delta t} B \\ 1 - p^{(j)}: & \text{(b):} \quad E_d = n S_d - e^{r_f^{(j)} \Delta t} B \end{cases}$$

Fig. 4.20. The concept of a replicating portfolio in the modified Cox-Ross-Rubinstein binomial tree model with a non-constant risk-free interest rate for time period $j, 1 \leq j \leq N$.

$$n = \frac{E_u - E_d}{S_u - S_d} \qquad E = \frac{p^{(j)} E_u + (1 - p^{(j)}) E_d}{e^{r_f^{(j)} \Delta t}}$$

$$B = \frac{E_u S_d - E_d S_u}{(S_u - S_d) e^{r_f^{(j)} \Delta t}} \qquad p^{(j)} = \frac{e^{r_f^{(j)} \Delta t} - d}{u - d}$$

Fig. 4.21. The solution of a replicating portfolio in the modified Cox-Ross-Rubinstein binomial tree model for time period $j, 1 \leq j \leq N$.

[96] See Cox, Ross & Rubinstein [36].

By chosing the replicating portfolio approach $E = nS - B$ (i.e., E is, as usual, replicated via n shares of stocks and the issue of a Δt-year Zero worth nominal B), the two equations (a) and (b) in Figure 4.20 have to be solved simultaneously to determine n and B. The solution to this problem is given in Figure 4.21. Figure 4.22 then gives the complete picture of real options pricing with a stochastic risk-free interest rate. It shows the complete lattice and the variables used for this method if a non-constant risk-free rate is applied.

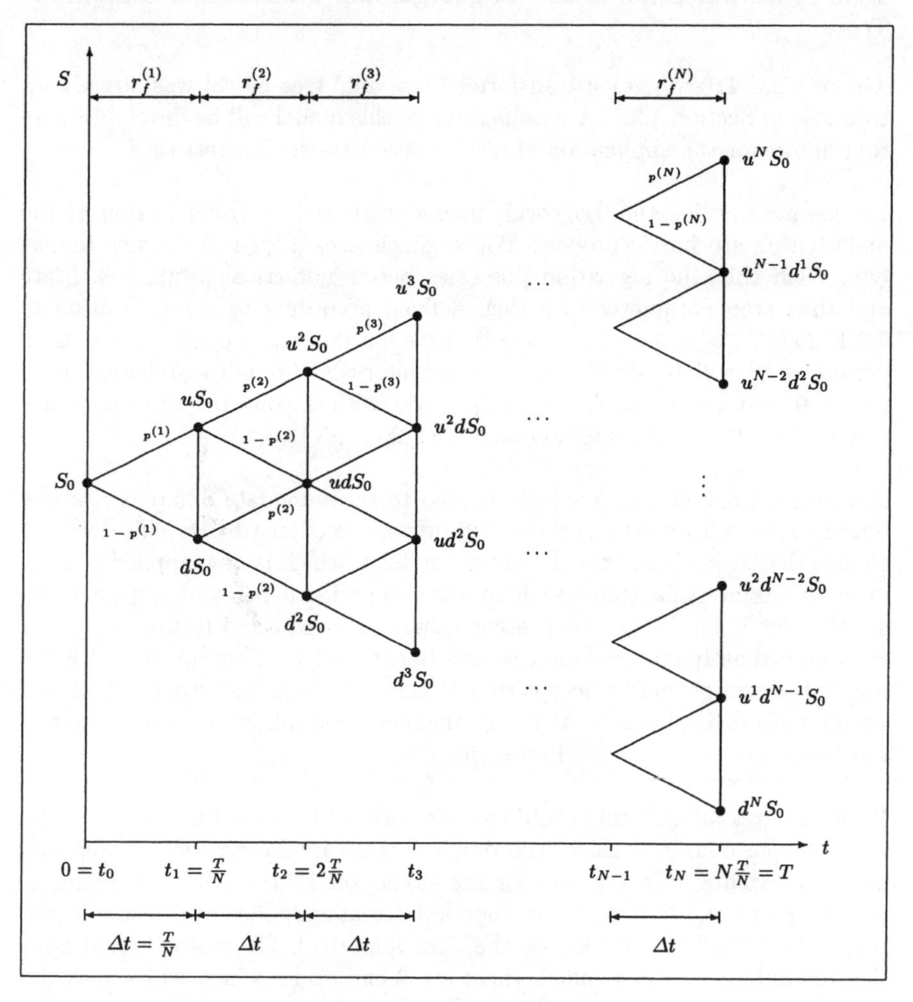

Fig. 4.22. Cox-Ross-Rubinstein binomial tree with update of the risk-free rate at each subinterval.

Also worth mentioning is that dividend adjustments to the Cox-Ross-Rubinstein binomial tree model can easily be made. However, since in Chapter 5 all analysis will be done for real options without dividend adjustments, these adjustments will not be described here. For more information on dividend adjustments in the cases of both a constant and a non-constant risk-free rate see Hull [62], Section 15.3.

4.4.3 A Modification of the Trigeorgis Log-Transformed Binomial Tree

The original Trigeorgis log-transformed binomial tree model was already introduced in Section 4.2.1. A modification of this model will be developed here that allows for the application of a stochastic risk-free interest rate.

Lattice models like the Trigeorgis model start with a discretization of the underlying's stochastic process. Trigeorgis chooses a log-transformed underlying such that the algorithm possesses better numerical properties. State and time step parameters are then defined according to the derivation in 4.2.1. Let $T > 0$ be the maturity time of the real option. Then, time period $[0, T]$ will be divided into N subintervals through equidistant time points $0 = t_0 < t_1 < \ldots < t_{N-1} < t_N$ with a constant time step size $t_j - t_{j-1} = \Delta t = \frac{T}{N}$ for subinterval $j \in \{1, 2, \ldots, N\}$.

The original model can now be extended to accommodate different risk-free interest rates within each of these N time periods of length Δt. This does not change the basic idea of the derivation in 4.2.1, which is now applied to each of the N subintervals. However, it results in a period-dependent state step. To see this, let V be the (gross) present value of the expected future cash flows from immediately undertaking the investment project. This concurs with the original Trigeorgis model accept that V is now considered "stepwise" on the subintervals defined above. Although the idea is sound, it can easily result in the *bushy tree problem* already described in Chapter 2.

If the risk-free interest rate is different on each subinterval, the results will be multi-dimensional trees since according to Figure 4.9 the state size H depends on the prevailing risk-free rate for the subperiod. This means, for example, that for $T = 1$ and $N = 4$, the number of different endpoints at time $t = T = 1$ is $2^4 = 16$ with $2^{4-1} = 8$ trees for the last subinterval. The reason is that after the first subinterval has ended, there are 2 end nodes which will act as the starting points for new trees (with a different H) for subinterval 2. For a choice of $N = 40$, this results in half a billion single trees for the last subinterval alone. This is a typical bushy tree that cannot be handled any more, especially when pricing an option to defer which requires multiple parallel runs.

A solution to this problem is to only update the risk-free interest rate at certain time points, e.g., to update the risk-free rate at the beginning of each quarter and then leave it constant over the quarter. This idea was already presented for the Cox-Ross-Rubinstein binomial tree in Section 4.4.2 above although it was not mandatory since the shape of the tree is independent from the risk-free interest rate. However, for the Trigeorgis model it is the only way to circumvent, or better to constrain, the bushy tree problem. Figure 4.23 below displays graphically how the idea of updating the risk-free rate each quarter shapes the bushy tree and makes it possible to handle. In the following, this idea will be described mathematically.

To model a quarterly update of the risk-free interest rate, it is first assumed that T is a multiple of 0.25 years[97]. Then, T is divided into $L = 4T$ subperiods with $T_l := 0.25 \cdot l$ for $l \in \{0, 1, \ldots, L\}$. Especially, $T_0 = 0$ and $T_L = t_N = T$. In the following, it will also be assumed that N is a multiple of L. On each of these L subperiods the risk-free interest rate is the same[98].

It is now possible to analyze the underlying process on each of these L subintervals separately. To restate, $l \in \{0, 1, \ldots, L\}$ is the index for the subintervals for changes of the risk-free rate and $i \in \{0, 1, \ldots, N\}$ is the index for the time steps of the valuation tree.

On each of the L subintervals the process $(V_t^{(l)})_{T_{l-1} \le t \le T_l}, 1 \le l \le L$, is assumed to follow a Geometric Brownian Motion:

$$dV_t^{(l)} = \alpha^{(l)} V_t^{(l)} dt + \sigma V_t^{(l)} dB_t, \quad T_{l-1} \le t \le T_l.$$

$V_0^{(1)} \in \mathbb{R}^+$ is given, and $V_{T_{l-1}}^{(l)} = V_{T_{l-1}}^{(l-1)}$ holds for $l = 2, \ldots, L$. Moreover, $\alpha^{(l)}$ is the instantaneous expected return on the project in subinterval l and σ is the instantaneous standard deviation which is assumed to be the same in each subinterval. Process $(V_t)_{0 \le t \le T}$ is then defined via

$$V_t = \sum_{l=1}^{L} 1_{(T_{l-1}, T_l]} V_t^{(l)}, \quad 0 < t \le T.$$

The specific choice of $\alpha^{(l)} = \alpha$ for all $l \in \{1, \ldots, L\}$ yields the same underlying process as in the original Trigeorgis model, see equation (4.5).

---

[97] It is easy to omit this assumption. However, the mathematical description becomes easier in the following if this restriction is made.

[98] It is also possible to generalize this idea by dividing $[0, T]$ in L subperiods with N as a multiple of L. To construct these subperiods the definition $T_l := l\frac{T}{L}, l \in \{0, 1, \ldots, L\}$, is needed. Within each of these L subintervals the risk-free interest rate is the same. However, in this thesis and in the computer simulation program the specification $L = 4T$ is chosen.

For $l \in \{1, \ldots, L\}$ the process $(Y_t^{(l)})_{T_{l-1} \leq t \leq T_l}$ on interval $[T_{l-1}, T_l]$ is defined via $Y_t^{(l)} := \ln(V_t^{(l)}), T_{l-1} \leq t \leq T_l$. In any infinitesimal time interval dt the process $Y^{(l)} := (Y_t^{(l)})_{t \geq 0}$ follows an Arithmetic Brownian Motion. Under risk-neutrality (i.e., $\alpha^{(l)} = r_f^{(l)}$ = risk-free rate in subperiod $[T_{l-1}, T_l]$) this gives

$$dY^{(l)} = \ln\left(\frac{V_{t+dt}^{(l)}}{V_t^{(l)}}\right) = \left(r_f^{(l)} - \frac{1}{2}\sigma^2\right) dt + \sigma dB_t, \; T_{l-1} \leq t \leq T_l,$$

for the infinitesimal time period dt. On each interval $[T_{l-1}, T_l]$ the original Trigeorgis model yields:

$$dY^{(l)} \overset{i.i.d.}{=} N\left(\left[r_f^{(l)} - \frac{1}{2}\sigma^2\right] dt, \sigma^2 dt\right)$$

with a different, from $r_f^{(l)}$ dependent mean. On each interval $[T_{l-1}, T_l]$ the transformation

$$K := \sigma^2 dt$$

is now applied which gives

$$dY^{(l)} \overset{i.i.d.}{=} N\left(\mu^{(l)} K, K\right), \quad \mu^{(l)} := \frac{r_f^{(l)}}{\sigma^2} - \frac{1}{2}.$$

Let for the following be $l \in \{1, 2, \ldots, L\}$. On subinterval l the log-transformed process $(Y_t^{(l)})_{T_{l-1} \leq t \leq T_l}$ can be approximated with a discrete-time version. Within each subinterval $[T_{l-1}, T_l]$ this discrete-time version is a Markov random walk that goes up by $\Delta Y^{(l)}$ (with probability $P^{(l)}$) and down by $-\Delta Y^{(l)}$ with probability $1 - P^{(l)}$.

Let $H^{(l)} := \Delta Y^{(l)}$ for the remainder of the thesis. According to Trigeorgis [132], page 321, it has to hold:

$$\begin{aligned} E(\Delta Y^{(l)}) &= 2P^{(l)} H^{(l)} - H^{(l)}, \\ Var(\Delta Y^{(l)}) &= (H^{(l)})^2 - (2P^{(l)} H^{(l)} - H^{(l)})^2. \end{aligned}$$

As done in the original Trigeorgis model, the two consistency criteria

(*i*) mean of the continuous process = mean of the discrete process,

(*ii*) variance of the continuous process = variance of the discrete process.

have to be applied. This yields:

$$2P^{(l)} H^{(l)} - H^{(l)} = \mu^{(l)} K \implies P^{(l)} = \tfrac{1}{2}\left(1 + \frac{\mu^{(l)} K}{H^{(l)}}\right),$$

$$(H^{(l)})^2 - (\mu^{(l)} K)^2 = K \implies H^{(l)} = \sqrt{K + \left(\mu^{(l)} K\right)^2} \geq \mu^{(l)} K.$$

The fact that $\Delta Y^{(l)}$ is different for each of the L subintervals requires, as already mentioned, multi-dimensional grids. In detail, this yields another grid on each final node of a subinterval and on each subinterval a different risk-free rate has to be applied. The idea not to update the risk-free interest rate for each of the N subintervals in $[0, T]$ but only for the L quarters in $[0, T]$ reduces the bushy tree to a managable construct[99]. For $T = 1$ and $N = 12$ with a different risk-free rate for each quarter, this means $(\frac{12}{4} + 1)^4 = 256$ final nodes and $(\frac{12}{4} + 1)^3 = 64$ trees at the last subinterval. Table 4.3 shows a comparison of a quarterly update of the risk-free rate with an update of the risk-free rate at each subinterval $j \in \{1, 2, \ldots, N\}$.

Table 4.3. Computational requirements of the modified Trigeorgis log-transformed binomial tree model for different update frequencies of the risk-free interest rate for a 1-year real option, measured by the number of trees for the last subinterval.

N	L (= 4 T)	Update each $j \in \{1, 2, \ldots, N\}$	Update each $l \in \{1, 2, \ldots, L\}$
12	4	$2^{12-1} \approx 2\,000$	$\left(\frac{12}{4} + 1\right)^3 = 64$
24	4	$2^{24-1} \approx 8\,000\,0000$	$\left(\frac{24}{4} + 1\right)^3 = 343$
48	4	$2^{48-1} \approx 100\,000\,000\,000\,000$	$\left(\frac{48}{4} + 1\right)^3 = 2\,197$

Source: own calculations based on the construction of the algorithm

Figure 4.23 graphically explains the ideas of this modified method. Thereby, $N = 3L$ (i.e., $T_1 = 3\Delta t$) was chosen which is, for example, a monthly time step size with a quarterly adjustment of the risk-free rate. The multi-dimensionality cannot be graphically displayed in all its details since even in such a simple example it becomes too complex to display.

As usual, on each of the nodes within the algorithm, adjustments have to be made according to the real options(s) embedded in the investment project. Since these adjustments are the same for the modified Trigeorgis model as for the original Trigeorgis model they will not be restated here.

[99] This holds true as long as there is no option to defer. Such an option would still result in extremely long-time calculations due to multiple parallel runs of the algorithm. For such a real option this algorithm is not suited at all as Chapter 5 will show.

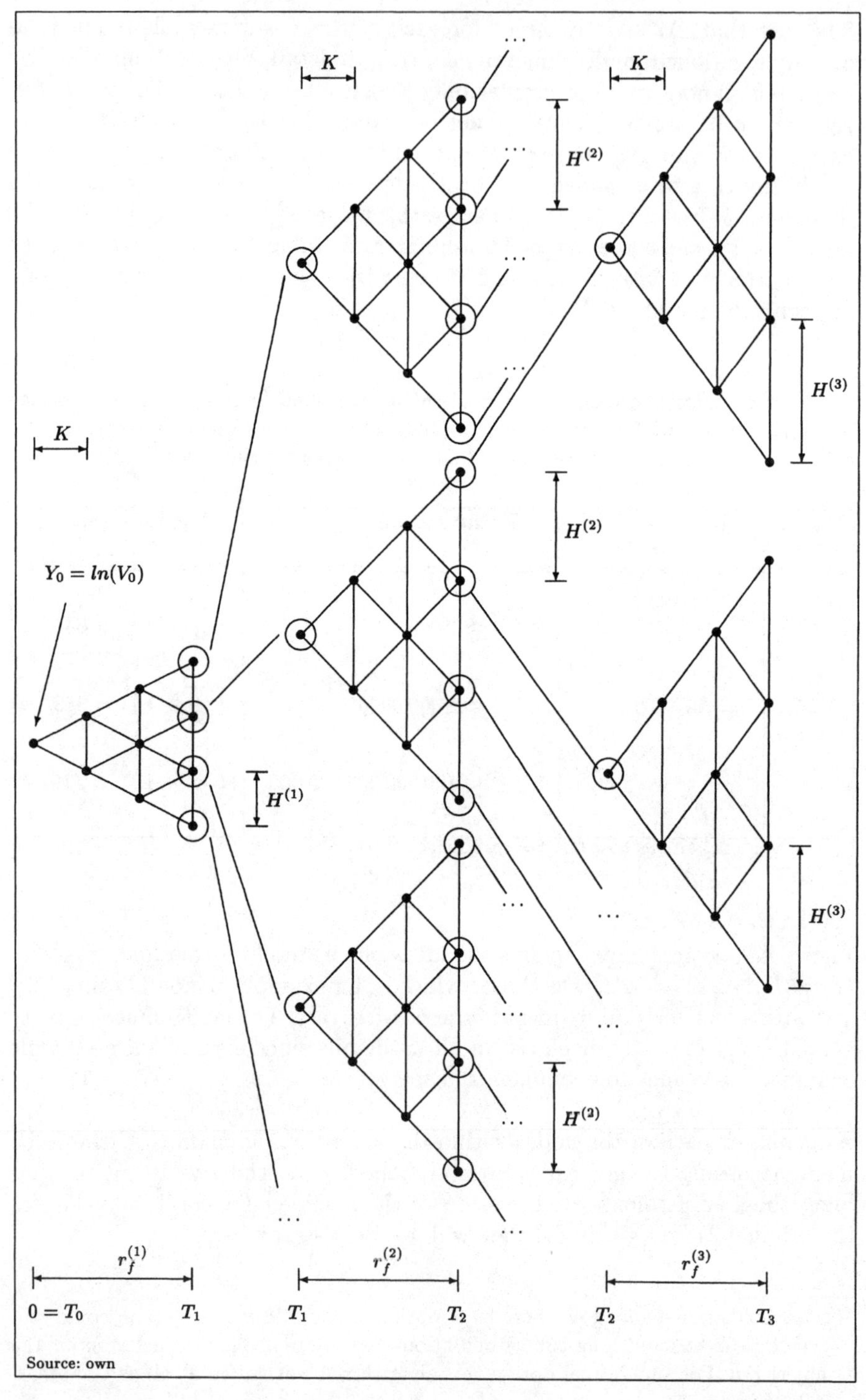

Fig. 4.23. Modified Trigeorgis log-transformed binomial tree approach for variable interest rates.

4.5 Summary

In this chapter various real options valuation tools were presented in detail. Besides explaining methods for real options pricing which are commonly used today, two modifications of classical valuation tools were introduced. These modified methods incorporate a non-constant risk-free interest rate which can be adjusted several times during the life of the real option.

According to the central idea of this thesis, the main focus was placed on the numerical implementation of each model. However, the necessary theoretical aspects and model properties were described as well. Especially, the criteria of stability and consistency were discussed since they are crucial for practical applications. The efficiency criterion of the algorithms will be an important aspect in Chapter 5 where different simple and complex real options are analyzed with a computer simulation program. The three main topics of Chapter 4 were:

- Numerical methods for real options pricing with constant interest rates (Section 4.2)
- Schwartz-Moon model (Section 4.3)
- Real options pricing with stochastic interest rates (Section 4.4)

The first topic, numerical methods, dealt with lattice methods that can be applied to various real options problems and finite difference methods that are restricted in the application. Lattice methods are computationally inefficient because with each run only one single real options value can be calculated for a particular starting value of the underlying. This holds true for both lattice schemes presented, i.e., for the classical Cox-Ross-Rubinstein binomial tree model and the Trigeorgis log-transformed binomial tree model. Their main advantage in Corporate Finance practice is that they can accommodate various real options types and even complex real options easily. It is important to note that in the Trigeorgis model a log-transformed underlying is used which yields positive features for the algorithm.

Log-transformed finite difference methods were the other area of Section 4.2. Compared with lattice methods, they have the big advantage that for each computer calculation the result is a whole array of real options values for different start values of the underlying in the finite difference grid. However, the use of these methods is restricted to cases where the partial differential equation is known which describes the development of the option value over time depending on the underlying. This was not necessary in the lattice methods since the idea there was to discretize the underlying process but not the partial differential equation. Two log-transformed finite difference methods were presented: the explicit and implicit finite difference methods.

The second area of Chapter 4 was the Schwartz-Moon model (Section 4.3), a recently developed real options pricing model for R&D investment in the pharmaceutical industry. While the model itself is mathematically very sophisticated, the NPV version of it is intuitive and easy to apply. Schwartz and Moon used the NPV solution as a benchmark for the stochastic solution that has to be derived via simulation programs which apply the NPV solution as the "start value". The model was slightly modified to accommodate a deferred project start when the risk-free interest rate is no longer constant.

The third and last area, Section 4.4, dealt with real options pricing when the risk-free interest rates are not constant any more. First, the Ingersoll-Ross model (4.4.1) was presented analytically. It is the first model that introduced the idea of a real option being created by stochastic interest rates even if the future cash flow structure is deterministic. This section also presented the two models that will be thoroughly analyzed in Chapter 5 via numerical simulation and historical backtesting. These two models are modifications of the Cox-Ross-Rubinstein binomial tree model (4.4.2) and of the Trigeorgis log-transformed binomial tree model (4.4.3). Both models incorporate a non-constant interest rate that can be calculated by applying the stochastic term structure models from Chapter 3. In the modified Trigeorgis log-transformed binomial tree model, the risk-free rate cannot be changed at each node of the tree due to the computational inefficiency of such an algorithm. Hence, it is updated quarterly.

All real options pricing tools presented in this chapter are implemented in the computer simulation program and, if theoretically possible, can be combined with each of the stochastic term stucture models from Chapter 3 for simulation and historical backtesting. This will be the topic of the following chapter.

5

Analysis of Various Real Options in Simulations and Backtesting

5.1 Introduction

In this chapter the real options valuation methods and stochastic term structure models introduced so far will be applied and numerically analyzed in five test situations for various real options, especially for the real options cases presented in Section 1.1. These cases are:

1. **Case 1:** Option to abandon the project at any time during the construction period for a salvage value X. Since such a real option is an American put option, the salvage value is the strike price of this option. The initial investment cost for the project is I_0.
2. **Case 2:** Option to abandon the project at any time during the construction period for a salvage value X (case 1) and option to expand the project once by expand factor e (e.g., expand project by $e = 30\%$) for an expand investment at the end of the construction period. The expand investment is assumed to be a fraction of the initial investment cost I_0.
3. **Case 3:** Complex real option in case 2 combined with an option to defer the project start by exactly one year. The project can start today or in exactly one year from today if the investment in one year has a positive NPV. The initial investment cost to start the project in one year is assumed to be I_0 as well. If, in one year from now, the NPV will be negative, the project will not be started at all.

The stochastic term structure models considered in this chapter are:

- **Equilibrium models:**
  Vasicek model and Cox-Ingersoll-Ross model.
- **No-arbitrage models:**
  Ho-Lee model, Hull-White one-factor model and Hull-White two-factor model.

To use a term structure model in practice, the coefficients of the short-rate model have to be determined. This is in itself a broad topic that will briefly be mentioned in Section 5.2. This so-called *calibration procedure* is presented for all equilibrium models introduced in this thesis as well as for the Ho-Lee no-arbitrage model: Section 5.2.1 briefly introduces a computational tool developed by Schulmerich[1] in 1997 to estimate the short-rate parameters for the Vasicek and the Cox-Ingersoll-Ross models (among others). Section 5.2.2 describes two calibration procedures for the Ho-Lee model.

In order to simulate all term structure models of Chapter 3, all real options pricing methods of Chapter 4 and the historical backtesting procedures, a computer simulation program[2] was developed for this thesis. However, not all real options pricing methods and not all stochastic term structure models will and can be combined with all of the three cases. One reason for this is the sheer amount of combinations that arise from three real options cases, historical backtesting, four real options valuation tools, and eight term structure models (including the simple case of a constant risk-free rate). More importantly, several real options valuation tools cannot be applied with a non-constant risk-free interest rate, e.g., the log-transformed implicit and explicit finite difference methods.

Another important aspect is the choice of the discretization parameters (for the time and/or the state axis). The number of simulated term structures for a stochastically modelled term structure has to be investigated. Additionally, the real options valuation tools presented have to be compared in order to identify the most suitable tool for analyzing the three real options cases. All these aspects were taken into account in a sound test strategy that comprises five consecutive stages referred to in the following as *test situations*:

1. Section 5.4: *Test situation 1: the Schwartz-Moon model with a deferred project start*
   Variable interest rates in the case of the Schwartz-Moon model with a deferred project start.

2. Section 5.5: *Test situation 2: preliminary tests for real options valuation*
   Preliminary tests to investigate the parameters of the real options valuation tools and to compare the valuation methods.

3. Section 5.6: *Test situation 3: real options valuation with a stochastic interest rate using equilibrium models*
   Influence of the salvage and expand factors in cases 1, 2 and 3 on the real options (Section 5.6.1); analysis of all cases for equilibrium models

[1] See Schulmerich [117].

[2] The computer program is written in Borland C++, version 5.02, for PCs.

(Vasicek model in Section 5.6.2 and Cox-Ingersoll-Ross model in Section 5.6.3).

4. Section 5.7: *Test situation 4: real options valuation with a stochastic interest rate using no-arbitrage models*

   Analysis of cases 1, 2 and 3 for the Ho-Lee model (Section 5.7.1), comparison of the Hull-White one-factor model with the Hull-White two-factor model (Section 5.7.2), and comparison of the Ho-Lee model with the Hull-White one-factor model (Section 5.7.3).

5. Section 5.8: *Test situation 5: real options valuation in historical backtesting*

   Historical backtesting for cases 1, 2 and 3 through a comparison of the real options pricing when using a stochastically modelled risk-free interest rate (Ho-Lee model), a constant rate, interest rates that equal the currently implied forward rates (Hull's approach), and the historical risk-free rates of the backtesting period.

The exact research strategy for each of these five test situations and its logical flow of ideas will be elaborated in Section 5.3. For test situation 1, an Excel spreadsheet was developed. For the remaining test situations 2 to 5 the developed computer simulation program is used. With the exception of the first one, each test situation is conducted by recapitulating of the test situation's main results. As usual, a summary of Chapter 5 is given in Section 5.9.

In the following, the underlying values, the calculated values and the corresponding standard deviations are all given in US $ (in short $) whereby the unit is omitted in most cases. All parameters with respect to the underlying are indicated by the subscript $_S$ (i.e., σ_S, T_S, and N_S) to distinguish them from parameters with respect to the term structure of interest rates, which are indicated by the subscript $_r$ (i.e., σ_r, T_r, and N_r).

5.2 Calibration of Stochastic Interest Rate Models

All the stochastic term structure models introduced in Chapter 3 have in common that some parameters have to be specified in advance, based on the data available in the capital markets. This process is called *calibration*. Calibration methods for the Vasicek and the Cox-Ingersoll-Ross models are introduced in 5.2.1. These methods estimate the short-rate parameters given a time series of short rates.

Term-structure-consistent models like the Ho-Lee model not only need one single parameter specified up-front but need the whole yield curve as an input parameter, i.e., the model has to be fit to the current yield curve. Only the calibration process for the Ho-Lee model will be described here in 5.2.2 since only this term structure model will be used in historical backtesting for reasons elaborated in the empirical analysis section[3].

5.2.1 Calibration Procedure for the Vasicek and the Cox-Ingersoll-Ross Models

In Schulmerich [117] several statistical methods are described in theory and practice on how the coefficients for various mean reversion models can be determined. Two of these are the Vasicek model and the Cox-Ingersoll-Ross model. His publication is accompanied by a proprietary computer program that estimates these coefficients for given time series. The main statistical tool is the martingale estimation function. For more information on this topic see Schulmerich [117], Chapter 6.

The way this computer tool is applied is such that the time series has to be loaded into the computer tool to estimate the short-rate parameters for the Vasicek and the Cox-Ingersoll-Ross models. The short-rate time series is the time series for instantaneous interest rates. An admissible approximation is to use a time series with a very short time to maturity (e.g., 1 month) instead of the instantaneous rate[4].

5.2.2 Calibration Procedure for the Ho-Lee Model

A short-rate process in the Ho-Lee model is given via the stochastic differential equation (see (3.17))

$$\begin{aligned} dr_t &= \theta(t)dt \ + \ \sigma dB_t, \ \ t \geq 0, \quad \text{where} \\ \theta(t) &= \frac{\partial f(0,t)}{\partial t} \ + \ \sigma^2 t \ \ \forall t \geq 0, \ \ \sigma \in \mathbb{R}^+. \end{aligned}$$

$\theta(\cdot)$ represents a time-dependent drift that contains the derivative of the initial instantaneous forward rate curve f and the volatility parameter of the short-rate process. For this model two parameters have to be specified: the volatility parameter σ and the function θ.

[3] For more information on calibration procedures for various term structure models see Brown & Schaefer [23], Chan, Karolyi, Longstaff & Sanders [30] (especially Part 2), and Wilmott [135], Section 34.

[4] The estimations for the 1-month Libor in Schulmerich [117], pages 187-188, Tables H.9 and H.10, with T given in years and parameter β given in percentage are similar to the values used in this thesis.

Estimation of volatility parameter σ: According to Clewlow and Strickland[5] there are two major methods to estimate the volatility parameter. The first involves interest rate options, the second involves caplets. For the first method, m actively traded pure discount bond put options need to be considered. Let p_i^{market}, with $i = 1, 2, \ldots, m$, denote the market price of m European put options on actively traded Zero bonds. The price of these options can also be calculated directly via a modified Black-Scholes formula[6]. Let s be the maturity time point of the European put option and let $T \geq s$ be the maturity time point of the underlying Zero. Then the price $p(s,T)$ of a European put option with strike X that matures at time s on a Zero that matures at time T is

$$\left.\begin{aligned} p(s,T) &= X\,P(0,s)\,N(-d_2) - P(0,T)N(-d_1), \;\; 0 \leq s \leq T, \text{ where} \\ d_1 &:= \tfrac{1}{\sigma_P} \log\left(\frac{P(0,T)}{X\,P(0,s)}\right) + \tfrac{\sigma_P}{2}, \\ d_2 &:= d_1 - \sigma_P, \\ \sigma_P &:= \sigma(T-s)\,\sqrt{s}. \end{aligned}\right\} \quad (5.1)$$

Here, $N(\cdot)$ is the cumulative standard normal distribution function. Using $p_i^{model}(\sigma)$ as the price of a European put option according to the model in (5.1) for one of the m put options, the volatility parameter $\sigma \in \mathbb{R}^+$ for the Ho-Lee model can be calculated as the solution σ^* of the minimalization problem

$$\min_{\sigma \in \mathbb{R}^+} \sqrt{\sum_{i=1}^{m} \left(\frac{p_i^{model}(\sigma) - p_i^{market}}{p_i^{market}}\right)^2}.$$

Clewlow and Strickland also elaborate on calibrating the volatility parameter σ to individual caplets, an idea that will not be presented here in detail. Caplets are, like caps, quoted on Libor. For these caplets the Black volatilities are calculated and published. Since caplets are quoted for specific Libor rates, using these caplets for calibrating results in parameters that, combined with a Libor curve as the initial yield curve for a no-arbitrage model, are perfectly suited to value swaps and swaptions. However, since U.S. Zero yields are used in this thesis, parameter calibration using caplets results in a small systematic error for the volatility estimate in the Ho-Lee model. For more information on calibrating the volatility in the Ho-Lee model to individual caplets see Clewlow & Strickland [31], pages 212-215.

[5] See Clewlow & Strickland [31], pages 211-215.

[6] See Clewlow & Strickland [31], formula (7.6) on page 210, or Black & Scholes [9].

Calculation of the function θ: The function θ contains the partial derivative of the instantaneous forward rate. Therefore, the instantaneous forward rate first has to be calculated from the market data and then the partial derivative has to be determined. The instantaneous forward rate is given in equation (3.3) as

$$f(0,t) = -\frac{\partial}{\partial t} \ln P(0,t), \quad t \geq 0, \tag{5.2}$$

where $P(0,t)$ is the current price of a Zero bond that matures at time t. For the Zero, the following pricing relationship holds (see equation (3.1) of Chapter 3):

$$P(0,t) = e^{-R(0,t)t} \tag{5.3}$$

where $R(0,t)$ is the current continuously compounded yield (spot rate) of a Zero bond that matures at time t. Combining (5.2) and (5.3) yields

$$f(0,t) = \frac{\partial}{\partial t} (R(0,t)\,t).$$

Therefore, if the current spot rate curve $R(0,t)$ is given, the instantaneous forward rate curve $f(0,t)$ can easily be calculated. Consequently, function θ in the Ho-Lee model can also be calculated easily as

$$\frac{\partial f(0,t)}{\partial t} \approx \frac{f(0,t+\Delta\tau) - f(0,t)}{\Delta\tau} \tag{5.4}$$

with sufficiently small $\Delta\tau > 0$. This $\Delta\tau$ is exactly the $\Delta\tau$ from the cubic spline interpolation (see 3.2.2) and the stochastic process approximation (see 2.5), a specification that is set automatically in the computer simulation program.

5.3 Description of the Test Strategy

This section describes the test strategy followed in the empirical analysis of Chapter 5. The test strategy was developed to answer the following questions:

- How do the various real options influence the total value of the project, i.e., the project value including all real options?
- In the case of stochastically modelled risk-free interest rates, how does the choice of the term structure model and its parameters influence the real options value and accordingly the net present value of the project?
- Does a stochastically modelled risk-free rate better capture the interest rate volatility that could be observed especially between 1999 and 2002 in the capital markets compared with models with a constant risk-free rate (historical backtesting)?

To answer these questions three cases were analyzed. The real options included in these cases were chosen on the basis of recent empirical analysis done by Vollrath[7] among a sample of firms with headquarters in Germany. This survey was already explained in detail in Section 1.1. The combination of the real options with the three cases was based on the idea of showing how the project value gradually changes by including more and more real options. The final case 3 is a modified version of a natural-resource project of a major multinational oil company that was valued by applying the real options theory. This project, its embedded real options and a sensible choice of the real options parameters are described in Trigeorgis [129] and Trigeorgis [132], Section 11.5.

The test strategy is to first introduce the idea of interest rate uncertainty from a simple yet impressive point of view in order to create awareness of this issue. For this the Schwartz-Moon model with a modification to incorporate an option to defer will be used.

Second, preliminary tests have to be conducted to investigate the quality of the real options valuation tools and the appropriate choice of discretization parameters. Here, the real options value (ROV), total project value (TPV), and net present value (NPV) of the project will be analyzed. The ROV is the present value of the real option(s). The TPV is the total project's present value, the present value of the project including the ROV but without considering any initial investment cost to start the project. Finally, NPV is the project's net present value including the ROV and the initial investment cost I_0. With $S = S_0$ as the underlying value, the present value of the estimated project's future cash flow (including all future costs but not I_0), the following relationships hold[8]:

$$\begin{aligned} NPV &= ROV + S_0 - I_0, \\ TPV &= ROV + S_0, \\ NPV &= TPV - I_0. \end{aligned}$$

The relationship between these three variables and the initial investment cost I_0 as well as boundaries are graphically explained for a European put option and an American put option in Figures 5.1 and 5.2, respectively. The maturity of the option is T, the constant risk-free rate is r_f, and the strike is X. A European put option is a real option to abandon the investment project at time T for a salvage value X. An American put option is a real option to abandon the investment project at any time in interval $[0, T]$ for a salvage value X.

[7] See Vollrath [134].

[8] See Section 2.6.1 and Trigeorgis [128], page 161: *Under the expanded [real options] framework, the total economic desirability of an investment opportunity is explicitly seen as the sum of its static NPV of directly measurable expected cash flows and of the option premium reflecting the value of managerial operating flexibility and strategic interactions.*

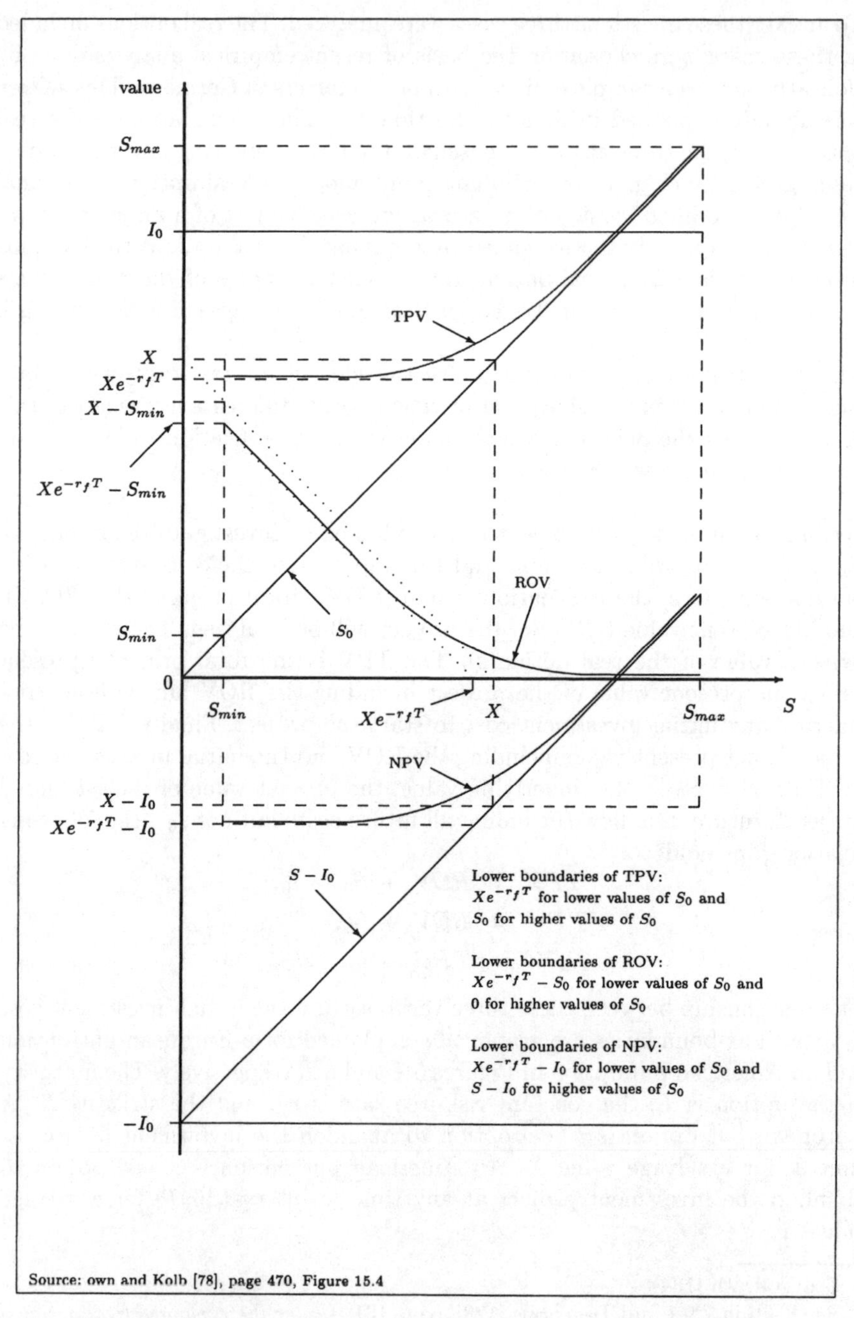

Source: own and Kolb [78], page 470, Figure 15.4

Fig. 5.1. General relationships and boundaries for a European put.

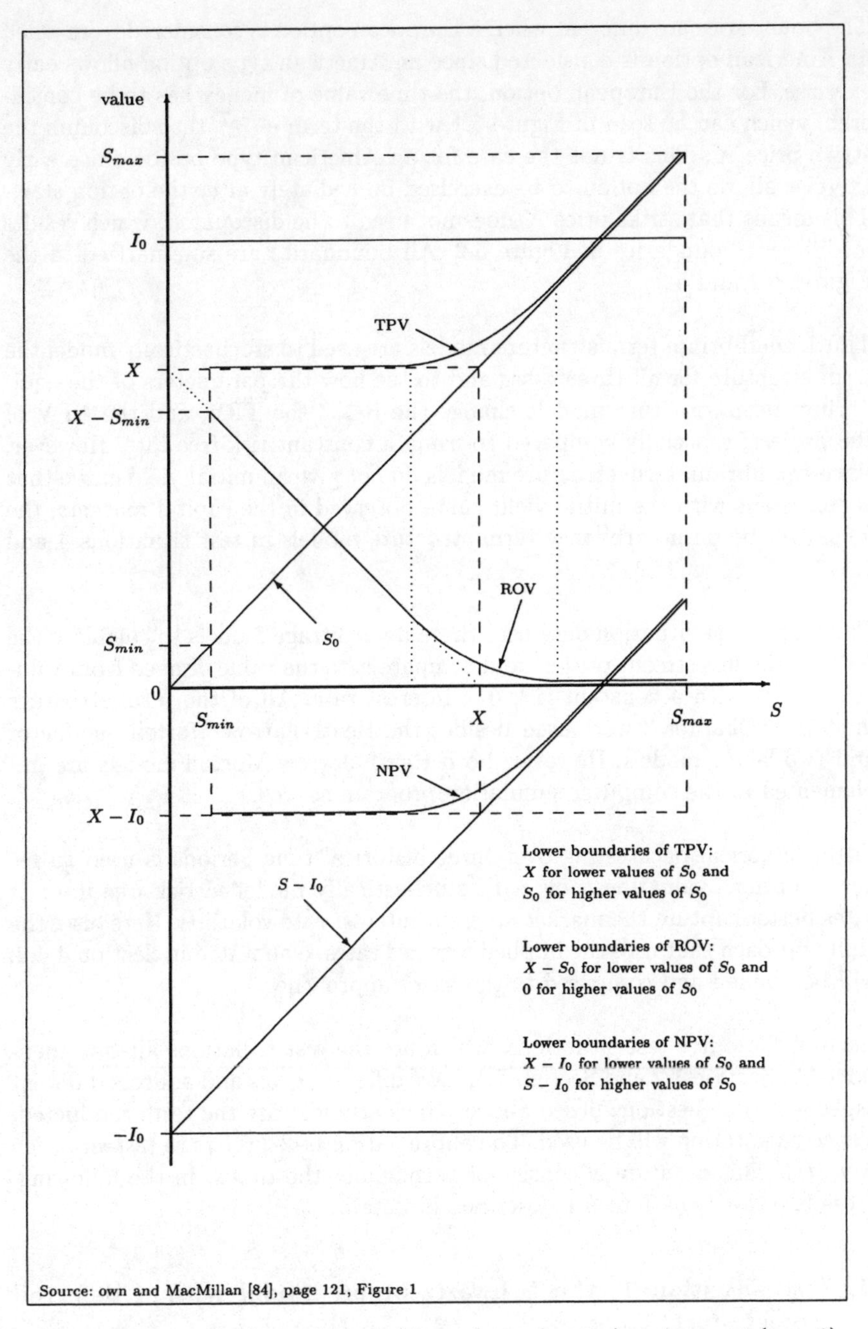

Fig. 5.2. General relationships and boundaries for an American put (case 1).

The boundaries are different when a European option is considered from when an American option is considered since an American type option allows early exercise. For the European option, the time value of money has to be considered, which can be seen in Figure 5.1 with the term $e^{-r_f T}$ that discounts the strike price X. This is not the case for an American type option since early exercise allows the option to be exercised immediately after the option start. This means that strike price X does not need to be discounted, which results in different boundaries in Figure 5.2. All boundaries are summarized in the Figures 5.1 and 5.2.

Third, equilibrium term structure models are used to stochastically model the term structure for all three cases and to see how the parameters of the equilibrium term structure models change the ROV, the TPV, and the NPV of the project, especially compared to using a constant risk-free rate. However, since equilibrium term structure models do not give an initial yield curve that is consistent with the initial yield curve observed in the capital markets, the focus will be on no-arbitrage term structure models in test situations 4 and 5.

The fourth test situation only uses these no-arbitrage models to calculate the NPV of the investment project and compare it to the value derived from valuation tools with a constant risk-free interest rate. All of the term structure models of Chapter 3 were used besides the Heath-Jarrow-Morton one-factor and two-factor models. However, both Heath-Jarrow-Morton models are implemented in the computer simulation program as well.

Fifth, historical backtesting over three historical time periods is used to see if real options valuation tools with stochastically modelled risk-free interest rates better capture the market inherent interest rate volatility. Here also, the Hull approach that uses the implied forward rates, elaborated in Section 4.4.2, will be applied and compared to the other approaches.

Each of these five test situations, which are the test situations already mentioned in Section 1.2 and Section 5.1, have different goals and address different aspects of the questions posed above. To clearly identify the tests conducted, a specific notation will be used: The abbreviation *ts-2-4* refers to *test situation 2, test 4*. This notation is consistent throughout the thesis. In the following, these five test situations are described in detail.

1. **Test situation 1: the Schwartz-Moon model with a deferred project start**

   The Schwartz-Moon model will be used to include a deferred project start. The assumptions are very strict so that the model is rather abstract. It will be shown how the development of the risk-free rate influences the value of an investment project and the manager's investment decision itself. Three scenarios are analyzed: First, the basic cases of a constant and flat term structure without a defer option (ts-1-1) and with a defer option (ts-1-2) are analyzed. Second, an initially flat yield curve shifts upwards by 50 bps per quarter (ts-1-3) in one example and shifts downwards by −50 bps per quarter in another example (ts-1-5). Then, the project's NPV is calculated depending on when the project starts. In the third scenario an initially flat yield curve shifts upwards by 100 bps per quarter in one example (ts-1-4) and shifts downwards by −100 bps per quarter in another example (ts-1-6). The example with a downward move of −100 bps per quarter is particularly realistic as could be seen between the end of 2000 and the end of 2001 for the 1-mos. U.S. Zero yields, see Figure 4.16.

The result in the downward shifting case will indicate that it might be preferable to wait with the project start due to the interest rate movement. Therefore, in *test situation 4*, only a downward sloping initial yield curve will be considered since such a yield curve indicates, according to the pure expectations theory, that the yields decrease in the future. Note that the Schwartz-Moon example is programmed in Microsoft Excel and is not part of the computer simulation program.

2. **Test situation 2: preliminary tests for real options valuation**

   Before the actual testing can start, the choice of the discretization parameters has to be investigated. Parameters N_S and M for the time and state axis, respectively, have to be chosen for the log-transformed explicit and implicit finite difference methods. The number N_S of time steps has to be specified for the Cox-Ross-Rubinstein binomial tree method and the Trigeorgis log-transformed binomial tree method. Moreover, if the term structure is stochastically modelled, the appropriate number of simulated paths has to be figured. The latter is especially important for the Cox-Ross-Rubinstein model since this is the main tool for real options valuation with a stochastic risk-free interest rate.

   On the other hand, the Trigeorgis model will only be used with a stochastic interest rate for test purposes since the computational time even with the fastest PCs available nowadays[9] will be very long[10]. In this case, the problem called the bushy tree problem shows up and restricts any practical use tremendously.

[9] The PCs used for this thesis were solely Pentium 4 computers with tact frequency of between 1.8 MHz. and 2.4 MHz., the fastest personal computers available in early 2003.

[10] In case 3 this means that it can take weeks to get a reasonable result.

The results obtained by different choices of discretization parameters have to be compared to a benchmark. As European put options can be priced with the Black-Scholes formula, which will be used as a benchmark to the numerical methods[11], the option type of a European put option was chosen to investigate an option to abandon a project at a certain time point T_S for a salvage value X. The Black-Scholes formula[12] calculates the current price p of a European put option (assuming no dividend payments) as

$$\left.\begin{aligned} p &= -S_0\, N(-d_1) \;+\; X\, e^{-r_f T_S} N(-d_2), \quad \text{where} \\ d_1 &:= \frac{\ln\left(\frac{S_0}{X}\right) + \left(r_f + \frac{1}{2}\sigma^2\right)}{\sigma\sqrt{T_S}} \;=\; d_2 \;+\; \sigma\sqrt{T_S}, \\ S_0 &= \text{current price of the underlying}, \\ T_S &= \text{time to maturity of option (assuming } t = 0 \text{ today)}, \\ X &= \text{strike price}, \\ r_f &= \text{risk-free interest rate.} \end{aligned}\right\} \tag{5.5}$$

Equation (5.5) serves as a benchmark when evaluating how suitable a presented algorithm is for pricing a real option and how the discretization parameters of the various real options valuation tools have to be chosen. The goal is to choose the most appropriate real options valuation tool for Corporate Finance practice, which will then be used in the remaining test situations. In order to achieve this goal, several tests will be conducted in test situation 2. Table 5.1 displays all tests of this test situation.

Table 5.1. Combinations of real options valuation tools, term structure models and test ideas for test situation 2.

No.	Real options valuation tool	Risk-free rate	Test idea
$2-1$	Black-Scholes formula	constant	benchmark
$2-2$	Log-transformed explicit finite differences	constant	$N_S = 720 \cdot T_S$ and $M = 100$
$2-3$	Log-transformed explicit finite differences	constant	$N_S = 1080 \cdot T_S$ and $M = 100$
$2-4$	Log-transformed explicit finite differences	constant	$N_S = 1440 \cdot T_S$ and $M = 100$

[11] To price a European option instead of an American option as in case 1, the computer simulation program has to be modified only slightly at node (i, j) with $j < N$ for each numerical method since the comparison of the prevailing option value with the difference of the strike price and the corresponding underlying value is not necessary for a European option which can only be exercised at maturity T_S, i.e., at $j = N$.

[12] See Trigeorgis [132], page 92.

Table 5.1 continued.

No.	Real options valuation tool	Risk-free rate	Test idea
2 – 5	Log-transformed explicit finite differences	constant	$N_S = 1800 \cdot T_S$ and $M = 100$
2 – 6	Log-transformed explicit finite differences	constant	$N_S = 2160 \cdot T_S$ and $M = 100$
2 – 7	Log-transformed explicit finite differences	constant	$N_S = 2160 \cdot T_S$ and $M = 50$
2 – 8	Log-transformed explicit finite differences	constant	$N_S = 2160 \cdot T_S$ and $M = 150$
2 – 9	Log-transformed explicit finite differences	constant	$N_S = 2160 \cdot T_S$ and $M = 200$
2 – 10	Log-transformed implicit finite differences	constant	$N_S = 720 \cdot T_S$ and $M = 100$
2 – 11	Log-transformed implicit finite differences	constant	$N_S = 1800 \cdot T_S$ and $M = 100$
2 – 12	Log-transformed implicit finite differences	constant	$N_S = 2160 \cdot T_S$ and $M = 50$
2 – 13	Cox-Ross-Rubinstein binomial tree	constant	$N_S = 60 \cdot T_S$ and $M = 100$
2 – 14	Cox-Ross-Rubinstein binomial tree	constant	$N_S = 120 \cdot T_S$ and $M = 100$
2 – 15	Cox-Ross-Rubinstein binomial tree	constant	$N_S = 240 \cdot T_S$ and $M = 100$
2 – 16	Cox-Ross-Rubinstein binomial tree	constant	$N_S = 360 \cdot T_S$ and $M = 100$
2 – 17	Cox-Ross-Rubinstein binomial tree	constant	$N_S = 720 \cdot T_S$ and $M = 100$
2 – 18	Cox-Ross-Rubinstein binomial tree	constant	$N_S = 1080 \cdot T_S$ and $M = 100$
2 – 19	Trigeorgis log-transformed binomial tree	constant	$N_S = 720 \cdot T_S$ and $M = 100$ (TPV)
2 – 20	Trigeorgis log-transformed binomial tree	constant	$N_S = 720 \cdot T_S$ and $M = 100$ (ROV)
2 – 21	Trigeorgis log-transformed binomial tree	constant	$N_S = 360 \cdot T_S$ and $M = 100$ (TPV)
2 – 22	Trigeorgis log-transformed binomial tree	constant	$N_S = 360 \cdot T_S$ and $M = 100$ (ROV)
2 – 23	Trigeorgis log-transformed binomial tree	constant	$N_S = 180 \cdot T_S$ and $M = 100$ (TPV)
2 – 24	Trigeorgis log-transformed binomial tree	constant	$N_S = 180 \cdot T_S$ and $M = 100$ (ROV)
2 – 25	Trigeorgis log-transformed binomial tree	constant	$N_S = 60 \cdot T_S$ and $M = 100$ (TPV)
2 – 26	Trigeorgis log-transformed binomial tree	constant	$N_S = 60 \cdot T_S$ and $M = 100$ (ROV)
2 – 27	Cox-Ross-Rubinstein binomial tree	Vasicek	$T_S = 3$, $N_S = 1080$, $M = 50$ with 25, 100, and 400 simulated future interest rate paths in the Vasicek model ($\alpha = 0.6$, $\beta = 0.018$)

Table 5.1 continued.

No.	Real options valuation tool	Risk-free rate	Test idea
2 – 28	Cox-Ross-Rubinstein binomial tree	Vasicek	$T_S = 3$, $N_S = 1080$, $M = 50$ with 25, 100, and 400 simulated future interest rate paths in the Vasicek model ($\alpha = 0.6$, $\beta = 0.024$)
2 – 29	Cox-Ross-Rubinstein binomial tree	Vasicek	$T_S = 3$, $N_S = 1080$, $M = 50$ with 25, 100, and 400 simulated future interest rate paths in the Vasicek model ($\alpha = 0.6$, $\beta = 0.03$)
2 – 30	Cox-Ross-Rubinstein binomial tree and Trigeorgis log-transformed binomial tree	constant	case 1 with $T_S = 3$ and salvage factor of 0
2 – 31	Cox-Ross-Rubinstein binomial tree and Trigeorgis log-transformed binomial tree	constant	case 1 with $T_S = 1$ and salvage factor of 0
2 – 32	Cox-Ross-Rubinstein binomial tree and Trigeorgis log-transformed binomial tree	Vasicek	case 1 with $T_S = 1$ and salvage factor of 0
2 – 33	Trigeorgis log-transformed binomial tree	Vasicek	case 1 with $T_S = 1, N_S = 36$ and $T_S = 1, N_S = 48$
2 – 34	Trigeorgis log-transformed binomial tree	Vasicek	case 2 with $T_S = 1, N_S = 36$ and $T_S = 1, N_S = 48$
2 – 35	Trigeorgis log-transformed binomial tree	Cox-Ingersoll-Ross	case 1 with $T_S = 1, N_S = 36$ and $T_S = 1, N_S = 48$
2 – 36	Trigeorgis log-transformed binomial tree	Cox-Ingersoll-Ross	case 2 with $T_S = 1, N_S = 36$ and $T_S = 1, N_S = 48$
2 – 37	Trigeorgis log-transformed binomial tree	Ho-Lee	case 1 with $T_S = 1, N_S = 24$ and $T_S = 1, N_S = 40$
2 – 38	Trigeorgis log-transformed binomial tree	Ho-Lee	case 2 with $T_S = 1, N_S = 24$ and $T_S = 1, N_S = 40$
2 – 39	Trigeorgis log-transformed binomial tree	Hull-White one-factor	case 1 with $T_S = 1, N_S = 24$ and $T_S = 1, N_S = 40$
2 – 40	Trigeorgis log-transformed binomial tree	Hull-White one-factor	case 2 with $T_S = 1, N_S = 24$ and $T_S = 1, N_S = 40$

Table 5.1 continued.

No.	Real options valuation tool	Risk-free rate	Test idea
2 – 41	Trigeorgis log-transformed binomial tree	Hull-White two-factor	case 1 with $T_S = 1, N_S = 24$ and $T_S = 1, N_S = 40$
2 – 42	Trigeorgis log-transformed binomial tree	Hull-White two-factor	case 2 with $T_S = 1, N_S = 24$ and $T_S = 1, N_S = 40$
2 – 43	Cox-Ross-Rubinstein binomial tree and Trigeorgis log-transformed binomial tree	constant	case 1 with $T_S = 3, N_S = 1080$ and $T_S = 3, N_S = 360$
2 – 44	Cox-Ross-Rubinstein binomial tree and Trigeorgis log-transformed binomial tree	constant	case 2 with $T_S = 3, N_S = 1080$ and $T_S = 3, N_S = 360$
2 – 45	Cox-Ross-Rubinstein binomial tree and Trigeorgis log-transformed binomial tree	constant	case 3 with $T_S = 3, N_S = 1080$ and $T_S = 3, N_S = 360$
2 – 46	Cox-Ross-Rubinstein binomial tree	Cox-Ingersoll-Ross	cases 1 and 2 with $T_S = 1, N_S = 36$ and $T_S = 1, N_S = 48$
2 – 47	Cox-Ross-Rubinstein binomial tree	Ho-Lee	cases 1 and 2 with $T_S = 1, N_S = 24$ and $T_S = 1, N_S = 40$

Source: own

3. **Test situation 3: real options valuation with a stochastic interest rate using equilibrium models**

   Here, the idea of the previous test situation will be made concrete for the real options cases 1, 2 and 3. The goal is to analyze the influence of a non-constant risk-free rate on the real options value. The two equilibrium models presented in Chapter 3, the Vasicek model and the Cox-Ingersoll-Ross model, are used to achieve this goal.

   It will be interesting to see how the volatility parameter and the long-term interest rate level drive the value of the three real options cases, especially since the Vasicek model has a constant volatility and the Cox-Ingersoll-Ross model has a volatility that depends on the level of the short rate. The valuation results will be compared to the real options values obtained when applying a constant risk-free rate. The valuation tool used for test situation 3 is exclusively the Cox-Ross-Rubinstein binomial tree method, which - as a result of test situation 2 - turns out to be the most appropriate real options valuation tool in practice.

   The resulting combinations of real options valuation tools, term structure models, and real options cases for test situation 3 are listed in Table 5.2.

Table 5.2. Combinations of real options valuation tools, term structure models and real options cases for test situation 3.

No.	Real options valuation tool	Risk-free rate	Real options case
$3-1$	Cox-Ross-Rubinstein binomial tree	constant	case 1 with variable salvage factor
$3-2$	Cox-Ross-Rubinstein binomial tree	constant	case 1
$3-3$	Cox-Ross-Rubinstein binomial tree	constant	case 2
$3-4$	Cox-Ross-Rubinstein binomial tree	constant	case 3
$3-5$	Cox-Ross-Rubinstein binomial tree	Vasicek	case 1
$3-6$	Cox-Ross-Rubinstein binomial tree	Vasicek	case 2
$3-7$	Cox-Ross-Rubinstein binomial tree	Vasicek	case 3
$3-8$	Cox-Ross-Rubinstein binomial tree	Cox-Ingersoll-Ross	case 1
$3-9$	Cox-Ross-Rubinstein binomial tree	Cox-Ingersoll-Ross	case 2
$3-10$	Cox-Ross-Rubinstein binomial tree	Cox-Ingersoll-Ross	case 3
$3-11$	Cox-Ross-Rubinstein binomial tree	Cox-Ingersoll-Ross	case 2 with variable expand factor
$3-12$	Cox-Ross-Rubinstein binomial tree	Cox-Ingersoll-Ross	case 3 with variable expand factor
$3-13$	Cox-Ross-Rubinstein binomial tree	Cox-Ingersoll-Ross	case 1 with variable salvage factor

Source: own

4. **Test situation 4: real options valuation with a stochastic interest rate using no-arbitrage models**

   This test situation (Section 5.7) extends the previous test situation to no-arbitrage term structure models. The initial yield curve for such a model will be an inverted yield curve (according to the results from the Schwartz-Moon model). Here, the inverted yield curve of December 1, 2000 was chosen. This test situation is structured in the following way:

   a) Ho-Lee model: Section 5.7.1

   b) Hull-White one-factor vs. Hull-White two-factor model: Section 5.7.2

   c) Ho-Lee model versus Hull-White one-factor model: Section 5.7.3

   The Heath-Jarrow-Morton one-factor and two-factor models will not be used here since the focus will be on those stochastic term structure models that can be generated using the discretization methods of Section 2.5. Nevertheless, in the computer simulation program real options pricing can also be done with the Heath-Jarrow-Morton one- and two-factor models for all three real options cases.

Of course, in this test situation too, the results will be compared to the results with a constant risk-free interest rate. For the no-arbitrage models the volatility parameters will be varied systematically to see the impact of the volatility parameter on the project for cases 1, 2 and 3. The resulting combinations of real options valuation tools, term structure models, and real options cases are listed in Table 5.3.

Table 5.3. Combinations of real options valuation tools, term structure models and real options cases for test situation 4.

No.	Real options valuation tool	Risk-free rate	Real options case
4 – 1	Cox-Ross-Rubinstein binomial tree	Ho-Lee	case 1 with variable salvage factor
4 – 2			case 2 with variable expand factor
4 – 3			case 3 with variable expand factor
4 – 4	Cox-Ross-Rubinstein binomial tree	Ho-Lee	case 1
4 – 5			case 2
4 – 6			case 3
4 – 7	Cox-Ross-Rubinstein binomial tree	Hull-White one-factor	case 1
4 – 8			case 2
4 – 9			case 3
4 – 10	Cox-Ross-Rubinstein binomial tree	Hull-White two-factor	case 1
4 – 11			case 2
4 – 12			case 3
4 – 13	Cox-Ross-Rubinstein binomial tree	Hull-White two-factor	case 2 with variable expand factor
4 – 14			case 3 with variable expand factor
4 – 15	Cox-Ross-Rubinstein binomial tree	constant risk-free rate	cases 1, 2 and 3

Source: own

5. **Test situation 5: real options valuation in historical backtesting**

This test situation is devoted to historical backtesting of all three real options cases with a construction stage of 3 years during three different historical time periods. All of these historical time periods include the time between the end of 2000 and the end of 2001 when the term structure of interest rates dramatically moved downwards at the short end by approximately 100 bps per quarter. The results from historical backtesting will then be compared to the results obtained when applying a constant risk-free rate, the Ho-Lee model, and Hull's implied forward rates approach described in Section 4.4.

The only real options valuation tool that will be applied here is the Cox-Ross-Rubinstein binomial tree since the Trigeorgis method takes too much computational time[13] in the case of non-constant risk-free interest rates. However, it is generally possible to do historical backtesting with the Trigeorgis model in the computer simulation program as well[14]. The resulting combinations of real options valuation tools, real options cases, and initial term structures are listed in Table 5.4.

Table 5.4. Combinations of real options valuation tools, real options cases, and initial term structures for test situation 5.

No.	Real options valuation tool	Real options	Shape of initial term structure
$5-1$	Cox-Ross-Rubinstein binomial tree	case 1	inverted
$5-2$	Cox-Ross-Rubinstein binomial tree	case 1	flat
$5-3$	Cox-Ross-Rubinstein binomial tree	case 1	normal
$5-4$	Cox-Ross-Rubinstein binomial tree	case 2	inverted
$5-5$	Cox-Ross-Rubinstein binomial tree	case 2	flat
$5-6$	Cox-Ross-Rubinstein binomial tree	case 2	normal
$5-7$	Cox-Ross-Rubinstein binomial tree	case 3	inverted
$5-8$	Cox-Ross-Rubinstein binomial tree	case 3	flat
$5-9$	Cox-Ross-Rubinstein binomial tree	case 3	normal

Source: own

Of course, it would not be wise to use the Vasicek model or the Cox-Ingersoll-Ross model for backtesting since they do not reproduce the currently prevailing term structure. Apart from the Ho-Lee model, no other no-arbitrage term structure models will be used as a result of test sit-

[13] This will be one result from test situation 2.

[14] This holds true for all cases if the risk-free rate is constant, but is only possible for cases 1 and 2 when the interest rate is stochastic due to the bushy tree problem that results in an exploding computational time if an option to defer is embedded in an investment project, as in case 3. Moreover, cases 1 and 2 are only implemented for $T_S = 1$ year when priced with the Trigeorgis log-transformed binomial tree method due to this bushy tree problem.

uation 4. The research goal for the final test situation and the thesis in general is to see whether it is better in a volatile interest rate environment to use a real options valuation tool with a stochastically modelled risk-free interest rate or if a tool that uses a constant risk-free rate is sufficient or if Hull's implied forward rates approach is more appropriate.

The overall goal of this test strategy is to conduct a systematic and thorough analysis of different complex real options with various stochastic term structure models that are applied through simulations, and using historical backtesting for evaluation. Well-known numerical valuation methods are applied and modified to accommodate for variable interest rates. Although many publications analyzed investments with multiple real options, most of these publications assumed a constant risk-free interest rate. Consequently, a systematic analysis and a comparison with methods that simulate these non-constant interest rates stochastically for various real options cases including historical backtesting has not yet been done.

5.4 Test Situation 1: The Schwartz-Moon Model with a Deferred Project Start

The deterministic solution of the Schwartz-Moon model is the central point of the analysis in this section. An option to defer the project start is then included in this model. It is assumed that the model "starts" at defer time $t = T_d > 0$ in the case of a deferred project start and that all parameter specifications besides the risk-free interest rate do not change. As already mentioned in Section 4.3, the following parameter specifications are needed for the deterministic NPV solution: T as the time to completion of the asset, V as the estimated value of the asset to be obtained at time T when the asset is completed, K as the deterministic, not discounted cost of completion of that asset, and I_m as the constant investment rate during the construction period. This gives $I_m = \frac{K}{T}$.

For discounting, the following interest rates and premia are needed: r_f as the risk-free interest rate and μ as the instantaneous drift of the present value process $(V_t)_{t \geq 0}$ in the stochastic Schwartz-Moon model. Moreover, η is the risk premium on the asset return such that $(\mu - \eta)$ is the risk-adjusted drift for the value of the asset. λ is the probability of a catastrophic event and can be interpreted as an annual "tax rate" on the value of the project. The appropriate discount rate for the asset value is $r_f + \lambda - \mu + \eta$ while the appropriate discount rate for the expected cost to completion is $r_f + \lambda$.

According to equation (4.38), the present value at time $t = 0$ of the asset value V_T is:

$$PV_0(V_T) \quad = \quad V_T\, e^{-(r_f+\lambda-\mu+\eta)T}. \tag{5.6}$$

According to equation (4.37), the present value at time $t = 0$ of the cost to completion K of the asset value is:

$$PV_0(K) \quad = \quad \frac{K}{T(r_f+\lambda)} \left[1 - e^{-(r_f+\lambda)T}\right]. \tag{5.7}$$

Since $T \cdot I_m = K$, for the NPV at time $t = 0$ of the project holds according to (5.6) and (5.7):

$$\left.\begin{aligned} NPV = NPV_0 \quad &= \quad PV_0(V_T) - PV_0(K) \\ &= V_T\, e^{-(r_f+\lambda-\mu+\eta)T} \quad - \quad \frac{I_m}{r_f+\lambda}\left[1 - e^{-(r_f+\lambda)T}\right]. \end{aligned}\right\} \tag{5.8}$$

Table 5.5 shows the results of an investment project with parameters $V_T = 150$, $K = 35$, $I_m = 20$, $\lambda = 0.1$, $\mu = 0$, and $\eta = 0.08$ without an option to defer.

Table 5.5. Schwartz-Moon model with a constant risk-free rate and no option to defer ($V_T = 150$, $K = 35$, $I_m = 20$, $\lambda = 0.1$, $\mu = 0$, $\eta = 0.08$).

r_f	$PV_0(V_T)$	$PV_0(K)$	NPV	r_f	$PV_0(V_T)$	$PV_0(K)$	NPV
0.0100	107.57	31.84	75.73	0.0600	98.56	30.53	68.03
0.0125	107.10	31.77	75.33	0.0625	98.13	30.46	67.66
0.0150	106.63	31.70	74.93	0.0650	97.70	30.40	67.30
0.0175	106.17	31.64	74.53	0.0675	97.27	30.34	66.94
0.0200	105.70	31.57	74.13	0.0700	96.85	30.27	66.57
0.0225	105.24	31.50	73.74	0.0725	96.42	30.21	66.21
0.0250	104.78	31.44	73.35	0.0750	96.00	30.15	65.86
0.0275	104.32	31.37	72.95	0.0775	95.58	30.09	65.50
0.0300	103.87	31.30	72.57	0.0800	95.17	30.02	65.14
0.0325	103.42	31.24	72.18	0.0825	94.75	29.96	64.79
0.0350	102.96	31.17	71.79	0.0850	94.34	29.90	64.44
0.0375	102.52	31.11	71.41	0.0875	93.93	29.84	64.09
0.0400	102.07	31.04	71.03	0.0900	93.52	29.78	63.74
0.0425	101.62	30.98	70.64	0.0925	93.11	29.71	63.39
0.0450	101.18	30.91	70.27	0.0950	92.70	29.65	63.05
0.0475	100.74	30.85	69.89	0.0975	92.30	29.59	62.70
0.0500	100.30	30.78	69.51	0.1000	91.89	29.53	62.36
0.0525	99.86	30.72	69.14	-	-	-	-
0.0550	99.42	30.65	68.77	-	-	-	-
0.0575	98.99	30.59	68.40	-	-	-	-

Source: ts-1-1

The modified Schwartz-Moon model with an option to defer has two different risk-free interest rates. The first interest rate $r_f^{(1)}$ is the risk-free rate of the equations above, which now prevails at time $t = T^d$. The second interest rate $r_f^{(2)}$ is the risk-free rate that prevails today with a time to maturity of T^d. While $r_f^{(2)}$ can be observed in the current market, the risk-free rate $r_f^{(1)}$ is a future interest rate that is unknown today. This situation was graphically explained in Figure 4.14 of Section 4.3.

This gives the following NPV formulas for the NPV at time T_d (named NPV_{T_d}) and for today (named $NPV = NPV_0$):

$$NPV_{T^d} = V_{T+T^d}\, e^{-(r_f^{(1)}+\lambda-\mu+\eta)T} - \frac{I_m}{r_f^{(1)}+\lambda}\left[1 - e^{-(r_f^{(1)}+\lambda)T}\right] \tag{5.9}$$

and, therefore

$$NPV = e^{-r_f^{(2)} \cdot T_d}\, NPV_{T^d}. \tag{5.10}$$

The risk-free interest rate $r_f^{(1)}$ will be given in the context of three scenarios. These three scenarios are:

- Flat term structure of interest rates stays constant over time.
- Flat term structure of interest rates shifts upwards and downwards by 50 bps per quarter.
- Flat term structure of interest rates shifts upwards and downwards by 100 bps per quarter.

If the project start is deferred to time point $T_d > 0$, the model is assumed to start at this time with the same parameter specification except for the risk-free interest rate. Applying formula (5.9) gives NPV_{T_d} as the net present value of the project at time point T_d whereby the risk-free rate that prevails at time T_d has to be used in the Schwartz-Moon model. To discount NPV_{T_d} back to the present, i.e., back to $NPV_0 = NPV$, the current risk-free rate has to be used. The calculation of the NPV is according to (5.10).

All simulation results are put together in three tables: Table 5.6 displays the results for a constant risk-free interest rate over time with an option to defer. Table 5.7 gives the results for shifts (up and down) by 50 bps per quarter with an option to defer and Table 5.8 for shifts (up and down) by 100 bps per quarter with an option to defer.

Table 5.6. NPV in the Schwartz-Moon model with a constant risk-free interest rate and an option to defer ($V_{T+T^d} = 150$, $K = 35$, $I_m = 20$, $\lambda = 0.1$, $\mu = 0$, $\eta = 0.08$).

	Constant risk-free interest rate $r_f^{(1)} = r_f^{(2)} =$			
T_d	0.02	0.04	0.06	0.08
0.00	74.13	71.03	68.03	65.14
0.25	73.76	70.32	67.02	63.85
0.50	73.40	69.62	66.02	62.59
0.75	73.03	68.93	65.04	61.35
1.00	72.67	68.24	64.07	60.14
1.25	72.30	67.56	63.11	58.94
1.50	71.94	66.89	62.17	57.78
1.75	71.58	66.22	61.25	56.63
2.00	71.23	65.56	60.34	55.51
2.25	70.87	64.91	59.44	54.41
2.50	70.52	64.27	58.55	53.34
2.75	70.17	63.63	57.68	52.28
3.00	69.82	62.99	56.82	51.24

Source: ts-1-1 and ts-1-2

Table 5.7. NPV in the Schwartz-Moon model with a non-constant risk-free interest rate (parallel shifts by 50 bps/qtr.) and an option to defer ($V_{T+T^d} = 150$, $K = 35$, $I_m = 20$, $\lambda = 0.1$, $\mu = 0$, $\eta = 0.08$).

	Flat term structure +50 bps/qtr. up-shift: initial risk-free rate $r_f^{(2)}$ of				Flat term structure −50 bps/qtr. down-shift: initial risk-free rate $r_f^{(2)}$ of			
T_d	0.01	0.02	0.03	0.04	0.10	0.09	0.08	0.07
0.00	75.73	74.13	72.57	71.03	62.36	63.74	65.14	66.57
0.25	74.74	72.98	71.26	69.57	61.49	63.01	64.55	66.13
0.50	73.76	71.84	69.97	68.14	60.63	62.28	63.96	65.69
0.75	72.80	70.72	68.70	66.74	59.78	61.56	63.38	65.25
1.00	71.84	69.62	67.46	65.36	58.94	60.84	62.80	64.81
1.25	70.90	68.53	66.24	64.02	58.12	60.14	62.22	64.38
1.50	69.97	67.46	65.04	62.70	57.30	59.44	61.65	63.95
1.75	69.05	66.40	63.86	61.40	56.49	58.75	61.09	63.51
2.00	68.14	65.36	62.70	60.14	55.70	58.06	60.52	63.09
2.25	67.24	64.34	61.56	58.89	54.91	57.39	59.97	62.66
2.50	66.35	63.33	60.44	57.67	54.14	56.71	59.41	62.23
2.75	65.47	62.33	59.34	56.48	53.37	56.05	58.86	61.81
3.00	64.61	61.35	58.25	55.31	52.62	55.39	58.32	61.39

Source: ts-1-3 and ts-1-5

These results for the deferred project start are displayed graphically below: Figure 5.3 shows the results for a constant risk-free rate, Figure 5.4 for an upward shift by 50 bps per quarter and Figure 5.5 for a downward shift by −50 bps per quarter.

Table 5.8. NPV in the Schwartz-Moon model with a non-constant risk-free interest rate (parallel shifts by 100 bps/qtr.) and an option to defer ($V_{T+T^d} = 150$, $K = 35$, $I_m = 20$, $\lambda = 0.1$, $\mu = 0$, $\eta = 0.08$).

	Flat term structure +100 bps/qtr. up-shift: initial risk-free rate $r_f^{(2)}$ of				Flat term structure −100 bps/qtr. down-shift: initial risk-free rate $r_f^{(2)}$ of			
T_d	0.01	0.02	0.03	0.04	0.10	0.09	0.08	0.07
0.00	75.73	74.13	72.57	71.03	62.36	63.74	65.14	66.57
0.25	73.95	72.20	70.49	68.82	62.17	63.69	65.26	66.85
0.50	72.20	70.32	68.48	66.68	61.97	63.64	65.36	67.12
0.75	70.49	68.48	66.52	64.61	61.76	63.59	65.47	67.39
1.00	68.82	66.68	64.61	62.59	61.56	63.53	65.56	67.66
1.25	67.18	64.93	62.75	60.63	61.35	63.47	65.66	67.92
1.50	65.58	63.22	60.94	58.73	61.13	63.40	65.75	68.18
1.75	64.01	61.55	59.17	56.89	60.92	63.33	65.84	-
2.00	62.48	59.92	57.46	55.09	60.70	63.26	-	-
2.25	60.98	58.33	55.79	53.36	60.47	-	-	-
2.50	59.50	56.77	54.16	51.67	-	-	-	-
2.75	58.06	55.26	52.58	50.03	-	-	-	-
3.00	56.66	53.78	51.04	48.44	-	-	-	-

Source: ts-1-4 and ts-1-6

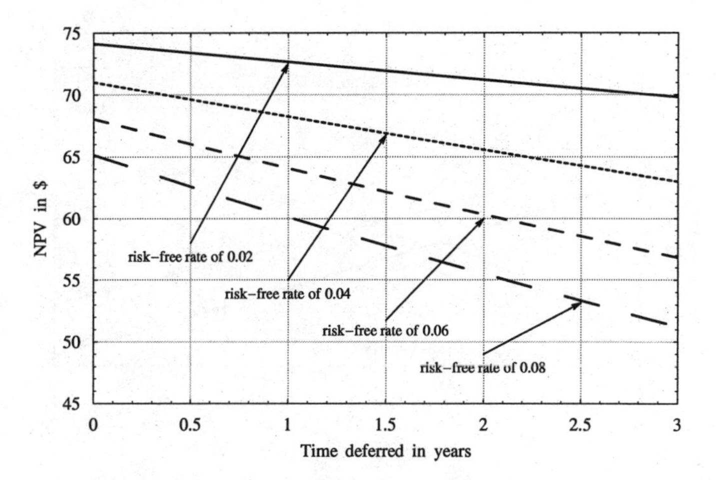

Fig. 5.3. NPV for a project with an option to defer, priced within the Schwartz-Moon model; flat term structure stays constant over time (ts-1-1 and ts-1-2; corresponding to Table 5.6; parameters: $V_{T+T^d} = 150$, $K = 35$, $I_m = 20$, $\lambda = 0.1$, $\mu = 0$, $\eta = 0.08$).

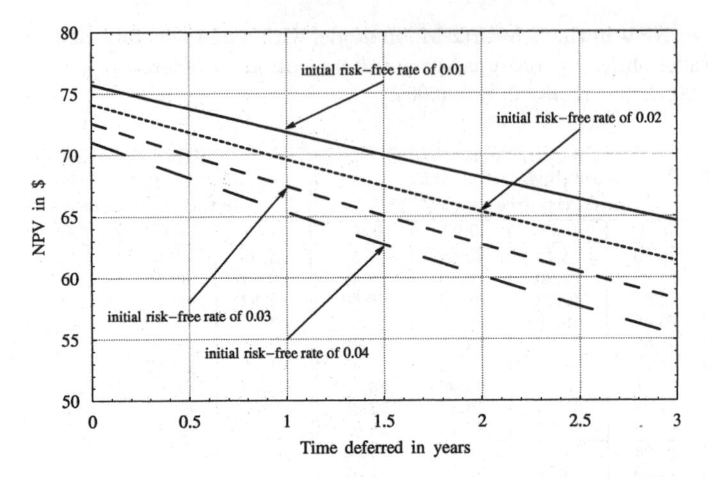

Fig. 5.4. NPV for a project with an option to defer, priced within the Schwartz-Moon model; flat initial term structure shifts upwards by 50 bps per quarter (ts-1-3; corresponding to Table 5.7; parameters: $V_{T+T^d} = 150$, $K = 35$, $I_m = 20$, $\lambda = 0.1$, $\mu = 0$, $\eta = 0.08$).

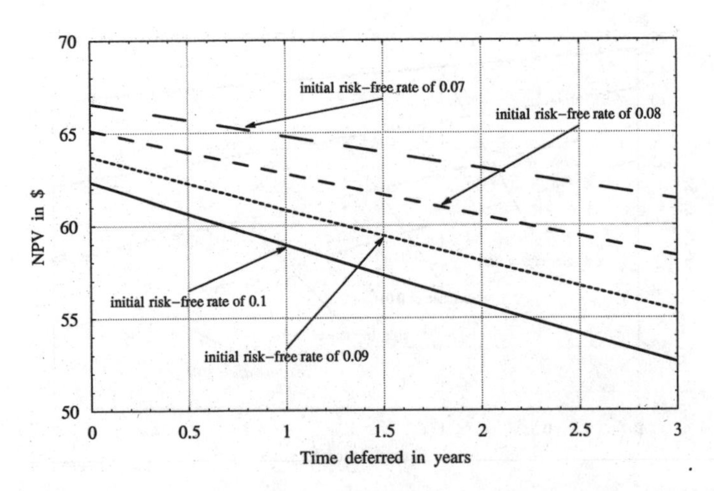

Fig. 5.5. NPV for a project with an option to defer, priced within the Schwartz-Moon model; flat initial term structure shifts downwards by −50 bps per quarter (ts-1-5; corresponding to Table 5.7; parameters: $V_{T+T^d} = 150$, $K = 35$, $I_m = 20$, $\lambda = 0.1$, $\mu = 0$, $\eta = 0.08$).

Figures 5.3, 5.4, and 5.5 show: With an increasing defer time T_d the current NPV_0 (denoted with NPV in these figures) decreases, regardless of initial risk-free rate of the flat term structure. For this investment example the conclusion can be drawn that given a low volatility and independent of the initial risk-free rate of the flat term structure it is optimal to start the investment immediately.

The following figures display data from Table 5.8 for a shift of 100 bps per quarter: Figures 5.6 and 5.7 show the results for an upward and downward shift by 100 bps per quarter, respectively.

When considering a higher volatility of the flat initial term structure the picture changes. Of course, as can be seen in Figure 5.6, a downward changing yield curve decreases the current net present value of the project if the project is deferred. If however, the flat term structure moves downwards by −100 bps per quarter as shown in Figure 5.7 (such a movement prevailed at the short end of the yield curve between December 2000 and December 2001 with a risk-free interest rate at the short end of approximately 7% in December 2000, see Figure 3.1 in Section 3.2.1), it can be optimal to defer the investment. The lower the initial flat term structure the more likely it is that deferring the investment generates value. This can be seen for an initially flat yield curve of 7% and 8% in Figure 5.7.

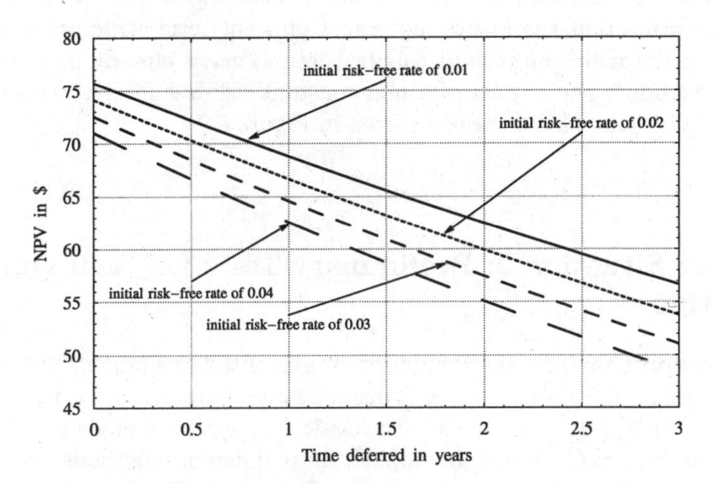

Fig. 5.6. NPV for a project with an option to defer, priced within the Schwartz-Moon model; flat initial term structure shifts upwards by 100 bps per quarter (ts-1-4; corresponding to Table 5.8; parameters: $V_{T+T^d} = 150$, $K = 35$, $I_m = 20$, $\lambda = 0.1$, $\mu = 0$, $\eta = 0.08$).

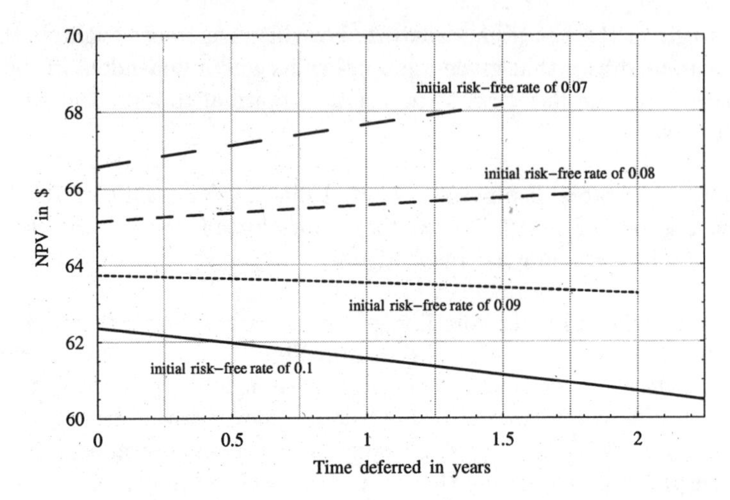

Fig. 5.7. NPV for a project with an option to defer, priced within the Schwartz-Moon model; flat initial term structure shifts downwards by −100 bps per quarter (ts-1-6; corresponding to Table 5.8; parameters: $V_{T+T^d} = 150$, $K = 35$, $I_m = 20$, $\lambda = 0.1$, $\mu = 0$, $\eta = 0.08$).

It needs to be stressed that this model is a simplified one. It is based on the assumption that the future movement of a flat term structure is known. Therefore, the main purpose of this test situation, as already mentioned, is only to introduce the impact of a non-constant risk-free interest rate on real options valuation. This is clearly shown in Figure 5.7.

5.5 Test Situation 2: Preliminary Tests for Real Options Valuation

Test situation 2 analyzes the simulation results of the preliminary tests. These tests examine the various choices of discretization parameters for real options valuation tools, but not for the stochastic interest rate models[15]. For the log-transformed explicit and the implicit finite difference methods, the underlying values and the time steps have to be discretized. The tests also provide a comparison between the log-transformed explicit and implicit finite difference methods.

Moreover, simulation results of the Cox-Ross-Rubinstein binomial tree method and the Trigeorgis log-transformed binomial tree method are displayed to investigate the differences in real options valuation for both models. The number

[15] The latter is topic of test situations 3 and 4.

of time steps, which is needed for the binomial tree in order to arrive at a stable real options value, is determined for both valuation tools.

These analyses are done in 11 consecutive steps. The test case in most of these steps is the European put option, i.e., case 1 modified from American to European style. The reason is that the value of a European option can also be calculated with the Black-Scholes formula (see equation (5.5)), which will work as the benchmark for all real options values calculated with the various real options valuation tools.

However, the Black-Scholes value may only be used as a benchmark if the risk-free interest rate is constant and not variable. If the interest rate is variable, only the Cox-Ross-Rubinstein and the Trigeorgis log-transformed binomial tree methods can be compared to each other without having a benchmark. Hereby, it has to be determined how many paths of the term structure of interest rates should be used to get a stable option value and a stable standard deviation within an appropriate computational time.

As already mentioned, ROV refers to the present value of the real option(s), TPV refers to the total project's present value as the present value of the project including the ROV but without considering any initial investment cost to start the project. Finally, NPV refers to the project's net present value including the ROV and the initial investment cost I_0. With S_0 as the current underlying value the following relationships hold:

$$\begin{aligned} NPV &= ROV + S_0 - I_0, \\ TPV &= ROV + S_0, \\ NPV &= TPV - I_0. \end{aligned}$$

The current underlying value S_0 is the present value of the project's estimated future cash inflow minus outflow, not considering any additional investment cost.

5.5.1 Analysis for Test Situation 2 in 11 Consecutive Steps

Step 1: Log-transformed explicit finite difference method

Step 1 answers the question: For given parameters T_S and M, how does the real options value change depending on parameter N_S? To answer this question, case 1 is modified to a European style real option allowing the Black-Scholes formula to be used as a benchmark.

Table 5.9. ROV of a European put option priced with the log-transformed explicit finite difference method ($S_{min} = 50$, $S_{max} = 150$, $M = 100$, $\sigma_S = 0.25$, $N_S = 720 \cdot T_S$, $X = 100$, $r_f = 0.04$).

S_0	— $T_S = 0.25$ — BS	LEFD	— $T_S = 1.00$ — BS	LEFD	— $T_S = 3.00$ — BS	LEFT	— $T_S = 5.00$ — BS	LEFT
50.000	49.005	49.005	46.103	46.130	39.920	41.173	35.497	39.146
55.806	43.199	43.199	40.369	40.374	35.070	35.731	31.479	33.852
62.287	36.718	36.719	34.116	34.117	30.113	30.455	27.460	29.030
69.519	29.493	29.492	27.496	27.495	25.191	25.384	23.529	24.649
77.592	21.519	21.518	20.829	20.828	20.463	20.618	19.772	20.719
86.603	13.242	13.241	14.592	14.592	16.090	16.297	16.271	17.276
96.659	6.138	6.138	9.299	9.302	12.209	12.567	13.094	14.373
107.883	1.885	1.884	5.306	5.318	8.914	9.558	10.292	12.064
120.411	0.343	0.342	2.673	2.722	6.245	7.369	7.890	10.397
134.394	0.034	0.034	1.175	1.346	4.188	6.061	5.894	9.404
150.000	0.002	0.004	0.446	0.952	2.682	5.655	4.285	9.097

Source: ts-2-1 and ts-2-2

Table 5.10. ROV of a European put option priced with the log-transformed explicit finite difference method ($S_{min} = 50$, $S_{max} = 150$, $M = 100$, $\sigma_S = 0.25$, $N_S = 1440 \cdot T_S$, $X = 100$, $r_f = 0.04$).

S_0	— $T_S = 0.25$ — BS	LEFD	— $T_S = 1.00$ — BS	LEFD	— $T_S = 3.00$ — BS	LEFT	— $T_S = 5.00$ — BS	LEFT
50.000	49.005	49.005	46.103	46.131	39.920	41.173	35.497	39.146
55.806	43.199	43.199	40.369	40.374	35.070	35.731	31.479	33.853
62.287	36.718	36.719	34.116	34.117	30.113	30.455	27.460	29.030
69.519	29.493	29.493	27.496	27.496	25.191	25.384	23.529	24.649
77.592	21.519	21.519	20.829	20.829	20.463	20.618	19.772	20.719
86.603	13.242	13.242	14.592	14.592	16.090	16.297	16.271	17.276
96.659	6.138	6.136	9.299	9.301	12.209	12.567	13.094	14.373
107.883	1.885	1.883	5.306	5.318	8.914	9.558	10.292	12.064
120.411	0.343	0.343	2.673	2.722	6.245	7.369	7.890	10.397
134.394	0.034	0.034	1.175	1.346	4.188	6.061	5.894	9.404
150.000	0.002	0.004	0.446	0.952	2.682	5.655	4.285	9.097

Source: ts-2-1 and ts-2-4

A comparison of Tables 5.9 and 5.10 shows that regardless of the different ratios $\frac{N_S}{T_S} = 720$ and $\frac{N_S}{T_S} = 1440$ there are virtually no differences for the real options value when priced with the log-transformed explicit finite difference method (LEFD). When comparing the real options value within Table 5.9 it can be seen that for low values of T_S the real options values derived from LEFD and the Black-Scholes formula (BS) are almost identical. This shows that LEFD is suited to price financial options which typically have a low maturity. If, however, T_S becomes larger, the difference between LEFD and the Black-Scholes formula increases for S_0 values that are close to S_{min} or S_{max}. The difference between BS and LEFD is not too big if S_0 is well in between S_{min} and S_{max}. This shows that LEFD is not suited to price longer term options, especially real options with long maturities. Figure 5.8 graphically displays one example for $T_S = 5$ and $N_S = 7200$.

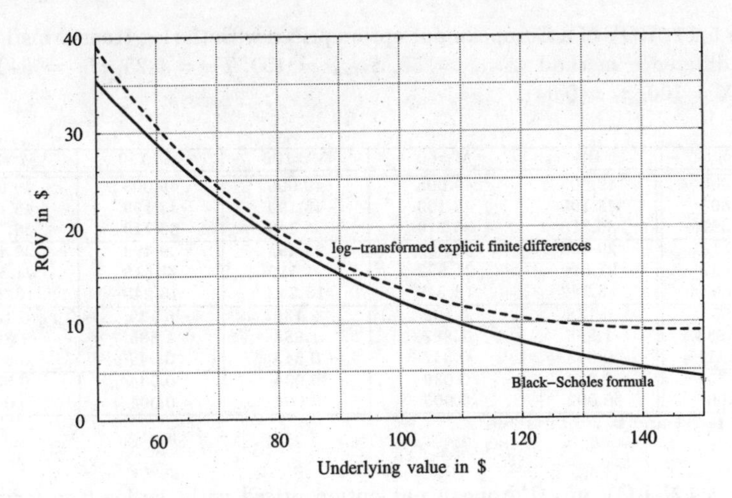

Fig. 5.8. Value of the real option to abandon the project for a salvage value X at the end of the construction period, priced with the log-transformed explicit finite difference method (ts-2-1 and ts-2-4; corresponding to Table 5.10; parameters: $S_{min} = 50$, $S_{max} = 150$, $M = 100$, $\sigma_S = 0.25$, $T_S = 5$, $N_S = 7200$, $X = 100$, $r_f = 0.04$).

Step 2: Log-transformed explicit finite difference method

Step 2 answers the question: For given parameters T_S and N_S, how does the real options value change depending on parameter M? To answer this question, case 1 is modified to a European style real option allowing the Black-Scholes formula to be used as a benchmark.

In Tables 5.11 to 5.14 the ROV for various values of M is displayed. The ratio $\frac{N_S}{T_S}$ is the same for each table ($\frac{N_S}{T_S} = 2160$) but time to maturity T_S varies. The real options value derived from the Black-Scholes formula is again used as a benchmark.

The real options value is almost the same for BS and for LEFD if T_S is small (Table 5.11). However, for larger values of T_S the deviation of the LEFD method from the BS formula increases, particularly if the option is out-of-the-money. For example, if $T_S = 5$ and S_0=150 the real options value is approximately twice as large for LEFD as for BS (see Table 5.14). More importantly, this result is almost unrelated to the choice of M: Regardless of whether $M = 50$ or $M = 200$, it hardly bears influence on the ROV. For further analysis, the choice of $M = 50$ is therefore sufficient for LEFD, independent of the choice of T_S.

Table 5.11. ROV of a European put option priced with the log-transformed explicit finite difference method ($S_{min} = 50$, $S_{max} = 150$, $T_S = 0.25$, $N_S = 540$, $\sigma_S = 0.25$, $X = 100$, $r_f = 0.04$).

S_0	BS	M=50	M=100	M=150	M=200
50.000	49.005	49.005	49.005	49.005	49.005
55.806	43.199	43.199	43.199	43.199	43.199
62.287	36.718	36.719	36.719	36.719	36.719
69.519	29.493	29.493	29.493	29.493	29.493
77.592	21.519	21.522	21.519	21.519	21.519
86.603	13.242	13.247	13.242	13.243	13.242
96.659	6.138	6.139	6.136	6.139	6.138
107.883	1.885	1.887	1.883	1.885	1.885
120.411	0.343	0.347	0.343	0.343	0.343
134.394	0.034	0.036	0.034	0.034	0.034
150.000	0.002	0.006	0.004	0.004	0.004

Source: ts-2-1 and ts-2-6 to ts-2-9

Table 5.12. ROV of a European put option priced with the log-transformed explicit finite difference method ($S_{min} = 50$, $S_{max} = 150$, $T_S = 1$, $N_S = 2160$, $\sigma_S = 0.25$, $X = 100$, $r_f = 0.04$).

S_0	BS	M=50	M=100	M=150	M=200
50.000	46.103	46.136	46.131	46.129	46.129
55.806	40.369	40.377	40.375	40.374	40.374
62.287	34.116	34.119	34.117	34.117	34.117
69.519	27.496	27.499	27.496	27.496	27.496
77.592	20.829	20.832	20.829	20.829	20.829
86.603	14.592	14.595	14.592	14.593	14.593
96.659	9.299	9.303	9.301	9.302	9.302
107.883	5.306	5.321	5.317	5.318	5.318
120.411	2.673	2.732	2.722	2.721	2.720
134.394	1.175	1.371	1.346	1.340	1.337
150.000	0.446	1.010	0.953	0.936	0.927

Source: ts-2-1 and ts-2-6 to ts-2-9

Table 5.13. ROV of a European put option priced with the log-transformed explicit finite difference method ($S_{min} = 50$, $S_{max} = 150$, $T_S = 3$, $N_S = 6480$, $\sigma_S = 0.25$, $X = 100$, $r_f = 0.04$).

S_0	BS	M=50	M=100	M=150	M=200
50.000	39.920	41.255	41.173	41.148	41.135
55.806	35.070	35.779	35.731	35.717	35.710
62.287	30.113	30.484	30.456	30.448	30.443
69.519	25.191	25.402	25.384	25.379	25.377
77.592	20.463	20.633	20.618	20.615	20.613
86.603	16.090	16.315	16.297	16.293	16.290
96.659	12.209	12.593	12.567	12.560	12.556
107.883	8.914	9.600	9.558	9.546	9.539
120.411	6.245	7.434	7.369	7.349	7.339
134.394	4.188	6.158	6.06	6.030	6.014
150.000	2.682	5.794	5.655	5.611	5.589

Source: ts-2-1 and ts-2-6 to ts-2-9

Table 5.14. ROV of a European put option priced with the log-transformed explicit finite difference method ($S_{min} = 50$, $S_{max} = 150$, $T_S = 5$, $N_S = 10800$, $\sigma_S = 0.25$, $X = 100$, $r_f = 0.04$).

S_0	BS	M=50	M=100	M=150	M=200
50.000	35.497	39.312	39.146	39.094	39.067
55.806	31.479	33.969	33.853	33.817	33.798
62.287	27.460	29.113	29.031	29.005	28.992
69.519	23.529	24.712	24.649	24.630	24.621
77.592	19.772	20.772	20.719	20.703	20.694
86.603	16.271	17.329	17.276	17.260	17.252
96.659	13.094	14.435	14.373	14.354	14.344
107.883	10.292	12.141	12.064	12.040	12.028
120.411	7.890	10.496	10.397	10.366	10.350
134.394	5.894	9.531	9.404	9.364	9.343
150.000	4.285	9.256	9.096	9.045	9.020

Source: ts-2-1 and ts-2-6 to ts-2-9

Step 3: Log-transformed explicit vs. implicit finite difference method

Step 3 answers the question: Do the log-transformed explicit and implicit methods yield different ROV results? To answer this question, case 1 is modified to a European style real option allowing the real options value of the log-transformed explicit and implicit finite difference methods to be compared with each other and with the values derived from the Black-Scholes formula as a benchmark.

Tables 5.15 and 5.16 display the ROV of the log-transformed explicit finite difference method (LEFD) and the log-transformed implicit finite difference method (LIFD) for various times to maturity and for various ratios $\frac{N_S}{T_S}$.

Table 5.15. ROV of a European put option priced with the log-transformed explicit and implicit finite difference methods ($S_{min} = 50$, $S_{max} = 150$, $N_S = 720 \cdot T_S$, $M = 100$, $\sigma_S = 0.25$, $X = 100$, $r_f = 0.04$).

	$T_S = 0.25$			$T_S = 1$			$T_S = 5$		
S_0	BS	LEFD	LIFD	BS	LEFD	LIFD	BS	LEFD	LIFD
50.000	49.005	49.005	49.005	46.103	46.130	46.131	35.497	39.146	39.148
55.806	43.199	43.199	43.199	40.369	40.374	40.375	31.479	33.852	33.854
62.287	36.718	36.719	36.719	34.116	34.117	34.118	27.460	29.030	29.032
69.519	29.493	29.492	29.493	27.496	27.495	27.497	23.529	24.649	24.650
77.592	21.519	21.518	21.521	20.829	20.828	20.829	19.772	20.719	20.719
86.603	13.242	13.241	13.243	14.592	14.592	14.591	16.271	17.276	17.276
96.659	6.138	6.138	6.132	9.299	9.302	9.299	13.094	14.373	14.373
107.883	1.885	1.884	1.880	5.306	5.318	5.316	10.292	12.064	12.064
120.411	0.343	0.342	0.344	2.673	2.722	2.722	7.890	10.397	10.396
134.394	0.034	0.034	0.035	1.175	1.346	1.347	5.894	9.404	9.403
150.000	0.002	0.004	0.005	0.446	0.952	0.954	4.285	9.097	9.096

Source: ts-2-1, ts-2-2, and ts-2-10

Table 5.16. ROV of a European put option priced with the log-transformed explicit and implicit finite difference methods ($S_{min} = 50$, $S_{max} = 150$, $N_S = 2160 \cdot T_S$, $M = 50$, $\sigma_S = 0.25$, $X = 100$, $r_f = 0.04$).

S_0	$T_S = 0.25$			$T_S = 1$			$T_S = 5$		
	BS	LEFD	LIFD	BS	LEFD	LIFD	BS	LEFD	LIFD
50.000	49.005	49.005	49.005	46.103	46.136	46.136	35.497	39.312	39.313
55.806	43.199	43.199	43.199	40.369	40.377	40.377	31.479	33.969	33.970
62.287	36.718	36.719	36.719	34.116	34.119	34.120	27.460	29.113	29.114
69.519	29.493	29.493	29.494	27.496	27.499	27.500	23.529	24.712	24.712
77.592	21.519	21.522	21.523	20.829	20.832	20.832	19.772	20.772	20.772
86.603	13.242	13.247	13.248	14.592	14.595	14.594	16.271	17.329	17.329
96.659	6.138	6.139	6.137	9.299	9.303	9.302	13.094	14.435	14.435
107.883	1.885	1.887	1.885	5.306	5.321	5.320	10.292	12.141	12.141
120.411	0.343	0.347	0.347	2.673	2.732	2.731	7.890	10.496	10.496
134.394	0.034	0.036	0.037	1.175	1.371	1.371	5.894	9.531	9.530
150.000	0.002	0.006	0.006	0.446	1.010	1.011	4.285	9.256	9.255

Source: ts-2-1, ts-2-7, and ts-2-12

Independent of these parameter choices, the results of LEFD and LIFD are always almost identical. However, since the stability and consistency restrictions of LIFD are weaker than for LEFD (see Chapter 4.2.2, especially Figure 4.12 and Figure 4.13), LIFD should be the preferred method of choice compared with LEFD.

On the other hand, both methods deliver false results for long times to maturity, which are typical for real options. In consequence, LEFD and LIFD will not be considered in the further analysis.

Step 4: Cox-Ross-Rubinstein binomial tree method for a constant risk-free interest rate

Step 4 answers the question: For a given parameter T_S, how does the real options value change depending on parameter N_S? To answer this question, case 1 is modified to a European style real option allowing the Black-Scholes formula to be used as a benchmark.

In the following, the Cox-Ross-Rubinstein binomial tree (and also later the Trigeorgis log-transformed binomial tree) is applied for various start values S_0 of the underlying. Analogous to the two log-transformed finite difference methods the lowest start value is named S_{min}, and the highest start value is named S_{max}. The number of intervals between the minimum and maximum start value is named M (analogous to the two log-transformed finite difference methods). The intervals between S_{min} and S_{max} are usually not of equal size in this test situation, whereas the intervals between $\ln(S_{min})$ and $\ln(S_{max})$ have the same length, especially when comparing the Cox-Ross-Rubinstein binomial tree method or the Trigeorgis log-transformed binomial tree method with one of the log-transformed finite difference methods.

Tables 5.17 to 5.19 display the real options values for $T_S = 0.25$, $T_S = 1$, and $T_S = 5$ years. Within each of these tables the ratio $\frac{N_S}{T_S}$ varies between 60 and 1080. Independent of the choice of T_S and independent of the ratio $\frac{N_S}{T_S}$ the result is always the same: The real options value calculated with the Cox-Ross-Rubinstein binomial tree method is almost always identical to the value delivered by the Black-Scholes formula. A low ratio of $\frac{N_S}{T_S} = 360$, which can be interpreted as one time step per day (assuming a 30/360 day count method), can therefore be chosen for further analysis.

Table 5.17. ROV of a European put option priced with the Cox-Ross-Rubinstein binomial tree method ($S_{min} = 50$, $S_{max} = 150$, $M = 100$, $\sigma_S = 0.25$, $T_S = 0.25$, $X = 100$, $r_f = 0.04$).

		$N_S/T_S =$					
S_0	BS	60	120	240	360	720	1080
50.000	49.005	49.005	49.005	49.005	49.005	49.005	49.005
55.806	43.199	43.199	43.199	43.199	43.199	43.199	43.199
62.287	36.718	36.718	36.718	36.718	36.718	36.718	36.718
69.519	29.493	29.489	29.491	29.492	29.493	29.493	29.493
77.592	21.519	21.512	21.516	21.515	21.518	21.519	21.519
86.603	13.242	13.251	13.230	13.247	13.246	13.243	13.243
96.659	6.138	6.071	6.158	6.125	6.146	6.139	6.140
107.883	1.885	1.935	1.910	1.897	1.883	1.882	1.881
120.411	0.343	0.345	0.327	0.343	0.337	0.340	0.343
134.394	0.034	0.023	0.032	0.032	0.033	0.033	0.034
150.000	0.002	0.001	0.001	0.002	0.002	0.002	0.002

Source: ts-2-1 and ts-2-13 to ts-2-18

Table 5.18. ROV of a European put option priced with the Cox-Ross-Rubinstein binomial tree method ($S_{min} = 50$, $S_{max} = 150$, $M = 100$, $\sigma_S = 0.25$, $T_S = 1$, $X = 100$, $r_f = 0.04$).

		$N_S/T_S =$					
S_0	BS	60	120	240	360	720	1080
50.000	46.103	46.101	46.102	46.103	46.103	46.103	46.103
55.806	40.369	40.361	40.367	40.367	40.368	40.369	40.369
62.287	34.116	34.110	34.113	34.115	34.113	34.115	34.115
69.519	27.496	27.497	27.485	27.495	27.495	27.497	27.496
77.592	20.829	20.806	20.835	20.826	20.831	20.830	20.830
86.603	14.592	14.598	14.589	14.599	14.596	14.593	14.593
96.659	9.299	9.337	9.309	9.293	9.303	9.300	9.300
107.883	5.306	5.310	5.321	5.314	5.306	5.305	5.305
120.411	2.673	2.667	2.664	2.676	2.669	2.671	2.674
134.394	1.175	1.186	1.180	1.173	1.174	1.174	1.175
150.000	0.446	0.443	0.441	0.446	0.446	0.446	0.446

Source: ts-2-1 and ts-2-13 to ts-2-18

Table 5.19. ROV of a European put option priced with the Cox-Ross-Rubinstein binomial tree method ($S_{min} = 50$, $S_{max} = 150$, $M = 100$, $\sigma_S = 0.25$, $T_S = 5$, $X = 100$, $r_f = 0.04$).

S_0	BS	$N_S/T_S =$ 60	120	240	360	720	1080
50.000	35.497	35.494	35.494	35.497	35.497	35.497	35.497
55.806	31.479	31.464	31.477	31.476	31.477	31.479	31.479
62.287	27.460	27.462	27.462	27.461	27.458	27.460	27.460
69.519	23.529	23.535	23.522	23.530	23.529	23.530	23.529
77.592	19.772	19.761	19.777	19.771	19.774	19.773	19.773
86.603	16.271	16.273	16.269	16.273	16.272	16.271	16.271
96.659	13.094	13.107	13.097	13.091	13.096	13.094	13.095
107.883	10.292	10.292	10.297	10.295	10.291	10.291	10.291
120.411	7.890	7.887	7.886	7.892	7.889	7.889	7.891
134.394	5.894	5.904	5.899	5.893	5.894	5.893	5.894
150.000	4.285	4.290	4.283	4.287	4.286	4.285	4.285

Source: ts-2-1 and ts-2-13 to ts-2-18

Step 5: Trigeorgis log-transformed binomial tree method for a constant risk-free interest rate

Step 5 answers the question: For a given parameter T_S, how does the real options value change depending on parameter N_S? To answer this question, case 1 is modified to a European style real option allowing the Black-Scholes formula to be used as a benchmark.

The results from this step are identical to the results from step 4 for the Cox-Ross-Rubinstein binomial tree method: Tables 5.20 to 5.22 show that independent of ratio $\frac{N_S}{T_S}$ and independent of the choice of T_S the real options value derived from the Trigeorgis log-transformed binomial tree method is the same as delivered by the Black-Scholes formula.

Table 5.20. ROV of a European put option priced with the Trigeorgis log-transformed binomial tree method ($S_{min} = 50$, $S_{max} = 150$, $M = 100$, $\sigma_S = 0.25$, $T_S = 0.25$, $X = 100$, $r_f = 0.04$).

S_0	BS	$N_S/T_S =$ 60	180	360	720
50.000	49.005	49.005	49.005	49.005	49.005
55.806	43.199	43.199	43.199	43.199	43.199
62.287	36.718	36.718	36.718	36.718	36.718
69.519	29.493	29.489	29.492	29.493	29.493
77.592	21.519	21.512	21.516	21.518	21.519
86.603	13.242	13.251	13.246	13.246	13.243
96.659	6.138	6.071	6.161	6.146	6.139
107.883	1.885	1.935	1.905	1.883	1.882
120.411	0.343	0.345	0.346	0.337	0.340
134.394	0.034	0.022	0.032	0.033	0.033
150.000	0.002	0.000	0.001	0.002	0.002

Source: ts-2-1, ts-2-20, ts-2-22, ts-2-24, and ts-2-26

Table 5.21. ROV of a European put option priced with the Trigeorgis log-transformed binomial tree method ($S_{min} = 50$, $S_{max} = 150$, $M = 100$, $\sigma_S = 0.25$, $T_S = 1$, $X = 100$, $r_f = 0.04$).

S_0	BS	$N_S/T_S =$ 60	180	360	720
50.000	46.103	46.101	46.103	46.103	46.103
55.806	40.369	40.361	40.368	40.368	40.369
62.287	34.116	34.110	34.114	34.113	34.115
69.519	27.496	27.497	27.495	27.495	27.497
77.592	20.829	20.806	20.828	20.831	20.830
86.603	14.592	14.598	14.589	14.596	14.593
96.659	9.299	9.336	9.294	9.303	9.300
107.883	5.306	5.310	5.297	5.305	5.305
120.411	2.673	2.666	2.663	2.669	2.671
134.394	1.175	1.185	1.169	1.174	1.174
150.000	0.446	0.442	0.443	0.446	0.445

Source: ts-2-1, ts-2-20, ts-2-22, ts-2-24, and ts-2-26

Table 5.22. ROV of a European put option priced with the Trigeorgis log-transformed binomial tree method ($S_{min} = 50$, $S_{max} = 150$, $M = 100$, $\sigma_S = 0.25$, $T_S = 5$, $X = 100$, $r_f = 0.04$).

S_0	BS	$N_S/T_S =$ 60	180	360	720
50.000	35.497	35.493	35.497	35.497	35.496
55.806	31.479	31.464	31.479	31.477	31.479
62.287	27.460	27.461	27.460	27.457	27.460
69.519	23.529	23.534	23.529	23.529	23.530
77.592	19.772	19.759	19.771	19.773	19.773
86.603	16.271	16.271	16.268	16.272	16.271
96.659	13.094	13.104	13.091	13.095	13.094
107.883	10.292	10.289	10.287	10.291	10.291
120.411	7.890	7.883	7.884	7.888	7.889
134.394	5.894	5.899	5.889	5.893	5.893
150.000	4.285	4.284	4.282	4.285	4.285

Source: ts-2-1, ts-2-20, ts-2-22, ts-2-24, and ts-2-26

Step 6: Trigeorgis log-transformed binomial tree method vs. Cox-Ross-Rubinstein binomial tree method for a constant risk-free interest rate

Step 6 answers the question: Do the Cox-Ross-Rubinstein and the Trigeorgis log-transformed binomial tree methods lead to different results for the real options value? To answer this question, case 1 is modified to a European style real option allowing the real options values to be compared with each other and with the Black-Scholes value as a benchmark.

Tables 5.23 to 5.25 display the ROVs derived from these two methods and from the Black-Scholes formula.

Table 5.23. ROV of a European put option priced with the Cox-Ross-Rubinstein binomial tree and Trigeorgis log-transformed binomial tree methods ($S_{min} = 50$, $S_{max} = 150$, $M = 100$, $\sigma_S = 0.25$, $T_S = 0.25$, $X = 100$, $r_f = 0.04$).

		$N_S/T_S = 60$		$N_S/T_S = 360$		$N_S/T_S = 720$	
S_0	BS	CRR	Trigeorgis	CRR	Trigeorgis	CRR	Trigeorgis
50.000	49.005	49.005	49.005	49.005	49.005	49.005	49.005
55.806	43.199	43.199	43.199	43.199	43.199	43.199	43.199
62.287	36.718	36.718	36.718	36.718	36.718	36.718	36.718
69.519	29.493	29.489	29.489	29.493	29.493	29.493	29.493
77.592	21.519	21.512	21.512	21.518	21.518	21.519	21.519
86.603	13.242	13.251	13.251	13.246	13.246	13.243	13.243
96.659	6.138	6.071	6.071	6.146	6.146	6.139	6.139
107.883	1.885	1.935	1.935	1.883	1.883	1.882	1.882
120.411	0.343	0.345	0.345	0.337	0.337	0.340	0.340
134.394	0.034	0.023	0.022	0.033	0.033	0.033	0.033
150.000	0.002	0.001	0.000	0.002	0.002	0.002	0.002

Source: ts-2-1, ts-2-13, ts-2-16, ts-2-17, ts-2-20, ts-2-22, and ts-2-26

Table 5.24. ROV of a European put option priced with the Cox-Ross-Rubinstein binomial tree and Trigeorgis log-transformed binomial tree methods ($S_{min} = 50$, $S_{max} = 150$, $M = 100$, $\sigma_S = 0.25$, $T_S = 1$, $X = 100$, $r_f = 0.04$).

		$N_S/T_S = 60$		$N_S/T_S = 360$		$N_S/T_S = 720$	
S_0	BS	CRR	Trigeorgis	CRR	Trigeorgis	CRR	Trigeorgis
50.000	46.103	46.101	46.101	46.103	46.103	46.103	46.103
55.806	40.369	40.361	40.361	40.368	40.368	40.369	40.369
62.287	34.116	34.110	34.110	34.113	34.113	34.115	34.115
69.519	27.496	27.497	27.497	27.495	27.495	27.497	27.497
77.592	20.829	20.806	20.806	20.831	20.831	20.830	20.830
86.603	14.592	14.598	14.598	14.596	14.596	14.593	14.593
96.659	9.299	9.337	9.336	9.303	9.303	9.300	9.300
107.883	5.306	5.310	5.310	5.306	5.305	5.305	5.305
120.411	2.673	2.667	2.666	2.669	2.669	2.671	2.671
134.394	1.175	1.186	1.185	1.174	1.174	1.174	1.174
150.000	0.446	0.443	0.442	0.446	0.446	0.446	0.445

Source: ts-2-1, ts-2-13, ts-2-16, ts-2-17, ts-2-20, ts-2-22, and ts-2-26

Table 5.25. ROV of a European put option priced with the Cox-Ross-Rubinstein binomial tree and Trigeorgis log-transformed binomial tree methods ($S_{min} = 50$, $S_{max} = 150$, $M = 100$, $\sigma_S = 0.25$, $T_S = 5$, $X = 100$, $r_f = 0.04$).

		$N_S/T_S = 60$		$N_S/T_S = 360$		$N_S/T_S = 720$	
S_0	BS	CRR	Trigeorgis	CRR	Trigeorgis	CRR	Trigeorgis
50.000	35.497	35.494	35.493	35.497	35.497	35.497	35.496
55.806	31.479	31.464	31.464	31.477	31.477	31.479	31.479
62.287	27.460	27.462	27.461	27.458	27.457	27.460	27.460
69.519	23.529	23.535	23.534	23.529	23.529	23.530	23.530
77.592	19.772	19.761	19.759	19.774	19.773	19.773	19.773
86.603	16.271	16.273	16.271	16.272	16.272	16.271	16.271
96.659	13.094	13.107	13.104	13.096	13.095	13.094	13.094
107.883	10.292	10.292	10.289	10.291	10.291	10.291	10.291
120.411	7.890	7.887	7.883	7.889	7.888	7.889	7.889
134.394	5.894	5.904	5.899	5.894	5.893	5.893	5.893
150.000	4.285	4.290	4.284	4.286	4.285	4.285	4.285

Source: ts-2-1, ts-2-13, ts-2-16, ts-2-17, ts-2-20, ts-2-22, and ts-2-26

This step provides the first comparison between the Cox-Ross-Rubinstein (CRR) and the Trigeorgis log-transformed binomial tree methods. The ROVs are almost identical for all choices of N_S and $\frac{N_S}{T_S}$. Therefore, only one method will be used for the further analysis in this thesis if the risk-free rate is constant.

Step 7: Cox-Ross-Rubinstein binomial tree method with a stochastically simulated risk-free rate using the Vasicek model for different numbers of simulated short-rate paths

Step 7 answers the question: How does the number of simulated future interest rate paths influence the real options value and its standard deviation? To answer this question, case 1 is modified to a European style real option.

Step 7 is the first step that includes a stochastically modelled risk-free interest rate instead of a constant one. Tables 5.26 and 5.27 display the ROV for an interest rate that is simulated with the Vasicek model. Table 5.26 applies the Vasicek model with a long-term interest rate mean of $\frac{\beta}{\alpha} = \frac{0.018}{0.6} = 0.03$ and Table 5.27 applies $\frac{\beta}{\alpha} = \frac{0.03}{0.6} = 0.05$. In both tables the ROV for 25, 100, and 400 simulated future term structures of interest rates are displayed together with the standard deviation (Std., given as an absolute value figure) of the ROV for the corresponding start value of the underlying.

Table 5.26. European put option priced with the Cox-Ross-Rubinstein binomial tree method with a stochastic interest rate using the Vasicek model (parameters for Vasicek model: $\alpha = 0.6$, $\beta = 0.018$, $\sigma_r = 0.02$, $T_r = 3$, $N_r = 10800$, $r_0 = 0.03$, Taylor 1.5 simulation scheme; parameters for Cox-Ross-Rubinstein binomial tree: $S_{min} = 50$, $S_{max} = 150$, $M = 50$, $\sigma_S = 0.25$, $T_S = 3$, $N_S = 1080$, $X = 100$).

	—— 25 paths ——		—— 100 paths ——		—— 400 paths ——	
S_0	ROV	Std.	ROV	Std.	ROV	Std.
50	41.771	3.306	42.299	2.717	42.372	2.715
60	33.538	3.084	34.025	2.536	34.093	2.532
70	26.461	2.788	26.895	2.293	26.957	2.288
80	20.595	2.449	20.970	2.016	21.024	2.009
90	15.860	2.101	16.176	1.729	16.222	1.722
100	12.129	1.767	12.390	1.455	12.429	1.448
110	9.237	1.464	9.450	1.206	9.482	1.198
120	7.005	1.198	7.176	0.987	7.202	0.980
130	5.305	0.972	5.442	0.800	5.463	0.794
140	4.020	0.783	4.128	0.645	4.145	0.639
150	3.047	0.628	3.132	0.517	3.145	0.512

Source: ts-2-27

Table 5.27. European put option priced with the Cox-Ross-Rubinstein binomial tree method with a stochastic interest rate using the Vasicek model (parameters for Vasicek model: $\alpha = 0.6$, $\beta = 0.03$, $\sigma_r = 0.02$, $T_r = 3$, $N_r = 10800$, $r_0 = 0.05$, Taylor 1.5 simulation scheme; parameters for Cox-Ross-Rubinstein binomial tree: $S_{min} = 50$, $S_{max} = 150$, $M = 50$, $\sigma_S = 0.25$, $T_S = 3$, $N_S = 1080$, $X = 100$).

	—— 25 paths ——		—— 100 paths ——		—— 400 paths ——	
S_0	ROV	Std.	ROV	Std.	ROV	Std.
50	36.834	3.058	37.321	2.514	37.388	2.511
60	28.978	2.808	29.420	2.309	29.482	2.305
70	22.390	2.491	22.775	2.049	22.830	2.044
80	17.065	2.144	17.391	1.765	17.438	1.758
90	12.874	1.801	13.143	1.482	13.182	1.476
100	9.651	1.484	9.869	1.222	9.901	1.215
110	7.211	1.205	7.385	0.992	7.411	0.986
120	5.369	0.967	5.505	0.796	5.526	0.790
130	3.995	0.770	4.102	0.634	4.119	0.628
140	2.977	0.609	3.060	0.502	3.073	0.497
150	2.220	0.480	2.284	0.395	2.295	0.391

Source: ts-2-29

For this step and all remaining tests in this thesis, the intervals between S_{min} and S_{max} are of equal size for the Cox-Ross-Rubinstein binomial tree method and the Trigeorgis log-transformed binomial tree method since no comparison with a log-transformed finite difference method is necessary any more. For example, $S_{min} = 50$, $S_{max} = 150$, and $M = 50$ means that the pricing tree is (independently) applied with the start values $50, 52, 54, \ldots, 150$.

The real options values and the standard deviations are similar for 100 and 400 simulated interest rate paths. However, there is a significant difference between the results for 25 and 100 simulated short-rate paths. Therefore, using 100 simulated interest rate paths is a good choice since it leads to good results within a reasonable computational time. These results are evident in both tables, i.e., for two different values of the mean reversion level.

Step 8: Cox-Ross-Rubinstein and Trigeorgis log-transformed binomial tree methods with a constant risk-free rate for a zero salvage value

Step 8 answers the question: What does the algorithm calculate as the real options value in case 1 when the salvage value is zero, i.e., when the investment project does not contain any real option? To answer this question, case 1 was priced with the two binomial tree methods with the salvage factor set to zero.

For this test, case 1, an American put option, is valued with the strike price of the option set to zero. Obviously, such an option cannot have any value, irrespective of the start value chosen for the underlying. The real options values for a constant interest rate are displayed in Table 5.28 (for $T_S = 3$) and in Table 5.29 (for $T_S = 1$). For both tables the risk-free interest rate is 0.03 and constant. The real options values for a stochastically simulated interest rate (Cox-Ingersoll-Ross model) are displayed in Table 5.30 (for $N_S = 48$) and in Table 5.31 (for $N_S = 24$), both with $T_S = 1$.

Table 5.28. ROV in case 1 with a zero salvage value priced with the Cox-Ross-Rubinstein binomial tree method and the Trigeorgis log-transformed binomial tree method with a constant interest rate ($r_f = 0.03$, $S_{min} = 50$, $S_{max} = 150$, $M = 50$, $\sigma_S = 0.25$, $T_S = 3$ with a salvage factor of 0).

	$N_S = 1080$		$N_S = 360$	
S_0	CRR	Trigeorgis	CRR	Trigeorgis
50	0.000000	-0.000125	0.000000	-0.000374
60	0.000000	-0.000150	0.000000	-0.000449
70	0.000000	-0.000175	0.000000	-0.000524
80	0.000000	-0.000200	0.000000	-0.000599
90	0.000000	-0.000225	0.000000	-0.000674
100	0.000000	-0.000250	0.000000	-0.000749
110	0.000000	-0.000275	0.000000	-0.000823
120	0.000000	-0.000299	0.000000	-0.000898
130	0.000000	-0.000324	0.000000	-0.000973
140	0.000000	-0.000349	0.000000	-0.001048
150	0.000000	-0.000374	0.000000	-0.001123

Source: ts-2-30

Table 5.29. ROV in case 1 with a zero salvage value priced with the Cox-Ross-Rubinstein binomial tree method and the Trigeorgis log-transformed binomial tree method with a constant interest rate ($r_f = 0.03$, $S_{min} = 50$, $S_{max} = 150$, $M = 50$, $\sigma_S = 0.25$, $T_S = 1$ with a salvage factor of 0).

	$N_S = 48$		$N_S = 24$	
S_0	CRR	Trigeorgis	CRR	Trigeorgis
50	0.000000	-0.000312	0.000000	-0.000623
60	0.000000	-0.000374	0.000000	-0.000748
70	0.000000	-0.000437	0.000000	-0.000873
80	0.000000	-0.000499	0.000000	-0.000998
90	0.000000	-0.000561	0.000000	-0.001122
100	0.000000	-0.000624	0.000000	-0.001247
110	0.000000	-0.000686	0.000000	-0.001372
120	0.000000	-0.000748	0.000000	-0.001496
130	0.000000	-0.000811	0.000000	-0.001621
140	0.000000	-0.000873	0.000000	-0.001746
150	0.000000	-0.000936	0.000000	-0.001870

Source: ts-2-31

Table 5.30. Real options case 1 with a zero salvage value priced with the Cox-Ross-Rubinstein binomial tree method and the Trigeorgis log-transformed binomial tree method and a stochastic interest rate simulated with the Cox-Ingersoll-Ross model (parameters for the Cox-Ingersoll-Ross model: $\alpha = 0.6$, $\beta = 0.018$, $\sigma_r = 0.06$, $T_r = 2$, $N_r = 3600$, $r_0 = 0.03$, Taylor 1.5 simulation scheme with 100 simulated paths; parameters for the option pricing models: $S_{min} = 50$, $S_{max} = 150$, $M = 50$, $\sigma_S = 0.25$, $T_S = 1$, $N_S = 48$ with a salvage factor of 0).

	Cox-Ross-Rubinstein		Trigeorgis	
S_0	ROV	Std.	ROV	Std.
50	0.000000	0.000000	-0.000317	0.000083
60	0.000000	0.000000	-0.000380	0.000099
70	0.000000	0.000000	-0.000444	0.000116
80	0.000000	0.000000	-0.000507	0.000132
90	0.000000	0.000000	-0.000571	0.000149
100	0.000000	0.000000	-0.000634	0.000165
110	0.000000	0.000000	-0.000697	0.000182
120	0.000000	0.000000	-0.000761	0.000198
130	0.000000	0.000000	-0.000824	0.000215
140	0.000000	0.000000	-0.000887	0.000231
150	0.000000	0.000000	-0.000951	0.000248

Source: ts-2-32

Table 5.31. Real options case 1 with a zero salvage value priced with the Cox-Ross-Rubinstein binomial tree method and the Trigeorgis log-transformed binomial tree method and a stochastic interest rate simulated with the Cox-Ingersoll-Ross model (parameters for the Cox-Ingersoll-Ross model: $\alpha = 0.6$, $\beta = 0.018$, $\sigma_r = 0.06$, $T_r = 2$, $N_r = 3600$, $r_0 = 0.03$, Taylor 1.5 simulation scheme with 100 simulated paths; parameters for the option pricing models: $S_{min} = 50$, $S_{max} = 150$, $M = 50$, $\sigma_S = 0.25$, $T_S = 1$, $N_S = 24$ with a salvage factor of 0).

	Cox-Ross-Rubinstein		Trigeorgis	
S_0	ROV	Std.	ROV	Std.
50	0.000000	0.000000	-0.000634	0.000164
60	0.000000	0.000000	-0.000760	0.000197
70	0.000000	0.000000	-0.000887	0.000230
80	0.000000	0.000000	-0.001014	0.000263
90	0.000000	0.000000	-0.001141	0.000295
100	0.000000	0.000000	-0.001267	0.000328
110	0.000000	0.000000	-0.001394	0.000361
120	0.000000	0.000000	-0.001521	0.000394
130	0.000000	0.000000	-0.001647	0.000427
140	0.000000	0.000000	-0.001774	0.000460
150	0.000000	0.000000	-0.001901	0.000492

Source: ts-2-32

When the Cox-Ross-Rubinstein binomial tree method is used, the real options value is (almost) identical to zero for all chosen start values of the underlying and for all chosen parameters N_S. This is not the case for the Trigeorgis log-transformed binomial tree method: The real options value is always slightly negative but very close to zero. The more out-of-the-money the option, the stronger the deviation from zero. The results are similar when considering a stochastic risk-free interest rate instead of a constant one. Both the real options values and the appropriate standard deviations are displayed in Tables 5.30 and 5.31.

Step 9: Trigeorgis log-transformed binomial tree method with a stochastic risk-free interest rate

Step 9 answers the question: How does parameter N_S influence the real options value of the Trigeorgis log-transformed binomial tree method if the risk-free rate is stochastic? To answer this question, various stochastic term structure models are applied to case 1 and case 2.

Cases 1 and 2 are analyzed here with three different stochastic interest rate models: the Vasicek model (Table 5.32), the Hull-White one-factor model (Table 5.33) and the Hull-White two-factor model (Table 5.34). T_S is always 1 year due to the fact that only this time to maturity was implemented in the computer simulation program for the Trigeorgis method because higher values of T_S lead to a dramatic increase of the computational time. This holds especially true for case 3 as will be shown in the two remaining steps. The values for N_S are very low since higher N_S values entail the bushy tree problem and cause excessive computational times.

Table 5.32. Trigeorgis log-transformed binomial tree method and a stochastic interest rate using the Vasicek model for cases 1 and 2 (parameters for the Vasicek model: $\alpha = 0.6$, $\beta = 0.018$, $\sigma_r = 0.01$, $T_r = 2, N_r = 7200$, $r_0 = 0.03$, Taylor 1.5 simulation scheme with 100 simulated paths; parameters for the Trigeorgis pricing method: $S_{min} = 50$, $S_{max} = 150$, $M = 50$, $\sigma_S = 0.25$, $T_S = 1$, salvage factor 0.75 of initial investment cost of 110, expand factor 0.3 of prevailing underlying value with expand investment of 0.2 of initial investment cost).

	Case 1				Case 2			
	$N_S = 36$		$N_S = 48$		$N_S = 36$		$N_S = 48$	
S_0	ROV	Std.	ROV	Std.	ROV	Std.	ROV	Std.
50	32.500	0.000	32.500	0.000	32.500	0.000	32.500	0.000
60	22.532	0.019	22.539	0.018	22.872	0.041	22.851	0.037
70	14.165	0.084	14.201	0.087	15.846	0.085	15.796	0.082
80	8.327	0.106	8.317	0.106	11.827	0.065	11.834	0.066
90	4.487	0.084	4.525	0.087	10.577	0.027	10.577	0.026
100	2.349	0.060	2.354	0.059	11.138	0.018	11.068	0.020
110	1.167	0.036	1.129	0.034	12.836	0.048	12.836	0.047
120	0.556	0.020	0.542	0.020	15.195	0.066	15.204	0.066
130	0.255	0.011	0.252	0.011	17.896	0.076	17.900	0.076
140	0.112	0.005	0.113	0.005	20.761	0.082	20.762	0.082
150	0.046	0.003	0.049	0.003	23.699	0.084	23.700	0.084

Source: ts-2-33 and ts-2-34

Table 5.33. Trigeorgis log-transformed binomial tree method and a stochastic interest rate using the Hull-White one-factor model for cases 1 and 2 (parameters for the Hull-White one-factor model: $\alpha = 0.3$, $\sigma_r = 0.015$, $T_r = 2$, $N_r = 3600$, Taylor 1.5 simulation scheme with U.S. Zero yield curve from December 1, 2000 as input parameter for each of the 100 simulated short-rate paths; parameters for the Trigeorgis pricing method: $S_{min} = 50$, $S_{max} = 150$, $M = 50$, $\sigma_S = 0.25$, $T_S = 1$, salvage factor 0.75 of initial investment cost of 110, expand factor 0.3 of prevailing underlying value with expand investment of 0.2 of initial investment cost).

	Case 1				Case 2			
	$N_S = 24$		$N_S = 40$		$N_S = 24$		$N_S = 40$	
S_0	ROV	Std.	ROV	Std.	ROV	Std.	ROV	Std.
50	32.500	0.000	32.500	0.000	32.500	0.000	32.500	0.000
60	22.500	0.000	22.500	0.000	22.500	0.000	22.500	0.000
70	13.195	0.070	13.190	0.062	14.862	0.075	14.821	0.073
80	7.222	0.114	7.204	0.113	11.111	0.058	11.122	0.060
90	3.702	0.098	3.670	0.098	10.272	0.014	10.341	0.015
100	1.801	0.066	1.821	0.068	11.254	0.046	11.297	0.046
110	0.850	0.039	0.858	0.040	13.241	0.083	13.261	0.083
120	0.390	0.022	0.386	0.021	15.788	0.104	15.793	0.104
130	0.167	0.011	0.169	0.011	18.591	0.114	18.586	0.115
140	0.061	0.005	0.071	0.006	21.495	0.119	21.495	0.120
150	0.023	0.003	0.028	0.003	24.450	0.122	24.459	0.123

Source: ts-2-39 and ts-2-40

Table 5.34. Trigeorgis log-transformed binomial tree method and a stochastic interest rate using the Hull-White two-factor model for cases 1 and 2 (parameters for the Hull-White two-factor model: $\alpha = 0.3$, $\sigma_{r,1} = 0.015$, $\xi = 1$, $\sigma_{r,2} = 0.01$, $\rho = -0.4$, $T_r = 2$, $N_r = 3600$, Euler simulation scheme with U.S. Zero yield curve from December 1, 2000 as input parameter for each of the 100 simulated short-rate paths; parameters for the Trigeorgis pricing method: $S_{min} = 50$, $S_{max} = 150$, $M = 50$, $\sigma_S = 0.25$, $T_S = 1$, salvage factor 0.75 of initial investment cost of 110, expand factor 0.3 of prevailing underlying value with expand investment of 0.2 of initial investment cost).

	Case 1				Case 2			
	$N_S = 24$		$N_S = 40$		$N_S = 24$		$N_S = 40$	
S_0	ROV	Std.	ROV	Std.	ROV	Std.	ROV	Std.
50	32.500	0.000	32.500	0.000	32.500	0.000	32.500	0.000
60	22.500	0.000	22.500	0.000	22.500	0.000	22.500	0.000
70	13.195	0.067	13.191	0.059	14.862	0.072	14.822	0.069
80	7.222	0.108	7.204	0.107	11.112	0.056	11.122	0.057
90	3.702	0.093	3.669	0.092	10.272	0.014	10.341	0.014
100	1.800	0.062	1.820	0.064	11.255	0.044	11.298	0.044
110	0.849	0.037	0.857	0.038	13.242	0.079	13.262	0.079
120	0.390	0.021	0.386	0.020	15.790	0.099	15.795	0.099
130	0.167	0.011	0.169	0.010	18.593	0.108	18.588	0.109
140	0.061	0.005	0.071	0.005	21.497	0.113	21.497	0.115
150	0.023	0.003	0.028	0.003	24.452	0.116	24.461	0.117

Source: ts-2-41 and ts-2-42

As can be seen in all tables, the real options values and the values for the corresponding standard deviation do not vary significantly for different values of N_S. This shows that the Trigeorgis log-transformed binomial tree model[16] can yield reliable results even for very low N_S values.

It needs to be mentioned that case 3 was not analyzed with this method for a non-constant interest rate. Tests with the Trigeorgis log-transformed binomial tree method for a non-constant interest rate revealed extremely long computational times (sometimes days for just one single short-rate path) due to the fact that the option to defer in case 3 requires multiple parallel runs of the algorithm to price real options with various underlying future start dates. For this very reason, the Trigeorgis log-transformed binomial tree method is not applied to case 3 for a stochastic interest rate but only for a constant interest rate.

Step 10: Trigeorgis log-transformed binomial tree method vs. Cox-Ross-Rubinstein log-transformed binomial tree method with a constant risk-free interest rate

Step 10 answers the question: How do the real options values differ between the Trigeorgis log-transformed and Cox-Ross-Rubinstein binomial tree methods when a constant risk-free interest rate is used and when the model parameters are identical? To answer this question, all cases 1, 2 and 3 are analyzed with the two real options pricing models.

Here, the first test with all three real options cases is conducted by applying the Cox-Ross-Rubinstein and the Trigeorgis log-transformed binomial tree methods as valuation tools with $T_S = 3$ and $N_S = 360$ as well as $N_S = 1080$. For all three cases the initial investment cost is $I_0 = 110$. It is assumed that if the investment project is abandoned the salvage value is 75% of the initial investment cost, i.e., the salvage value is 82.5 if the project is abandoned. The abandonment can take place at any time in time interval $[0, T_S]$.

For cases 2 and 3 the project can be expanded. The expand factor is 30% with expand cost of 20% of the initial investment cost, i.e., the underlying project value S can be increased to $(1 + 30\%)S$ by investing $20\% I_0 = 22$. However, the expansion can only be undertaken once at time T_S.

[16] This is also one of the findings (for a constant risk-free rate) Trigeorgis mentions when pricing options with his algorithm compared with the Cox-Ross-Rubinstein binomial tree algorithm, see Trigeorgis [132], pages 330-331, Tables 10.1 and 10.2.

The option to defer in case 3 allows to defer the start of the investment project by exactly 1 year. The project will only start in 1 year from today if the NPV is positive. Hereby it is assumed that the investment cost in 1 year from today is $I_0 = 110$ as well[17].

In all three Tables 5.35, 5.36, and 5.37 the ROV as well as the NPV of the investment project are displayed. Both terms were explained in the beginning of this section: ROV refers to the present value of the real option(s) and NPV refers to the project's net present value including the ROV and the initial investment cost I_0 (given as a present value figure). With S_0 as the underlying value, the present value of the estimated project's future cash flow (including all future costs but not I_0), the following relationship holds:

$$NPV \quad = \quad ROV + S_0 - I_0.$$

The sum $ROV + S_0$ is the total project value (TPV), i.e., the investment project's present value as the present value of the project including the ROV but without considering any initial investment to start the project. Finally, Figures 5.9 and 5.10 graphically display the ROV and the NPV, respectively, for all three cases in comparison.

Table 5.35. Cox-Ross-Rubinstein binomial tree and Trigeorgis log-transformed binomial tree methods with a constant interest rate for case 1 ($r_f = 0.03$, $S_{min} = 50$, $S_{max} = 150$, $M = 50$, $\sigma_S = 0.25$, $T_S = 3$, salvage factor 0.75 of initial investment cost of 110).

	$N_S = 360$				$N_S = 1080$			
	CRR		Trigeorgis		CRR		Trigeorgis	
S_0	ROV	NPV	ROV	NPV	ROV	NPV	ROV	NPV
50	32.500	-27.500	32.500	-27.500	32.500	-27.500	32.500	-27.500
60	23.555	-26.445	23.555	-26.445	23.558	-26.442	23.558	-26.442
70	16.946	-23.054	16.946	-23.054	16.943	-23.057	16.943	-23.057
80	12.113	-17.887	12.113	-17.887	12.110	-17.890	12.109	-17.891
90	8.618	-11.382	8.618	-11.382	8.620	-11.380	8.620	-11.380
100	6.123	-3.877	6.123	-3.877	6.123	-3.877	6.122	-3.878
110	4.349	4.349	4.349	4.349	4.345	4.345	4.345	4.345
120	3.088	13.088	3.088	13.088	3.089	13.089	3.088	13.088
130	2.195	22.195	2.195	22.195	2.199	22.199	2.199	22.199
140	1.572	31.572	1.572	31.572	1.569	31.569	1.569	31.569
150	1.121	41.121	1.121	41.121	1.126	41.126	1.125	41.125

Source: ts-2-43

[17] This assumption can easily be modified to, e.g., investment cost of $I_0(1 + r_f)$ for an investment start in 1 year from today as done in Trigeorgis [132], pages 157-158. However, according to Trigeorgis [132], page 157, this more complicated assumption is not crucial at all to the analysis, and will not be considered in this thesis.

Table 5.36. Cox-Ross-Rubinstein binomial tree and Trigeorgis log-transformed binomial tree methods with a constant interest rate for case 2 ($r_f = 0.03$, $S_{min} = 50$, $S_{max} = 150$, $M = 50$, $\sigma_S = 0.25$, $T_S = 3$, salvage factor 0.75 of initial investment cost of 110, expand factor 0.3 of prevailing underlying value with expand investment of 0.2 of initial investment cost).

	$N_S = 360$				$N_S = 1080$			
	CRR		Trigeorgis		CRR		Trigeorgis	
S_0	ROV	NPV	ROV	NPV	ROV	NPV	ROV	NPV
50	32.665	-27.335	32.665	-27.335	32.665	-27.335	32.665	-27.335
60	25.271	-24.729	25.271	-24.729	25.266	-24.734	25.266	-24.734
70	20.511	-19.489	20.511	-19.489	20.516	-19.484	20.516	-19.484
80	17.844	-12.156	17.844	-12.156	17.845	-12.155	17.845	-12.155
90	16.771	-3.229	16.770	-3.230	16.763	-3.237	16.762	-3.238
100	16.859	6.859	16.858	6.858	16.856	6.856	16.856	6.856
110	17.801	17.801	17.800	17.800	17.801	17.801	17.800	17.800
120	19.360	29.360	19.359	29.359	19.357	29.357	19.357	29.357
130	21.347	41.347	21.346	41.346	21.343	41.343	21.343	41.343
140	23.629	53.629	23.628	53.628	23.630	53.630	23.629	53.629
150	26.132	66.132	26.130	66.130	26.131	66.131	26.130	66.130

Source: ts-2-44

Table 5.37. Cox-Ross-Rubinstein binomial tree and Trigeorgis log-transformed binomial tree methods with a constant interest rate for case 3 ($r_f = 0.03$, $S_{min} = 50$, $S_{max} = 150$, $M = 50$, $\sigma_S = 0.25$, $T_S = 3$, salvage factor 0.75 of initial investment cost of 110, expand factor 0.3 of prevailing underlying value with expand investment of 0.2 of initial investment cost).

	$N_S = 360$				$N_S = 1080$			
	CRR		Trigeorgis		CRR		Trigeorgis	
S_0	ROV	NPV	ROV	NPV	ROV	NPV	ROV	NPV
50	60.051	0.051	60.051	0.051	60.051	0.051	60.051	0.051
60	50.407	0.407	50.407	0.407	50.408	0.408	50.408	0.408
70	41.691	1.691	41.691	1.691	41.688	1.688	41.688	1.688
80	34.648	4.648	34.648	4.648	34.638	4.638	34.638	4.638
90	29.702	9.702	29.701	9.701	29.700	9.700	29.699	9.699
100	26.862	16.862	26.861	16.861	26.848	16.848	26.848	16.848
110	25.762	25.762	25.761	25.761	25.750	25.750	25.750	25.750
120	25.983	35.983	25.981	35.981	25.975	35.975	25.974	35.974
130	27.115	47.115	27.114	47.114	27.113	47.113	27.112	47.112
140	28.855	58.855	28.853	58.853	28.859	58.859	28.859	58.859
150	31.010	71.010	31.008	71.008	31.007	71.007	31.006	71.006

Source: ts-2-45

The ROVs (and, therefore, the NPVs as well) derived from the Cox-Ross-Rubinstein and the Trigeorgis log-transformed binomial tree methods are almost identical in all three cases and for both $N_S = 360$ and $N_S = 1080$. However, the calculation time varies tremendously, especially if an option to defer is included into the investment project as done in case 3. The computational times are displayed in Table 5.38 for all three cases for a constant risk-free interest rate. This picture changes if a stochastically modelled interest rate is applied as in step 11, the last step of test situation 2. The corresponding tree of the Trigeorgis log-transformed binomial tree method gets very big ("bushy tree problem") for a non-constant interest rate, whereas the Cox-Ross-Rubinstein tree is not affected by the choice of the risk-free rate as elaborated in the previous chapter.

Therefore, case 3 cannot be handled properly by combining the Trigeorgis method with a non-constant risk-free rate. However, case 3 poses no problem, if a constant risk-free rate is used. Therefore, the computational times for case 3 valued with the Trigeorgis method are displayed in Table 5.38 (constant risk-free rate) but not in Table 5.43 (non-constant risk-free rate) of step 11.

Table 5.38. Computational time for cases 1, 2 and 3 priced with the Cox-Ross-Rubinstein binomial tree and the Trigeorgis log-transformed binomial tree methods for a constant interest rate ($T_S = 3$, $S_{min} = 50$, $S_{max} = 150$, $M = 50$, salvage factor 0.75 of initial investment cost of 110, expand factor 0.3 of prevailing underlying value with expand investment of 0.2 of initial investment cost).

Case for real options	Cox-Ross-Rubinstein tree $N_S = 360$	Cox-Ross-Rubinstein tree $N_S = 1080$	Trigeorgis log-transformed tree $N_S = 360$	Trigeorgis log-transformed tree $N_S = 1080$
case 1	< 1 sec.	22 sec.	< 1 sec.	15 sec.
case 2	< 1 sec.	22 sec.	< 1 sec.	15 sec.
case 3	5.3 min.	142 min.	3.6 min.	91 min.

Source: own calculation with Pentium 4, 2.4 MHz. personal computer

Figures 5.9 and 5.10 show an increase of ROV and NPV if more options are added to the investment project. The option to abandon works like a floor of $-(1 - 75\%)I_0 = -27.5$ for the NPV in cases 1 and 2 (see Figure 5.10). The option to defer the investment project in case 3 lifts the NPV with the result that it can never turn negative since the investment project can be deferred by exactly one year but does not need to be started then, if the NPV is not positive. The higher the underlying value S_0, the more valuable the option to expand the investment. The analysis of the different options (and their parameters) for cases 1-3 will be extended in test situation 3.

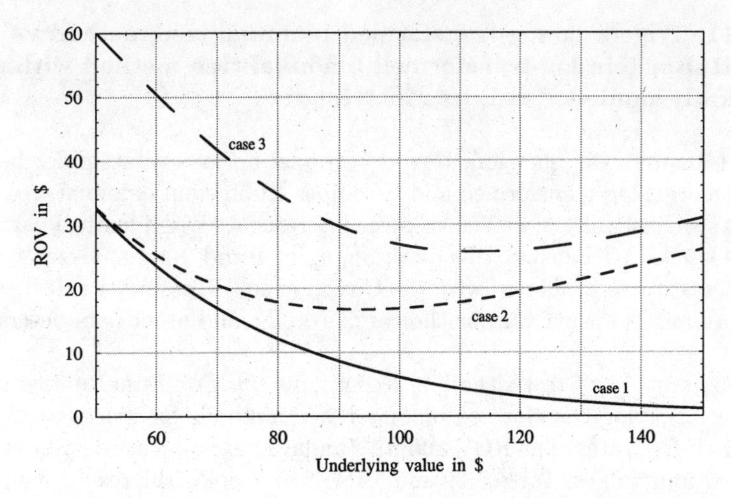

Fig. 5.9. ROV for real options cases 1, 2 and 3 priced with the Cox-Ross-Rubinstein binomial tree method and using a constant risk-free interest rate (ts-2-43, ts-2-44, and ts-2-45; parameters: $r_f = 0.03$, $S_{min} = 50$, $S_{max} = 150$, $M = 50$, $\sigma_S = 0.25$, $T_S = 3$, $N_S = 1080$, salvage factor 0.75 of initial investment cost of 110, expand factor 0.3 of prevailing underlying value with expand investment of 0.2 of initial investment cost).

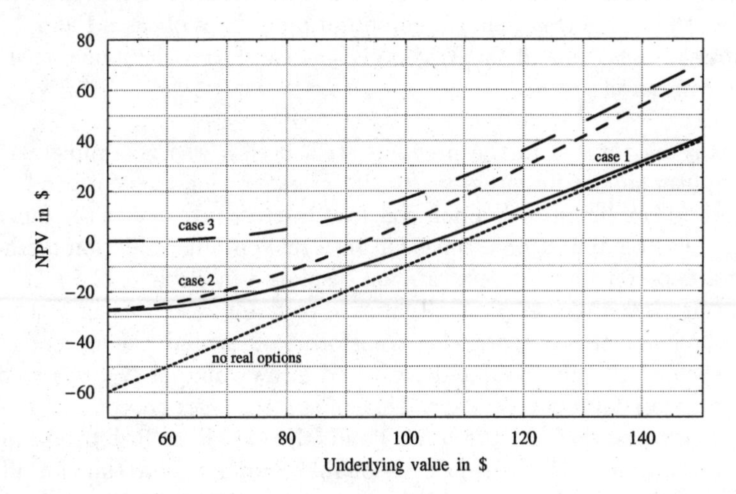

Fig. 5.10. NPV for real options cases 1, 2 and 3 priced with the Cox-Ross-Rubinstein binomial tree method and using a constant risk-free interest rate (ts-2-43, ts-2-44, and ts-2-45; parameters: $r_f = 0.03$, $S_{min} = 50$, $S_{max} = 150$, $M = 50$, $\sigma_S = 0.25$, $T_S = 3$, $N_S = 1080$, salvage factor 0.75 of initial investment cost of 110, expand factor 0.3 of prevailing underlying value with expand investment of 0.2 of initial investment cost).

Step 11: Trigeorgis log-transformed binomial tree method vs. Cox-Ross-Rubinstein log-transformed binomial tree method with a stochastically modelled risk-free interest rate

Step 11 answers the question: How do the real options values differ between the Trigeorgis log-transformed and Cox-Ross-Rubinstein binomial tree methods if a stochastically modelled risk-free interest rate with identical parameter choices is used? To answer this question, cases 1 and 2 are analyzed for risk-free interest rates simulated with the Cox-Ingersoll-Ross and the Ho-Lee term structure models using various choices of N_S, T_S and other parameters.

The final step 11 of test situation 2 compares the Cox-Ross-Rubinstein and the Trigeorgis log-transformed binomial tree methods for stochastically modelled risk-free rates. The ROV and its standard deviation are displayed in the following four tables. Table 5.39 and Table 5.40 display the results for cases 1 and 2, respectively, if the risk-free interest rate follows the Cox-Ingersoll-Ross model. Tables 5.41 and 5.42 show the results for cases 1 and 2, respectively, if the risk-free interest rate follows the Ho-Lee model. For all tables T_S was defined as 1 with very low values of N_S. A low value for N_S poses no problem due to the results of step 9 for both the Cox-Ross-Rubinstein and the Trigeorgis log-transformed binomial tree methods.

Case 3 was not analyzed here since the bushy tree problem of the Trigeorgis log-transformed binomial tree method results in extremely long computational times. At the end of this step, the computational times of cases 1 and 2 for the Cox-Ross-Rubinstein and the Trigeorgis log-transformed binomial tree methods are compared.

As already elaborated in the previous step, case 3 was not implemented in the computer simulation program for the Trigeorgis log-transformed binomial tree method combined with a non-constant interest rate due to the bushy tree problem. The computational times for both real options valuation methods in case of a non-constant interest rate are displayed in Table 5.43 for 100 simulated future interest rate paths. Tentative tests for case 3 using a stochastic interest rate and the Trigeorgis log-transformed binomial tree method showed that even for just one simulated path and a low valuc of N_S, the computational time can cover several days. This, of course, is not acceptable in Corporate Finance practice. On the other hand, the Cox-Ross-Rubinstein method yields real options values within a reasonable computational time for all three cases. Since there is virtually no difference in the ROV between the Cox-Ross-Rubinstein binomial tree method and the Trigeorgis log-transformed binomial tree method as shown in the previous tables, the Cox-Ross-Rubinstein method will be the only real options valuation tool used for the remaining test situations 3, 4 and 5.

Table 5.39. Real options case 1 priced with the Cox-Ross-Rubinstein binomial tree and Trigeorgis log-transformed binomial tree methods using a stochastic interest rate simulated with the Cox-Ingersoll-Ross model (parameters for Cox-Ingersoll-Ross model: $\alpha = 0.6$, $\beta = 0.018$, $\sigma_r = 0.06$, $T_r = 2$, $N_r = 7200$, $r_0 = 0.03$, Taylor 1.5 simulation scheme for 100 simulated short-rate paths; parameters for the Cox-Ross-Rubinstein binomial tree and the Trigeorgis log-transformed binomial tree: $S_{min} = 50$, $S_{max} = 150$, $M = 50$, $\sigma_S = 0.25$, $T_S = 1$, salvage factor 0.75 of initial investment cost of 110).

	$N_S = 36$				$N_S = 48$			
	CRR		Trigeorgis		CRR		Trigeorgis	
S_0	ROV	Std.	ROV	Std.	ROV	Std.	ROV	Std.
50	32.500	0.000	32.500	0.000	32.500	0.000	32.500	0.000
60	22.532	0.019	22.532	0.019	22.539	0.018	22.539	0.018
70	14.165	0.086	14.165	0.086	14.202	0.089	14.201	0.089
80	8.327	0.108	8.327	0.109	8.317	0.109	8.316	0.109
90	4.487	0.087	4.487	0.087	4.525	0.089	4.524	0.089
100	2.349	0.061	2.348	0.061	2.354	0.061	2.354	0.061
110	1.168	0.037	1.167	0.037	1.130	0.035	1.129	0.035
120	0.556	0.021	0.555	0.021	0.543	0.020	0.542	0.020
130	0.256	0.011	0.255	0.011	0.253	0.011	0.252	0.011
140	0.113	0.005	0.112	0.006	0.114	0.005	0.113	0.006
150	0.047	0.002	0.046	0.003	0.050	0.003	0.049	0.003

Source: ts-2-35 and ts-2-46

Table 5.40. Real options case 2 priced with the Cox-Ross-Rubinstein binomial tree and Trigeorgis log-transformed binomial tree methods using a stochastic interest rate simulated with the Cox-Ingersoll-Ross model (parameters for Cox-Ingersoll-Ross model: $\alpha = 0.6$, $\beta = 0.018$, $\sigma_r = 0.06$, $T_r = 2$, $N_r = 7200$, $r_0 = 0.03$, Taylor 1.5 simulation scheme for 100 simulated short-rate paths; parameters for the Cox-Ross-Rubinstein binomial tree and the Trigeorgis log-transformed binomial tree: $S_{min} = 50$, $S_{max} = 150$, $M = 50$, $\sigma_S = 0.25$, $T_S = 1$, salvage factor 0.75 of initial investment cost of 110, expand factor 0.3 of prevailing underlying value with expand investment of 0.2 of initial investment cost).

	$N_S = 36$				$N_S = 48$			
	CRR		Trigeorgis		CRR		Trigeorgis	
S_0	ROV	Std.	ROV	Std.	ROV	Std.	ROV	Std.
50	32.500	0.000	32.500	0.000	32.500	0.000	32.500	0.000
60	22.872	0.042	22.872	0.042	22.851	0.038	22.851	0.038
70	15.846	0.086	15.846	0.087	15.796	0.084	15.796	0.084
80	11.828	0.066	11.827	0.066	11.835	0.067	11.834	0.067
90	10.578	0.027	10.577	0.027	10.578	0.026	10.577	0.026
100	11.140	0.019	11.139	0.019	11.069	0.021	11.068	0.021
110	12.837	0.050	12.836	0.050	12.837	0.050	12.836	0.049
120	15.196	0.069	15.195	0.069	15.206	0.069	15.205	0.069
130	17.898	0.080	17.897	0.079	17.902	0.080	17.901	0.079
140	20.763	0.085	20.761	0.085	20.763	0.085	20.762	0.085
150	23.701	0.088	23.699	0.088	23.702	0.088	23.701	0.088

Source: ts-2-36 and ts-2-46

Table 5.41. Real options case 1 priced with the Cox-Ross-Rubinstein binomial tree and Trigeorgis log-transformed binomial tree methods using a stochastic interest rate simulated with the Ho-Lee model (parameters for Ho-Lee model: $\sigma_r = 0.01$, $T_r = 2$, $N_r = 3600$, Taylor 1.5 simulation scheme with U.S. Zero yield curve from December 1, 2000 as input parameter for each of the 100 simulated short-rate paths; parameters for the Cox-Ross-Rubinstein binomial tree and the Trigeorgis log-transformed binomial tree: $S_{min} = 50$, $S_{max} = 150$, $M = 50$, $\sigma_S = 0.25$, $T_S = 1$, salvage factor 0.75 of initial investment cost of 110).

	$N_S = 24$				$N_S = 40$			
	CRR		Trigeorgis		CRR		Trigeorgis	
S_0	ROV	Std.	ROV	Std.	ROV	Std.	ROV	Std.
50	32.500	0.000	32.500	0.000	32.500	0.000	32.500	0.000
60	22.500	0.000	22.500	0.000	22.500	0.000	22.500	0.000
70	13.198	0.054	13.199	0.054	13.194	0.048	13.195	0.048
80	7.228	0.088	7.230	0.088	7.211	0.087	7.212	0.087
90	3.708	0.076	3.709	0.075	3.676	0.075	3.676	0.075
100	1.806	0.051	1.805	0.050	1.826	0.052	1.825	0.052
110	0.855	0.030	0.852	0.030	0.862	0.031	0.860	0.031
120	0.395	0.016	0.391	0.017	0.390	0.016	0.388	0.016
130	0.173	0.008	0.168	0.009	0.172	0.008	0.169	0.008
140	0.068	0.004	0.062	0.004	0.075	0.004	0.072	0.004
150	0.029	0.002	0.023	0.002	0.032	0.002	0.028	0.002

Source: ts-2-37 and ts-2-47

Table 5.42. Real options case 2 priced with the Cox-Ross-Rubinstein binomial tree and Trigeorgis log-transformed binomial tree methods using a stochastic interest rate simulated with the Ho-Lee model (parameters for Ho-Lee model: $\sigma_r = 0.01$, $T_r = 2$, $N_r = 3600$, Taylor 1.5 simulation scheme with U.S. Zero yield curve from December 1, 2000 as input parameter for each of the 100 simulated short-rate paths; parameters for the Cox-Ross-Rubinstein binomial tree and the Trigeorgis log-transformed binomial tree: $S_{min} = 50$, $S_{max} = 150$, $M = 50$, $\sigma_S = 0.25$, $T_S = 1$, salvage factor 0.75 of initial investment cost of 110, expand factor 0.3 of prevailing underlying value with expand investment of 0.2 of initial investment cost).

	$N_S = 24$				$N_S = 40$			
	CRR		Trigeorgis		CRR		Trigeorgis	
S_0	ROV	Std.	ROV	Std.	ROV	Std.	ROV	Std.
50	32.500	0.000	32.500	0.000	32.500	0.000	32.500	0.000
60	22.500	0.000	22.500	0.000	22.500	0.000	22.500	0.000
70	14.865	0.059	14.867	0.058	14.825	0.055	14.826	0.055
80	11.115	0.045	11.116	0.044	11.125	0.046	11.126	0.045
90	10.273	0.010	10.272	0.009	10.342	0.010	10.341	0.010
100	11.254	0.035	11.251	0.036	11.295	0.035	11.294	0.035
110	13.241	0.065	13.236	0.064	13.258	0.064	13.256	0.064
120	15.788	0.081	15.783	0.080	15.791	0.080	15.788	0.080
130	18.592	0.088	18.585	0.088	18.584	0.089	18.580	0.088
140	21.496	0.093	21.488	0.092	21.493	0.093	21.488	0.003
150	24.452	0.095	24.443	0.094	24.457	0.095	24.452	0.095

Source: ts-2-38 and ts-2-47

Table 5.43. Computational time for cases 1, 2 and 3 priced with the Cox-Ross-Rubinstein binomial tree and the Trigeorgis log-transformed binomial tree methods for a non-constant interest rate (100 simulated short-rate paths with $T_r = T_S + 1$ and $N_r/T_r = 1800$; $S_{min} = 50$, $S_{max} = 150$, $M = 50$, salvage factor 0.75 of initial investment cost of 110, expand factor 0.3 of prevailing underlying value with expand investment of 0.2 of initial investment cost).

Case for real options	Cox-Ross-Rubinstein tree $T_S = 3$, $N_S = 360$	Cox-Ross-Rubinstein tree $T_S = 3$, $N_S = 1080$	Trigeorgis log-transformed tree $T_S = 1$, $N_S = 24$	Trigeorgis log-transformed tree $T_S = 1$, $N_S = 48$
case 1	7 min.	75 min.	33 min.	105 min.
case 2	7 min.	75 min.	33 min.	105 min.
case 3	15.5 hrs.	650 hrs.	-	-

Source: own calculation with Pentium 4, 2.4 MHz. personal computer

5.5.2 Recapitulation of the Main Results in Test Situation 2

Test situation 2 yielded interesting results that will be applied in the following three test situations. The main results and their implications are summarized as follows:

- The valuation results of both the log-transformed explicit and the implicit finite difference methods are almost identical. However, since the stability and consistency restrictions are more strict for the explicit than for the implicit method, the log-transformed implicit method should always be used.

- The log-transformed finite difference methods yield inaccurate results for longer times to maturity T_S, (e.g., $T_S > 1$ year) compared to a benchmark value derived from the Black-Scholes formula. Since real options are usually long-term options, the log-transformed finite difference methods are not well suited for real options pricing and will not be used in the following.

- The real options values derived from the Cox-Ingersoll-Ross method and the Trigeorgis log-transformed binomial tree method are almost identical for both a constant and a stochastic risk-free interest rate. However, the computational time for the Trigeorgis method combined with non-constant interest rates is extremely long while the Cox-Ross-Rubinstein binomial tree method offers quicker valuation results for both types of interest rates. Therefore, only the Cox-Ross-Rubinstein binomial tree method will be used in the following analysis.

- The Cox-Ross-Rubinstein method combined with a stochastically modelled risk-free rate yields stable accurate results if the number of simulated future interest rate paths is approximately 100 or higher. Due to computational efficiency, in the following, always 100 paths will be simulated for valuation in all three cases.

5.6 Test Situation 3: Real Options Valuation with a Stochastic Interest Rate Using Equilibrium Models

Test situation 3 is devoted to the analysis of all three cases applying a risk-free interest rate which is stochastically simulated by using an equilibrium model. This section first describes the influence of critical case parameters on the real options value and the net present value of the project (5.6.1). The second part (5.6.2) analyzes real options pricing when the Vasicek model is applied to derive the risk-free interest rate, while the third part (5.6.3) analyzes real options pricing when the Cox-Ingersoll-Ross model is applied to derive the risk-free interest rate. The most important results are summarized in the last section (5.6.4).

Before proceeding to analyze the three cases, the general notation used in the following will be repeated:

1. **Case 1:** Option to abandon the project at any time during the construction period for a salvage value.
2. **Case 2:** Option to abandon the project at any time during the construction period for a salvage value and option to expand the project once by an expand factor e (e.g., expand project by $e = 30\%$) for an expand investment (given as a fraction of the initial investment cost I_0) at the end of the construction period.
3. **Case 3:** Complex real option of case 2 combined with an option to defer the project start by exactly one year. This means that the project can start today or in exactly one year from today if the investment is advantageous then, i.e., if in one year from now the NPV is negative the project would not be started at all.

Furthermore, ROV refers to the present value of the real option(s), TPV refers to the total project's present value as the present value of the project including the ROV but without considering any initial investment cost to start the project. NPV refers to the project's net present value including the ROV and the initial investment cost I_0 (given as a present value figure). With S_0 as the current underlying value the following relationships hold:

$$\begin{aligned} NPV &= ROV + S_0 - I_0, \\ TPV &= ROV + S_0, \\ NPV &= TPV - I_0. \end{aligned}$$

The current underlying value S_0 is the present value of the project's estimated future cash inflow minus outflow, not considering any investment cost in addition to $I_0 = 110$. The choice of the parameters is based on a natural-resource project described in Trigeorgis [132], Section 11.5. The general idea is that the project's future cash inflow minus outflow, not considering any additional investment cost, is unknown and needs to be estimated. However, different estimated values for S_0 are usually obtained, depending on the method used for the estimation.

The choice of 0.2 for the investment factor and of 0.3 as the expand factor in the base case scenario are similar to the choice Trigeorgis describes in the natural-resource project. However, the choice of 0.75 as the salvage factor of the initial capital outlay I_0 is slightly higher than in Trigeorgis' natural-resource project (0.5 instead of 0.75). The investment outlay assumed here is only one single outlay as opposed to Trigeorgis' project that allows for several investment outlays (see Trigeorgis [132], page 358). The figures used and calculated in this thesis are dollar-value figures.

5.6.1 The Influence of the Salvage Factor and the Expand Factor

This section analyzes how the salvage factor influences the real options value in case 1 and how the expand factor influences the real options value in cases 2 and 3. Figures 5.11 and 5.12 display the ROV and NPV for $r_f = 0.03$, respectively.

The project's ROV is higher for lower values of S_0. This is plausible since the option to abandon a project at any time for a salvage value is of greater value when the estimated underlying value S_0 is low. A low value of S_0 makes it unlikely that the underlying value would increase by a large amount, resulting in a larger effect of the salvage value. The ROV is low if S_0 is high since in this situation the high value of S_0 makes it unlikely that, over time, this value moves below the salvage value.

As can be seen in Figures 5.11 and 5.12, the influence of the salvage factor on the real option is large. The higher the salvage factor, the stronger the influence on ROV and NPV for any given S_0. Hence, it is critical for a good real options analysis to estimate this factor as well as S_0 as accurately as possible. Again, the special case of a zero strike price results in a zero real options value, a scenario that was already analyzed in step 8 of the previous section.

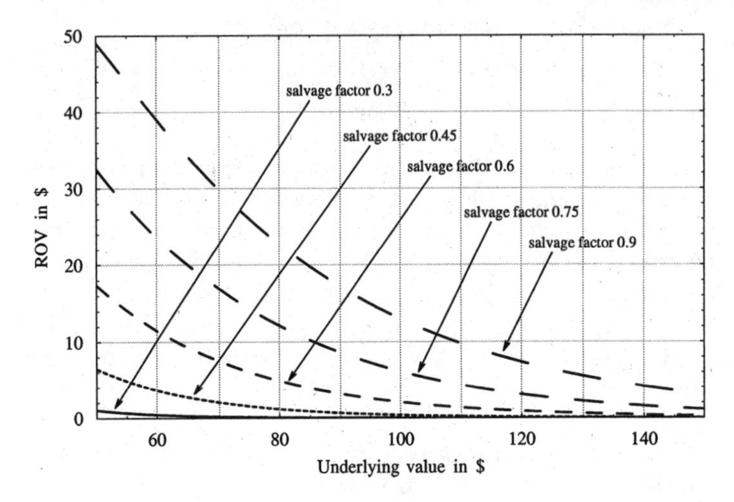

Fig. 5.11. ROV in case 1 with a variable salvage factor priced with the Cox-Ross-Rubinstein binomial tree method and using a constant risk-free interest rate (ts-3-1; parameters: $r_f = 0.03$, $S_{min} = 50$, $S_{max} = 150$, $M = 50$, $\sigma_S = 0.25$, $T_S = 3$, $N_S = 1080$).

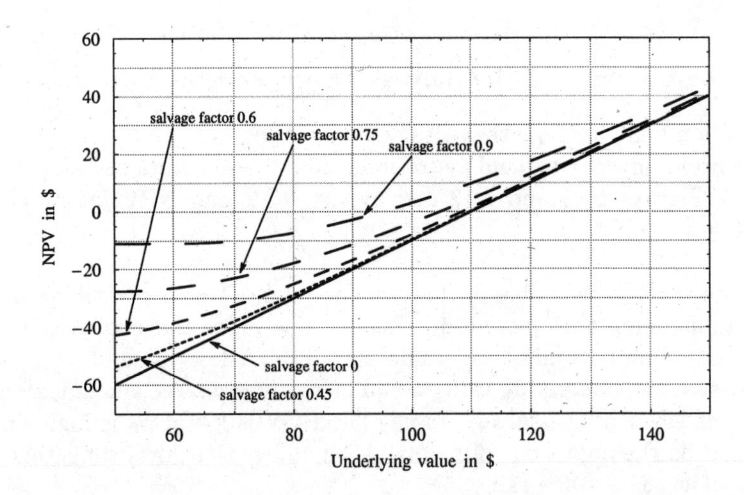

Fig. 5.12. NPV in case 1 with a variable salvage factor priced with the Cox-Ross-Rubinstein binomial tree method and using a constant risk-free interest rate (ts-3-1; parameters: $r_f = 0.03$, $S_{min} = 50$, $S_{max} = 150$, $M = 50$, $\sigma_S = 0.25$, $T_S = 3$, $N_S = 1080$).

Of all graphs in Figure 5.12 the graph that displays the project's NPV of $S_0 - I_0$ if no real options are present poses a lower boundary in general since the presence of real options can never reduce the project's NPV below $S_0 - I_0$. Moreover, for a larger salvage factor (e.g., 0.75 or 0.9), Figure 5.12 indicates another lower boundary which is $-(1 - \text{salvage factor})I_0$. This is the NPV for abandoning a project immediately after start when S_0 is far below the salvage value. This is clearly a boundary for lower values of S_0 in case of a larger salvage factor while $S_0 - I_0$ is a boundary in general and especially for higher values of S_0.

Figures 5.13 and 5.14 show the influence of the salvage factor for a higher risk-free interest rate of $r_f = 0.06$. Obviously, the influence of the increased risk-free rate on the project's ROV and NPV is less than the influence of the salvage factor (see Figures 5.13 and 5.11 for the ROV; see Figures 5.14 and 5.12 for the NPV).

Figures 5.15 and 5.16 display the ROV for real options cases 2 and 3 respectively with a given mean reversion level of the Cox-Ingersoll-Ross model of $\frac{\beta}{\alpha} = 0.03$. For case 2 the ROV graph is u-shaped whereby this u-shape is more conspicuous for higher expand factors. Economically, this makes sense since a higher expand factor has a higher influence (and results in a higher ROV) for higher underlying values S_0.

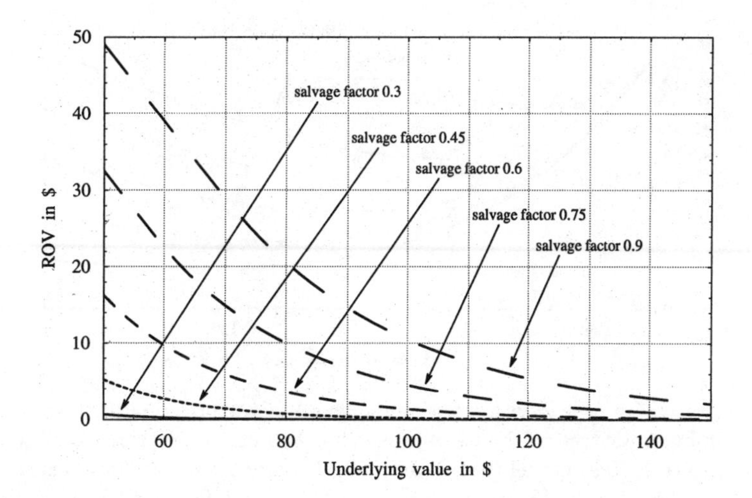

Fig. 5.13. ROV in case 1 with a variable salvage factor priced with the Cox-Ross-Rubinstein binomial tree method and using a constant risk-free interest rate (ts-3-1; parameters: $r_f = 0.06$, $S_{min} = 50$, $S_{max} = 150$, $M = 50$, $\sigma_S = 0.25$, $T_S = 3$, $N_S = 1080$).

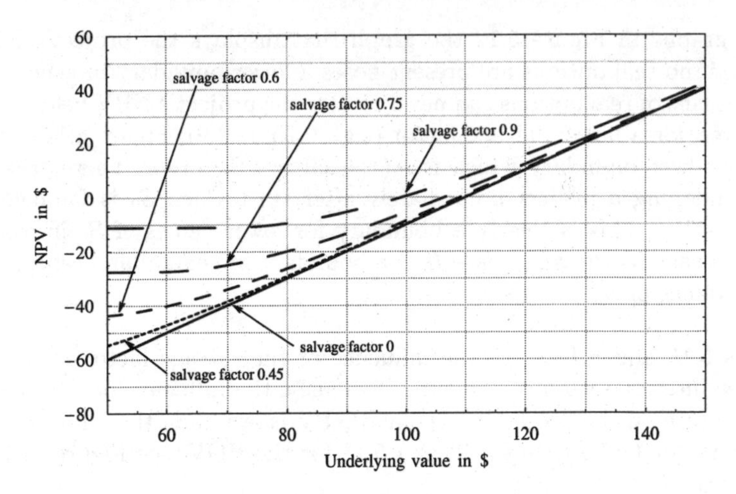

Fig. 5.14. NPV in case 1 with a variable salvage factor priced with the Cox-Ross-Rubinstein binomial tree method and using a constant risk-free interest rate (ts-3-1; parameters: $r_f = 0.06$, $S_{min} = 50$, $S_{max} = 150$, $M = 50$, $\sigma_S = 0.25$, $T_S = 3$, $N_S = 1080$).

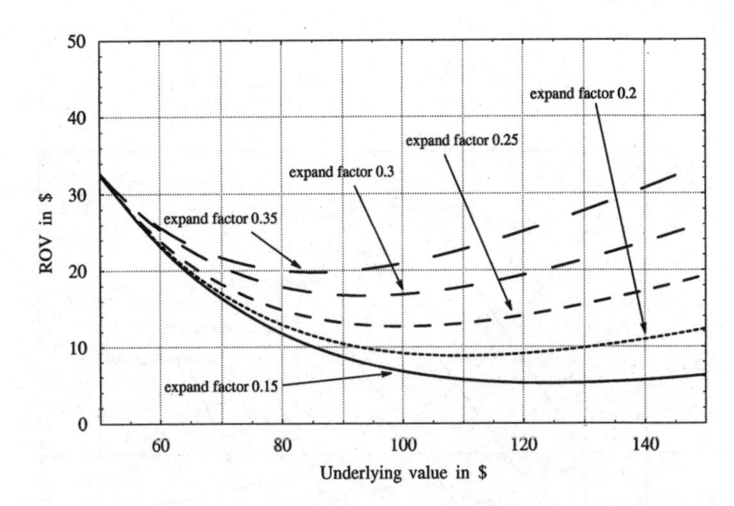

Fig. 5.15. ROV in case 2 with a variable expand factor priced with the Cox-Ross-Rubinstein binomial tree method and using a stochastic risk-free rate simulated with the Cox-Ingersoll-Ross model (ts-3-9 and ts-3-11; parameters for Cox-Ingersoll-Ross model: $\alpha = 0.6$, $\beta = 0.018$, $\sigma_r = 0.06$, $T_r = 4$, $N_r = 7200$, $r_0 = 0.03$, Taylor 1.5 simulation scheme for 100 simulated short-rate paths; parameters for Cox-Ross-Rubinstein binomial tree: $S_{min} = 50$, $S_{max} = 150$, $M = 50$, $\sigma_S = 0.25$, $T_S = 3$, $N_S = 1080$, salvage factor 0.75 of initial investment cost of 110, expand investment of 0.2 of initial investment cost).

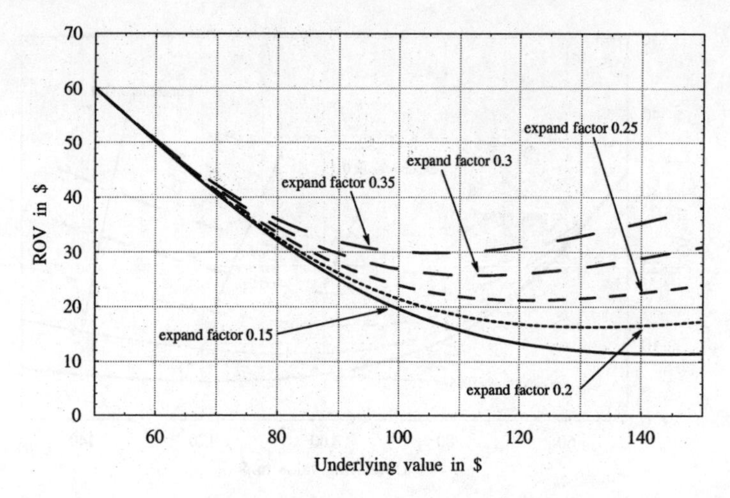

Fig. 5.16. ROV in case 3 with a variable expand factor priced with the Cox-Ross-Rubinstein binomial tree method and using a stochastic risk-free rate simulated with the Cox-Ingersoll-Ross model (ts-3-10 and ts-3-12; parameters for Cox-Ingersoll-Ross model: $\alpha = 0.6$, $\beta = 0.018$, $\sigma_r = 0.06$, $T_r = 4$, $N_r = 7200$, $r_0 = 0.03$, Taylor 1.5 simulation scheme for 100 simulated short-rate paths; parameters for Cox-Ross-Rubinstein binomial tree: $S_{min} = 50$, $S_{max} = 150$, $M = 50$, $\sigma_S = 0.25$, $T_S = 3$, $N_S = 300$, salvage factor 0.75 of initial investment cost of 110, expand investment of 0.2 of initial investment cost).

When adding an option to defer the project's start by one year, i.e., when considering case 3 versus case 2, the u-shape of the ROV remains. However, the ROV increases with $ROV = 60$ for $S_0 = 50$ since I_0 is 110 which gives $ROV = I_0 - S_0 = 60$ as the real options value for the lowest underlying value considered. This is an economically plausible result since an option to defer does not *require* to invest in one year but it *allows* to invest if in one year the NPV is positive. This means that the NPV has a lower boundary of zero since an investment will never be made for a negative NPV. On the other hand, the NPV is the sum of S_0 and the ROV minus the initial investment cost I_0. Both Figures 5.15 and 5.16 show that in case 2 and case 3 the influence of the expand factor is large. This implies that an accurate estimate of the expand factor is critical for an accurate valuation of an investment project with the real options approach.

Figures 5.17 and 5.18 display the influence of the expand factor on the real options value for case 2 and case 3, respectively, if the risk-free rate is stochastically simulated using the Cox-Ingersoll-Ross model with a mean reversion level of 0.06.

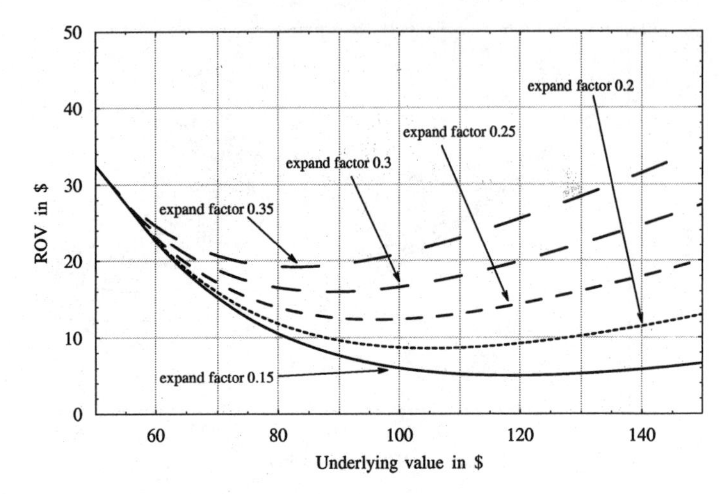

Fig. 5.17. ROV in case 2 with a variable expand factor priced with the Cox-Ross-Rubinstein binomial tree method and using a stochastic risk-free rate simulated with the Cox-Ingersoll-Ross model (ts-3-9 and ts-3-11; parameters for Cox-Ingersoll-Ross model: $\alpha = 0.6$, $\beta = 0.036$, $\sigma_r = 0.06$, $T_r = 4$, $N_r = 7200$, $r_0 = 0.06$, Taylor 1.5 simulation scheme for 100 simulated short-rate paths; parameters for Cox-Ross-Rubinstein binomial tree: $S_{min} = 50$, $S_{max} = 150$, $M = 50$, $\sigma_S = 0.25$, $T_S = 3$, $N_S = 1080$, salvage factor 0.75 of initial investment cost of 110, expand investment of 0.2 of initial investment cost).

The results are similar to those for a mean reversion level of 0.03 with only slight differences. This also means that the level of the risk-free interest rate is less important in a capital budgeting problem than the correct determination of the expand level (and, as seen above, of the salvage factor). It is important to mention that independent of the risk-free interest rate, the u-shape of the graphs in case 3 is less significant than in case 2. For example, the u-shape of the ROV graph in case 2 for an expand factor of 0.15 disappears completely for case 3. Only for higher values of the expand factor the u-shape can be observed for cases 2 and 3 likewise whereby the ROV in case 2 is much lower than in case 3 for the same S_0. In practice, this means that the option to defer is less important for higher values of S_0 and more important for lower values of S_0. Consequently, the ROV is much higher for lower values of S_0 in case 3 than in case 2.

Since the ROV graphs of all three cases are often shown in the section below, it is worth giving a brief summary here. The option to abandon (case 1) is important mainly for low values of S_0. The option to expand an investment is important only for higher values of S_0. Therefore, in case 3 the ROV is determined by the option to defer for low values of S_0, whereas it is dominated by the option to expand for high values of S_0. The option to defer the investment

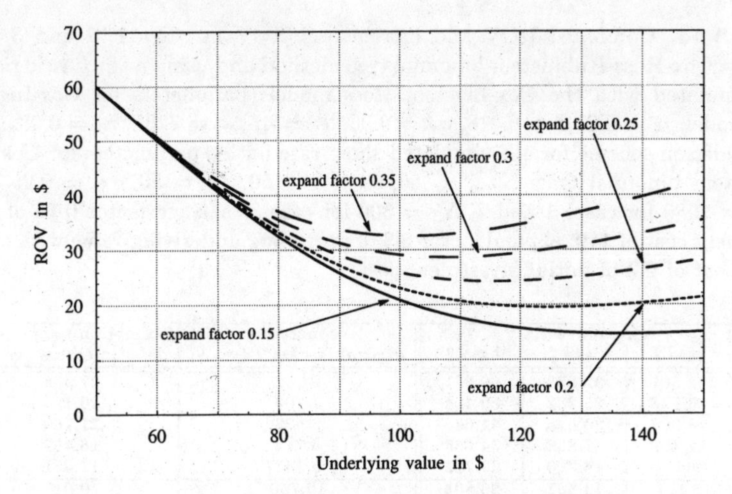

Fig. 5.18. ROV in case 3 with a variable expand factor priced with the Cox-Ross-Rubinstein binomial tree method and using a stochastic risk-free rate simulated with the Cox-Ingersoll-Ross model (ts-3-10 and ts-3-12; parameters for Cox-Ingersoll-Ross model: $\alpha = 0.6$, $\beta = 0.036$, $\sigma_r = 0.06$, $T_r = 4$, $N_r = 7200$, $r_0 = 0.06$, Taylor 1.5 simulation scheme for 100 simulated short-rate paths; parameters for Cox-Ross-Rubinstein binomial tree: $S_{min} = 50$, $S_{max} = 150$, $M = 50$, $\sigma_S = 0.25$, $T_S = 3$, $N_S = 300$, salvage factor 0.75 of initial investment cost of 110, expand investment of 0.2 of initial investment cost).

is significant for low values of S_0 but is of little importance for high values of S_0. Consequently, the option to defer shifts the ROV graph in case 2 upwards (compared to case 3) by a large amount for low values of S_0 but only by a small amount for high values of S_0. The combined values of the real options, their isolated values and their incremental values are shown in Table 5.44 and graphically displayed in Figures 5.19 and 5.20.

As already elaborated in Chapter 4, the additivity of real options values in a complex real option is usually not true. This is the case when the exercise ranges of two real options intersect[18]. Therefore, the incremental value of an option to expand in case 2 is not necessarily the value of the option to expand if no other real option were present. The same holds true for the incremental value of the option to defer in case 3. However, it is not the goal of this thesis to compare isolated (simple) real options values with bundled (complex) real options. Therefore, the simple real options value of expanding an investment and the simple real options value of deferring the investment are not investigated.

[18] See Trigeorgis [132], Chapter 7.

Table 5.44. Combined ROV and incremental ROV in cases 1, 2 and 3 priced with the Cox-Ross-Rubinstein binomial tree method and using a stochastic risk-free rate simulated with the Cox-Ingersoll-Ross model (parameters for Cox-Ingersoll-Ross model: $\alpha = 0.6$, $\beta = 0.018$, $\sigma_r = 0.06$, $T_r = 4$, $N_r = 7200$, $r_0 = 0.03$, Taylor 1.5 simulation scheme for 100 simulated short-rate paths; parameters for Cox-Ross-Rubinstein binomial tree: $S_{min} = 50$, $S_{max} = 150$, $M = 50$, $\sigma_S = 0.25$, $T_S = 3$, $N_S = 1080$ for cases 1 and 2, $N_S = 300$ for case 3, salvage factor 0.75 of initial investment cost of 110, expand factor 0.3 of prevailing underlying value with expand investment of 0.2 of initial investment cost).

S_0	Combined ROV			Incremental ROV of option to	
	case 1	case 2	case 3	expand in case 2 (vs. 1)	defer in case 3 (vs. 2)
50	32.501	32.683	60.051	0.182	27.368
60	23.578	25.282	50.404	1.703	25.123
70	16.952	20.525	41.694	3.573	21.169
80	12.109	17.852	34.635	5.742	16.783
90	8.616	16.770	29.721	8.155	12.951
100	6.117	16.867	26.884	10.750	10.017
110	4.340	17.814	25.785	13.474	7.970
120	3.084	19.373	25.982	16.289	6.609
130	2.196	21.362	27.150	19.165	5.789
140	1.567	23.649	28.884	22.082	5.235
150	1.124	26.152	31.045	25.028	4.893

Source: ts-3-8, ts-3-9, and ts-3-10

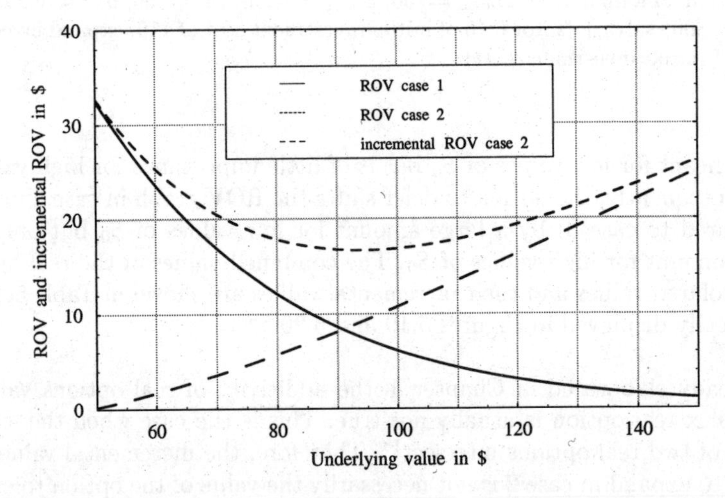

Fig. 5.19. Combined ROV and incremental ROV in cases 1 and 2 priced with the Cox-Ross-Rubinstein binomial tree method and using a stochastic risk-free rate simulated with the Cox-Ingersoll-Ross model (ts-3-8 and ts-3-9; corresponding to Table 5.44; parameters for Cox-Ingersoll-Ross model: $\alpha = 0.6$, $\beta = 0.018$, $\sigma_r = 0.06$, $T_r = 4$, $N_r = 7200$, $r_0 = 0.03$, Taylor 1.5 simulation scheme for 100 simulated short-rate paths; parameters for Cox-Ross-Rubinstein binomial tree: $S_{min} = 50$, $S_{max} = 150$, $M = 50$, $\sigma_S = 0.25$, $T_S = 3$, $N_S = 1080$, salvage factor 0.75 of initial investment cost of 110, expand factor 0.3 of prevailing underlying value with expand investment of 0.2 of initial investment cost).

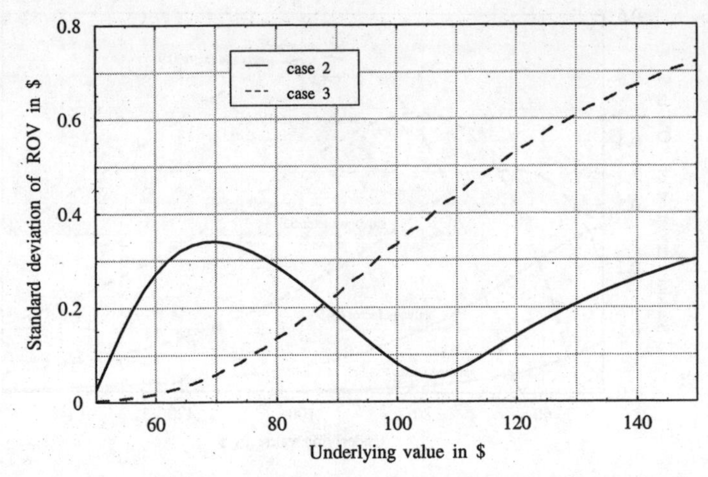

Fig. 5.20. Combined ROV and incremental ROV in cases 2 and 3 priced with the Cox-Ross-Rubinstein binomial tree method and using a stochastic risk-free rate simulated with the Cox-Ingersoll-Ross model (ts-3-9 and ts-3-10; corresponding to Table 5.44; parameters for Cox-Ingersoll-Ross model: $\alpha = 0.6$, $\beta = 0.018$, $\sigma_r = 0.06$, $T_r = 4$, $N_r = 7200$, $r_0 = 0.03$, Taylor 1.5 simulation scheme for 100 simulated short-rate paths; parameters for Cox-Ross-Rubinstein binomial tree: $S_{min} = 50$, $S_{max} = 150$, $M = 50$, $\sigma_S = 0.25$, $T_S = 3$, $N_S = 1080$ for case 2, $N_S = 300$ for case 3, salvage factor 0.75 of initial investment cost of 110, expand factor 0.3 of prevailing underlying value with expand investment of 0.2 of initial investment cost).

Finally, the standard deviation of the ROV (and, therefore, the TPV and the NPV) needs to be investigated. This will be done for a stochastic interest rate simulated with the Cox-Ingersoll-Ross model. In the following, the three cases will be analyzed separately at first. Then, cases 1 and 2 as well as cases 2 and 3 are combined in one graph to see how the addition of a new real option and, therefore, the incremental ROV determines the shape of the standard deviation graph. Figure 5.21 shows the standard deviation for case 1 with various salvage factors. Of course, the lower the salvage factor, the lower the standard deviation, irrespective of S_0. If, however, the salvage factor increases, the graph of the standard deviation rises immediately and peaks if the option is at-the-money (strike is the salvage factor times 110, the initial investment cost).

Figures 5.22 and 5.23 show the standard deviation for case 2 with a constant salvage factor of 0.75 and various expand factors. While the first figure displays the graphs corresponding to the expand factors 0.01, 0.05, 0.1, and 0.15, the second figure displays the graphs corresponding to the expand factors 0.2, 0.25, 0.3, and 0.35.

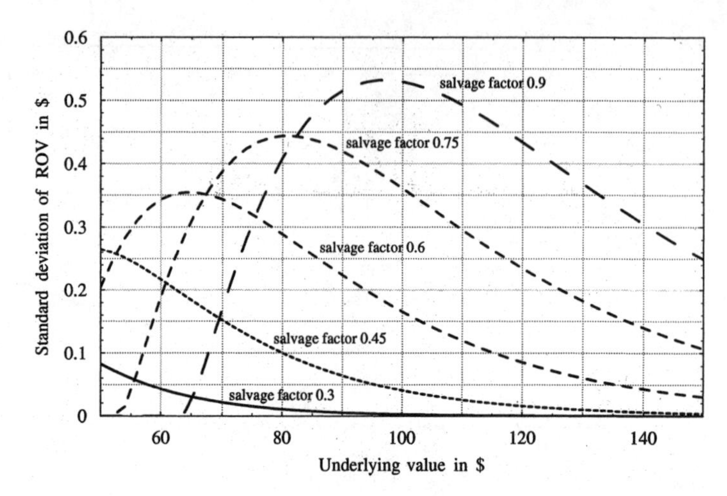

Fig. 5.21. Standard deviation of the ROV for case 1 and a variable salvage factor using a stochastic risk-free interest rate simulated with the Cox-Ingersoll-Ross model (ts-3-13; parameters for Cox-Ingersoll-Ross model: $\alpha = 0.6$, $\beta = 0.024$, $T_r = 4$, $N_r = 7200$, $r_0 = 0.04$, $\sigma_r = 0.06$, Taylor 1.5 simulation scheme for 100 simulated short-rate paths; parameters for Cox-Ross-Rubinstein binomial tree: $S_{min} = 50$, $S_{max} = 150$, $M = 50$, $\sigma_S = 0.25$, $T_S = 3$, $N_S = 1080$; initial investment cost of 110).

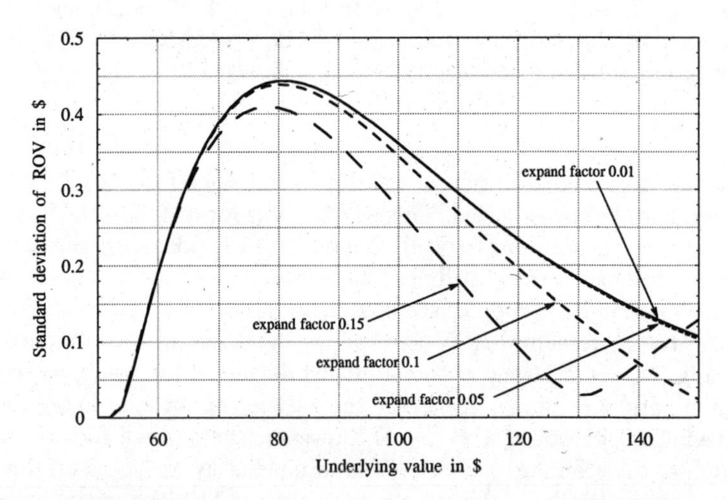

Fig. 5.22. Standard deviation of ROV for case 2 and a variable expand factor using a stochastic risk-free interest rate simulated with the Cox-Ingersoll-Ross model (diagram 1; ts-3-9 and ts-3-11; parameters for Cox-Ingersoll-Ross model: $\alpha = 0.6$, $\beta = 0.024$, $T_r = 4$, $N_r = 7200$, $r_0 = 0.04$, $\sigma_r = 0.06$, Taylor 1.5 simulation scheme for 100 simulated short-rate paths; parameters for Cox-Ross-Rubinstein binomial tree: $S_{min} = 50$, $S_{max} = 150$, $M = 50$, $\sigma_S = 0.25$, $T_S = 3$, $N_S = 1080$, salvage factor 0.75 of initial investment cost of 110, expand investment of 0.2 of initial investment cost).

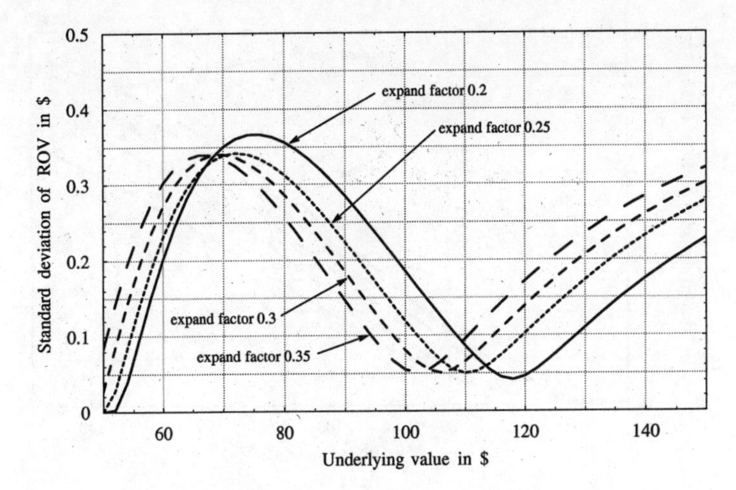

Fig. 5.23. Standard deviation of ROV for case 2 and a variable expand factor using a stochastic risk-free interest rate simulated with the Cox-Ingersoll-Ross model (diagram 2; ts-3-9 and ts-3-11; parameters for Cox-Ingersoll-Ross model: $\alpha = 0.6$, $\beta = 0.024$, $T_r = 4$, $N_r = 7200$, $r_0 = 0.04$, $\sigma_r = 0.06$, Taylor 1.5 simulation scheme for 100 simulated short-rate paths; parameters for Cox-Ross-Rubinstein binomial tree: $S_{min} = 50$, $S_{max} = 150$, $M = 50$, $\sigma_S = 0.25$, $T_S = 3$, $N_S = 1080$, salvage factor 0.75 of initial investment cost of 110, expand investment of 0.2 of initial investment cost).

The resulting graph of the standard deviation is very unusual and unexpected. The graph is devided into two areas. The first area displays a peak for low values of S_0 whereby the peak itself depends on the expand factor: the higher the expand factor, the lower the peak, and the earlier the peak is reached. But, the height of the peak does not change significantly for an expand factor above 0.25 with the result that the peak only moves to the left for higher expand factors. After the peak the graph reaches a local minimum whose value depends on the expand factor: the lower the expand factor, the lower the minimal value. Again, this minimum moves to the left for higher expand factors but does not change in height any more for an expand factor beyond 0.25. However, at the right of the minimum, the height of the graph depends on the expand factor again: the higher the expand factor, the higher the graph.

This phenomenom can also be observed for higher values of S_0 and higher expand factors in case 3. Figure 5.24 shows the standard deviation for case 3 with a constant salvage factor of 0.75 and various expand factors. However, the standard deviation for case 3 increases monotonically whereby the graph moves upwards for higher expand factors just like in case 2 for higher values of S_0.

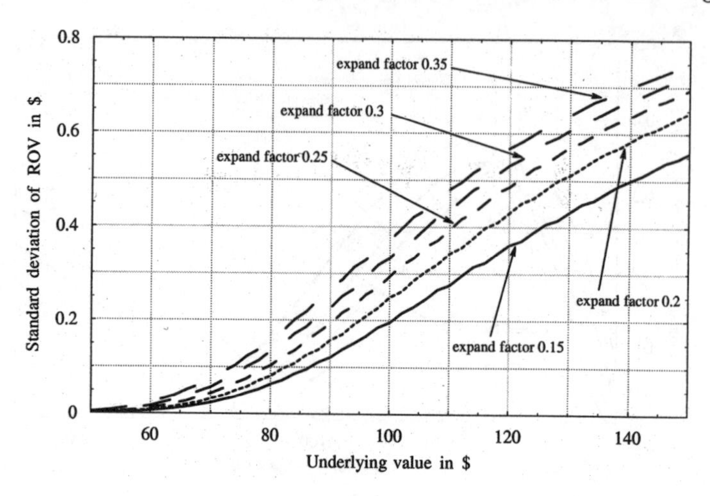

Fig. 5.24. Standard deviation of ROV for case 3 and a variable expand factor using a stochastic risk-free interest rate simulated with the Cox-Ingersoll-Ross model (ts-3-10 and ts-3-12; parameters for Cox-Ingersoll-Ross model: $\alpha = 0.6$, $\beta = 0.024$, $T_r = 4$, $N_r = 7200$, $r_0 = 0.04$, $\sigma_r = 0.06$, Taylor 1.5 simulation scheme for 100 simulated short-rate paths; parameters for Cox-Ross-Rubinstein binomial tree: $S_{min} = 50$, $S_{max} = 150$, $M = 50$, $\sigma_S = 0.25$, $T_S = 3$, $N_S = 300$, salvage factor 0.75 of initial investment cost of 110, expand investment of 0.2 of initial investment cost).

The question that needs to be answered now is: Why are the shapes of the graphs so different for the three cases? The answer to this question is given by the incremental ROV when adding a new real option to an investment project. For case 1 the start scenario is an investment project without any options. This, of course, means that the standard deviation of the ROV is zero. When adding an option to abandon to the project, the project's ROV changes. Since such an option is more valuable for lower S_0 values, the ROV changes considerably for lower values of S_0 but not for higher values of S_0 where such an option does not pose much value. This results in a standard deviation that varies significantly for lower values of S_0 but not for higher values of S_0. Figures 5.25 and 5.26 display graphically case 1 versus a situation without real options.

The difference between case 1 and case 2 consists in the option to expand the investment. This is graphically displayed in Figures 5.27 and 5.28. Since such an option is more valuable for higher underlying values of S_0, the incremental ROV of this option to expand is higher for higher values of S_0, influencing the shape of the standard deviation graph especially for these higher values.

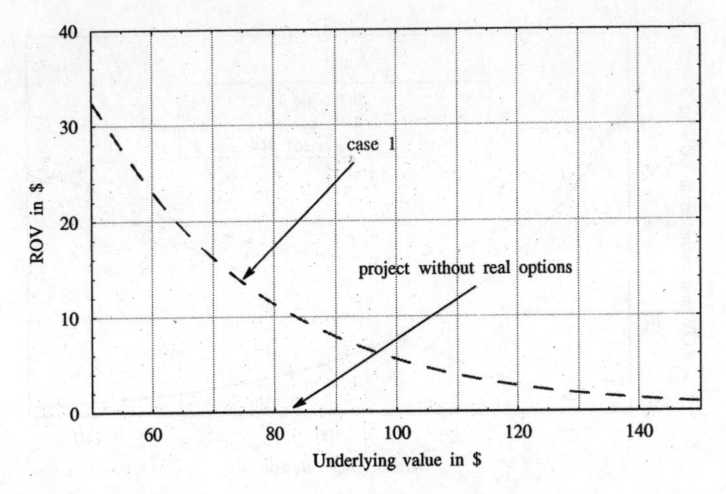

Fig. 5.25. ROV and incremental ROV for case 1 using a stochastic risk-free interest rate simulated with the Cox-Ingersoll-Ross model (ts-3-8; parameters for Cox-Ingersoll-Ross model: $\alpha = 0.6$, $\beta = 0.024$, $T_r = 4$, $N_r = 7200$, $r_0 = 0.04$, $\sigma_r = 0.06$, Taylor 1.5 simulation scheme for 100 simulated short-rate paths; parameters for Cox-Ross-Rubinstein binomial tree: $S_{min} = 50$, $S_{max} = 150$, $M = 50$, $\sigma_S = 0.25$, $T_S = 3$, $N_S = 1080$, salvage factor 0.75 of initial investment cost of 110).

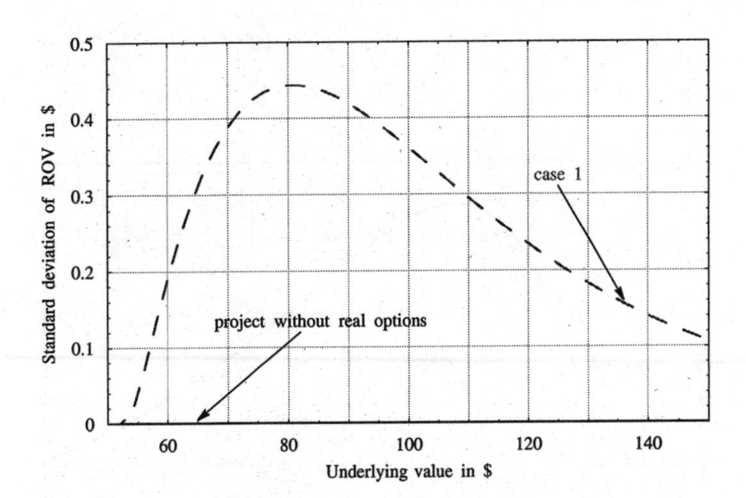

Fig. 5.26. Standard deviation of ROV for case 1 using a stochastic risk-free interest rate simulated with the Cox-Ingersoll-Ross model (ts-3-8; parameters for Cox-Ingersoll-Ross model: $\alpha = 0.6$, $\beta = 0.024$, $T_r = 4$, $N_r = 7200$, $r_0 = 0.04$, $\sigma_r = 0.06$, Taylor 1.5 simulation scheme for 100 simulated short-rate paths; parameters for Cox-Ross-Rubinstein binomial tree: $S_{min} = 50$, $S_{max} = 150$, $M = 50$, $\sigma_S = 0.25$, $T_S = 3$, $N_S = 1080$, salvage factor 0.75 of initial investment cost of 110).

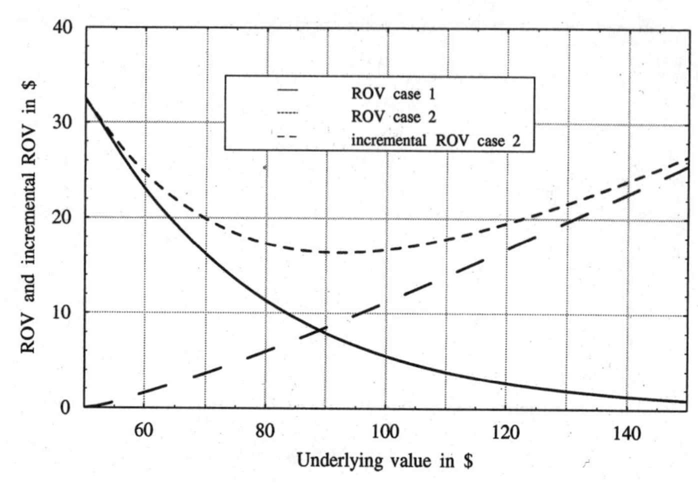

Fig. 5.27. ROV for cases 1 and 2 and incremental ROV for case 2 using a stochastic risk-free interest rate simulated with the Cox-Ingersoll-Ross model (ts-3-8 and ts-3-9; parameters for Cox-Ingersoll-Ross model: $\alpha = 0.6$, $\beta = 0.024$, $T_r = 4$, $N_r = 7200$, $r_0 = 0.04$, $\sigma_r = 0.06$, Taylor 1.5 simulation scheme for 100 simulated short-rate paths; parameters for Cox-Ross-Rubinstein binomial tree: $S_{min} = 50$, $S_{max} = 150$, $M = 50$, $\sigma_S = 0.25$, $T_S = 3$, $N_S = 1080$, salvage factor 0.75 of initial investment cost of 110, expand factor 0.3 of prevailing underlying value with expand investment of 0.2 of initial investment cost).

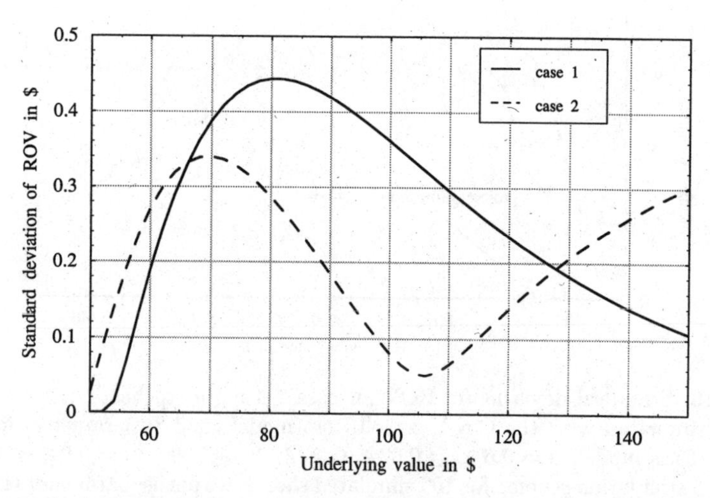

Fig. 5.28. Standard deviation of ROV for cases 1 and 2 using a stochastic risk-free interest rate simulated with the Cox-Ingersoll-Ross model (ts-3-8 and ts-3-9; corresponding to Figure 5.27).

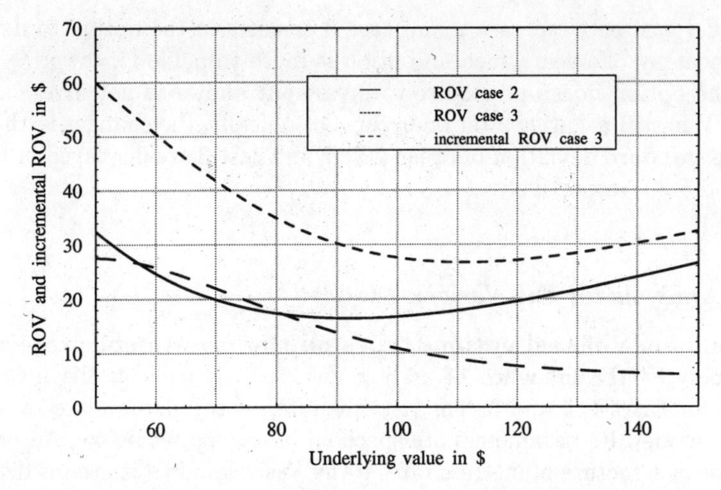

Fig. 5.29. ROV for cases 2 and 3 and incremental ROV for case 3 using a stochastic risk-free interest rate simulated with the Cox-Ingersoll-Ross model (ts-3-9 and ts-3-10; parameters for Cox-Ingersoll-Ross model: $\alpha = 0.6$, $\beta = 0.024$, $T_r = 4$, $N_r = 7200$, $r_0 = 0.04$, $\sigma_r = 0.06$, Taylor 1.5 simulation scheme for 100 simulated short-rate paths; parameters for Cox-Ross-Rubinstein binomial tree: $S_{min} = 50$, $S_{max} = 150$, $M = 50$, $\sigma_S = 0.25$, $T_S = 3$, $N_S = 1080$ for case 2, $N_S = 300$ for case 3, salvage factor 0.75 of initial investment cost of 110, expand factor 0.3 of prevailing underlying value with expand investment of 0.2 of initial investment cost).

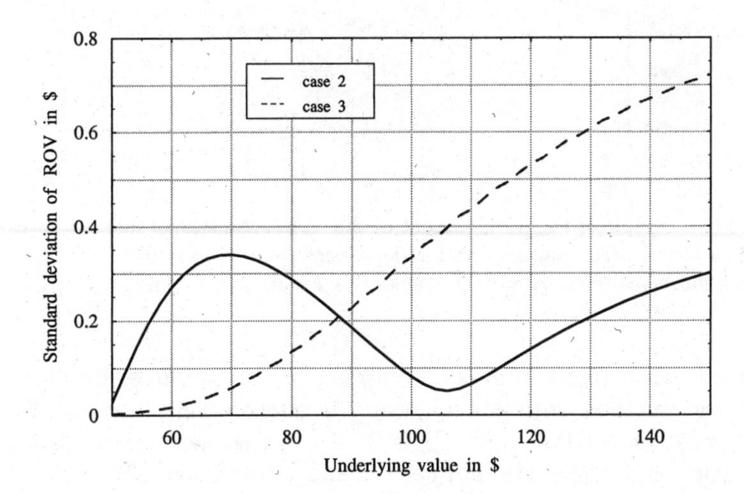

Fig. 5.30. Standard deviation of ROV for cases 2 and 3 using a stochastic risk-free interest rate simulated with the Cox-Ingersoll-Ross model (ts-3-9 and ts-3-10; corresponding to Figure 5.29).

The difference between case 2 and case 3 consists in the option to defer the investment by one year. Such an option is more valuable for lower S_0 values since the option does not require to invest but allows to not invest at all if the NPV is still negative after one year. Graphically, the changes in the ROV and the standard deviation between case 2 and case 3 are displayed in Figures 5.29 and 5.30, respectively.

5.6.2 Analysis for the Vasicek Model

The influence of real options types on the investment project. This part analyzes the influence of adding new real options on the investment project for cases 1, 2 and 3. The risk-free interest rate is simulated using the Vasicek model. Its parameters are specified in the figures below. An example of the term structure of interest rates if the Vasicek model is used is displayed in Figure 5.31.

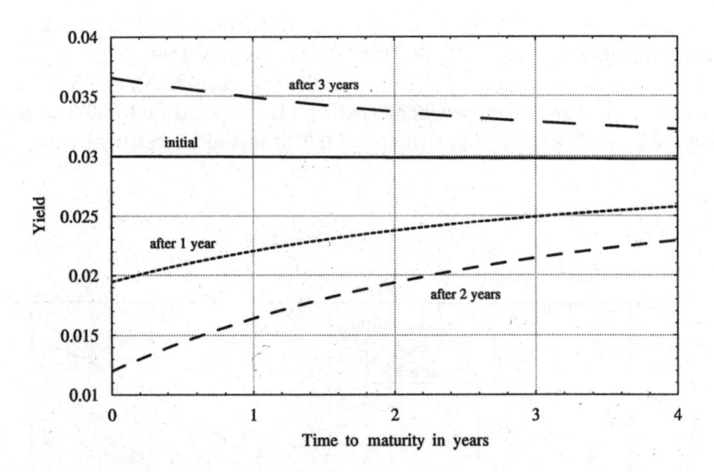

Fig. 5.31. Example of the development of the term structure of interest rates if the Vasicek model is used (parameters for the Vasicek model: $\alpha = 0.6$, $\beta = 0.018$, $T_r = 4$, $N_r = 7200$, $r_0 = 0.03$, $\sigma_r = 0.02$, Taylor 1.5 simulation scheme).

The investment project is priced according to the Cox-Ross-Rubinstein binomial tree method with the following parameters: $S_{min} = 50$, $S_{max} = 150$, $M = 50$, $\sigma_S = 0.25$, and $T_S = 3$; the salvage factor is 0.75 of initial investment cost of $I_0 = 110$. For cases 2 and 3 the expand factor is 0.3 with an expand investment of 0.2 of the initial investment cost I_0. $N_S = 1080$ for case 1 and case 2; $N_S = 300$ for case 3. In case 3 the defer period is exactly one year.

Figure 5.32 shows the ROV for this choice of parameters and for all three cases 1, 2 and 3. Figure 5.33 and Figure 5.34 display the corresponding TPV and NPV, respectively. The standard deviation is displayed in Figure 5.35. Unlike the terms ROV and NPV, which have already been analyzed in the previous section, the term TPV has only shortly been presented. The TPV is the NPV plus the initial investment cost I_0. As the shape of the ROV and NPV graphs are similar to those presented in the previous section, they will not be analyzed in this section. The graph of the TPV is just a linear translation of the NPV graph by I_0, so it does not need to be further explained either.

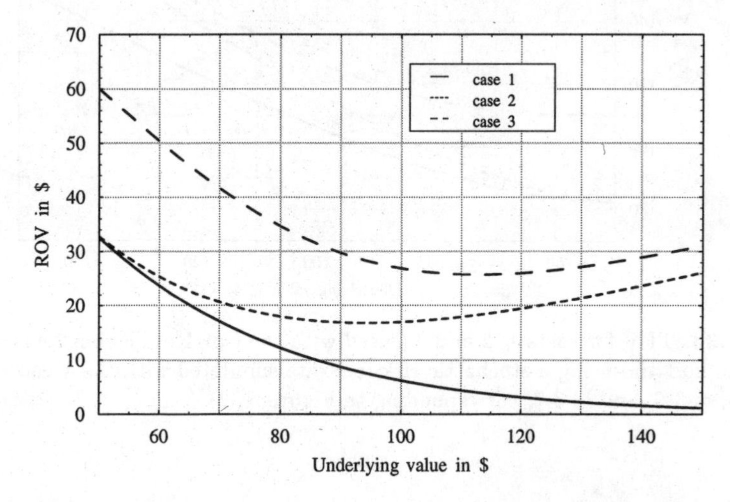

Fig. 5.32. ROV for cases 1, 2 and 3 priced with the Cox-Ross-Rubinstein binomial tree method and using a stochastic risk-free rate simulated with the Vasicek model (ts-3-5, ts-3-6, and ts-3-7; parameters for the Vasicek model: $\alpha = 0.6$, $\beta = 0.018$, $T_r = 4$, $N_r = 7200$, $r_0 = 0.03$, $\sigma_r = 0.02$, Taylor 1.5 simulation scheme for 100 simulated short-rate paths; parameters for the Cox-Ross-Rubinstein binomial tree: $S_{min} = 50$, $S_{max} = 150$, $M = 50$, $\sigma_S = 0.25$, $T_S = 3$, $N_S = 1080$ for case 1 and case 2, $N_S = 300$ for case 3, salvage factor 0.75 of initial investment cost of 110, expand factor 0.3 of prevailing underlying value with expand investment of 0.2 of initial investment cost).

The development of the ROV, which was already analyzed for cases 1, 2 and 3 in the previous section, is the basis for the development of the NPV which is displayed in Figure 5.34. There are two lower boundaries for every case. One boundary is $-0.25 \cdot I_0 = -27.5$ as the maximum loss in case the investment is undertaken but S_0 is far below the salvage value. The other boundary is $S_0 - I_0$ for each S_0, which is the NPV if the project has no embedded real options at all. In case 1 the NPV graph approaches these boundaries asymptotically. In case 2 the NPV is much higher for higher values of S_0 for which an expansion of the project has the highest impact. The additional option to defer in case 3 shifts the NPV graph upwards for the whole range of S_0 values, but this shift is much stronger for lower values of S_0 for which an option to defer is most valuable.

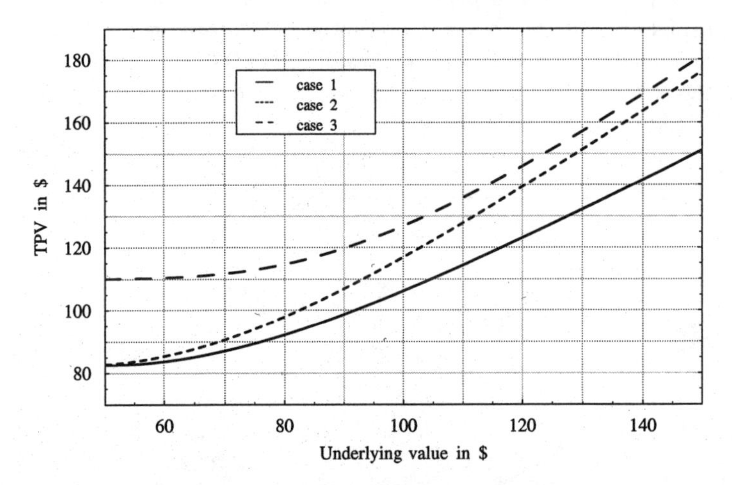

Fig. 5.33. TPV for cases 1, 2 and 3 priced with the Cox-Ross-Rubinstein binomial tree method and using a stochastic risk-free rate simulated with the Vasicek model (ts-3-5, ts-3-6, and ts-3-7; corresponding to Figure 5.32).

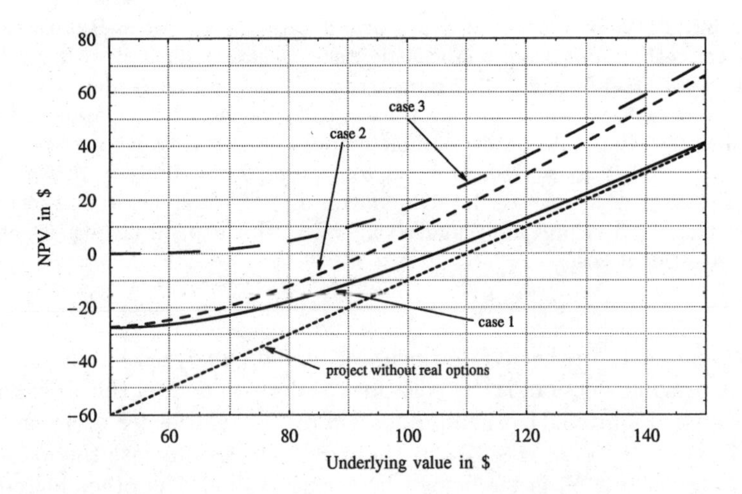

Fig. 5.34. NPV for cases 1, 2 and 3 priced with the Cox-Ross-Rubinstein binomial tree method and using a stochastic risk-free rate simulated with the Vasicek model (ts-3-5, ts-3-6, and ts-3-7; corresponding to Figure 5.32).

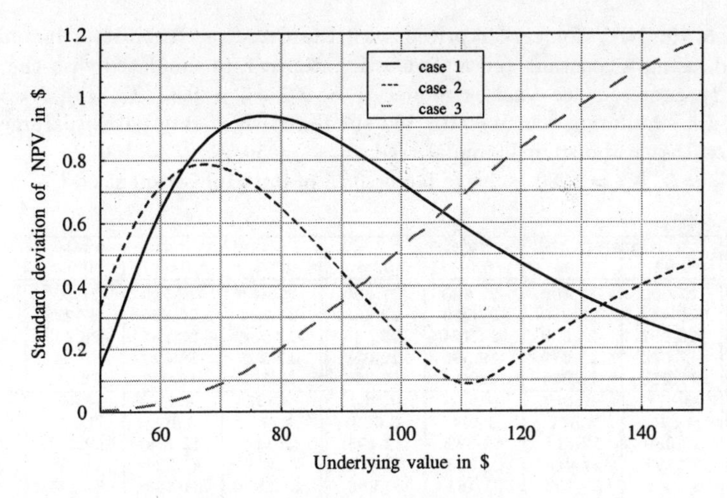

Fig. 5.35. Standard deviation of NPV for cases 1, 2 and 3 priced with the Cox-Ross-Rubinstein binomial tree method and using a stochastic risk-free rate simulated with the Vasicek model (ts-3-5, ts-3-6, and ts-3-7; corresponding to Figure 5.32).

The influence of the short-rate mean reversion level on the investment project. This part analyzes the influence of the mean reversion level on the ROV, the NPV, and the NPV's standard deviation. The risk-free interest rate is simulated using the Vasicek model. All parameters are specified in the tables below. In case 3 the defer period is exactly one year.

Table 5.45. ROV for case 1 priced with the Cox-Ross-Rubinstein binomial tree method, using a constant and a stochastic interest rate simulated with the Vasicek model (parameters for Vasicek model: $\alpha = 0.6$, $\sigma_r = 0.02$, $T_r = 4$, $N_r = 7200$, Taylor 1.5 simulation scheme with 100 simulated short-rate paths; parameters for the Cox-Ross-Rubinstein binomial tree: $S_{min} = 50$, $S_{max} = 150$, $M = 50$, $\sigma_S = 0.25$, $T_S = 3$, $N_S = 1080$, salvage factor 0.75 of initial investment cost of 110).

	Constant risk-free rate: $r_f =$				Stochastic risk-free rate: $\beta/\alpha =$			
S_0	0.03	0.04	0.05	0.06	0.03	0.04	0.05	0.06
50	32.500	32.500	32.500	32.500	32.549	32.500	32.500	32.500
60	23.558	23.050	22.717	22.537	23.706	23.136	22.771	22.573
70	16.943	16.170	15.506	14.934	17.062	16.249	15.563	14.978
80	12.110	11.321	10.615	9.982	12.194	11.378	10.659	10.016
90	8.620	7.916	7.283	6.711	8.679	7.958	7.315	6.737
100	6.123	5.535	5.009	4.536	6.165	5.566	5.033	4.556
110	4.345	3.874	3.454	3.080	4.377	3.898	3.473	3.096
120	3.089	2.718	2.392	2.103	3.114	2.738	2.407	2.116
130	2.199	1.913	1.663	1.444	2.219	1.929	1.676	1.455
140	1.569	1.350	1.161	0.996	1.585	1.363	1.171	1.005
150	1.126	0.959	0.815	0.692	1.138	0.969	0.824	0.700

Source: ts-3-2 and ts-3-5

Table 5.46. NPV for case 1 priced with the Cox-Ross-Rubinstein binomial tree method, using a constant and a stochastic interest rate simulated with the Vasicek model (parameters for Vasicek model: $\alpha = 0.6$, $\sigma_r = 0.02$, $T_r = 4$, $N_r = 7200$, Taylor 1.5 simulation scheme with 100 simulated short-rate paths; parameters for the Cox-Ross-Rubinstein binomial tree: $S_{min} = 50$, $S_{max} = 150$, $M = 50$, $\sigma_S = 0.25$, $T_S = 3$, $N_S = 1080$, salvage factor 0.75 of initial investment cost of 110).

	Constant risk-free rate: $r_f =$				Stochastic risk-free rate: $\beta/\alpha =$			
S_0	0.03	0.04	0.05	0.06	0.03	0.04	0.05	0.06
50	-27.500	-27.500	-27.500	-27.500	-27.451	-27.500	-27.500	-27.500
60	-26.442	-26.950	-27.283	-27.463	-26.294	-26.864	-27.229	-27.427
70	-23.057	-23.830	-24.494	-25.066	-22.938	-23.751	-24.437	-25.022
80	-17.890	-18.679	-19.385	-20.018	-17.806	-18.622	-19.341	-19.984
90	-11.380	-12.084	-12.717	-13.289	-11.321	-12.042	-12.685	-13.263
100	-3.877	-4.465	-4.991	-5.464	-3.835	-4.434	-4.967	-5.444
110	4.345	3.874	3.454	3.080	4.377	3.898	3.473	3.096
120	13.089	12.718	12.392	12.103	13.114	12.738	12.407	12.116
130	22.199	21.913	21.663	21.444	22.219	21.929	21.676	21.455
140	31.569	31.350	31.161	30.996	31.585	31.363	31.171	31.005
150	41.126	40.959	40.815	40.692	41.138	40.969	40.824	40.700

Source: ts-3-2 and ts-3-5

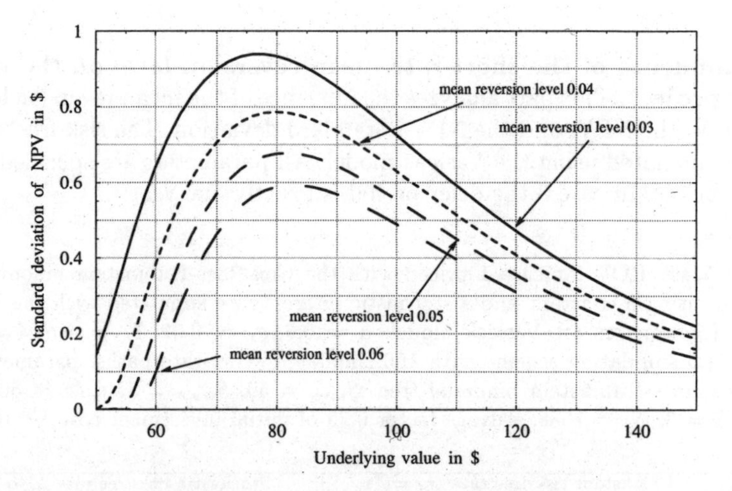

Fig. 5.36. Standard deviation of NPV for case 1 priced with the Cox-Ross-Rubinstein binomial tree method and using a stochastic risk-free rate simulated with the Vasicek model (ts-3-5; corresponding to Tables 5.45 and 5.46).

Tables 5.45 and 5.46 display the ROV and the NPV for case 1, respectively, using a constant and a stochastically modelled risk-free rate. The corresponding standard deviation of the NPV (and, therefore, the ROV) is displayed in Figure 5.36.

One of the most important results of this thesis is given in Table 5.46: There is no significant difference between the NPV (as the mean for 100 simulated short-rate paths) priced by using a simulated risk-free rate according to the Cox-Ingersoll-Ross model and the NPV priced by taking the mean reversion level of the Cox-Ingersoll-Ross model as the constant risk-free interest rate in the Cox-Ross-Rubinstein binomial tree algorithm. This result is obtained for every stochastic interest rate model and every real options case analyzed. The analysis of cases 2 and 3 below supports this finding.

Another important point derived from case 1 is that the NPV (for any chosen S_0) decreases with an increasing risk-free rate (or mean reversion level for a stochastic term structure model). This seems to be intuitively correct. However, the following analysis of cases 2 and 3 will show that this intuition is misleading if more options are added to the investment project. Moreover, Figure 5.36 reveals: the lower the mean reversion level, the higher the standard deviation of ROV and NPV. This makes sense since the influence of relative stochastic changes in the interest rate on the NPV is stronger for a low interest rate level than for a higher one. The peak of all graphs is hereby attained when the option to abandon is at- or in-the-money.

Tables 5.47 and 5.48 display the ROV and the NPV for case 2, respectively, using a constant and a stochastically modelled risk-free rate. The corresponding standard deviation of the NPV (and, therefore, the ROV) is displayed in Figure 5.37.

Table 5.47. ROV for case 2 priced with the Cox-Ross-Rubinstein binomial tree method, using a constant and a stochastic interest rate simulated with the Vasicek model (parameters for Vasicek model: $\alpha = 0.6$, $\sigma_r = 0.02$, $T_r = 4$, $N_r = 7200$, Taylor 1.5 simulation scheme with 100 simulated short-rate paths; parameters for the Cox-Ross-Rubinstein binomial tree: $S_{min} = 50$, $S_{max} = 150$, $M = 50$, $\sigma_S = 0.25$, $T_S = 3$, $N_S = 1080$, salvage factor 0.75 of initial investment cost of 110, expand factor 0.3 of prevailing underlying value with expand investment of 0.2 of initial investment cost).

	Constant risk-free rate: $r_f =$				Stochastic risk-free rate: $\beta/\alpha =$			
S_0	0.03	0.04	0.05	0.06	0.03	0.04	0.05	0.06
50	32.665	32.507	32.500	32.500	32.782	32.550	32.502	32.500
60	25.266	24.642	24.133	23.720	25.399	24.723	24.189	23.759
70	20.516	19.850	19.274	18.773	20.621	19.922	19.328	18.816
80	17.845	17.312	16.857	16.470	17.928	17.373	16.906	16.512
90	16.763	16.417	16.141	15.925	16.831	16.471	16.186	15.965
100	16.856	16.699	16.600	16.551	16.915	16.748	16.643	16.590
110	17.801	17.809	17.864	17.960	17.853	17.854	17.905	17.997
120	19.357	19.502	19.683	19.894	19.405	19.544	19.721	19.929
130	21.343	21.596	21.875	22.176	21.387	21.635	21.911	22.209
140	23.630	23.967	24.321	24.690	23.670	24.003	24.355	24.721
150	26.131	26.532	26.943	27.361	26.168	26.565	26.974	27.391

Source: ts-3-3 and ts-3-6

Table 5.48. NPV for case 2 priced with the Cox-Ross-Rubinstein binomial tree method, using a constant and a stochastic interest rate simulated with the Vasicek model (parameters for Vasicek model: $\alpha = 0.6$, $\sigma_r = 0.02$, $T_r = 4$, $N_r = 7200$, Taylor 1.5 simulation scheme with 100 simulated short-rate paths; parameters for the Cox-Ross-Rubinstein binomial tree: $S_{min} = 50$, $S_{max} = 150$, $M = 50$, $\sigma_S = 0.25$, $T_S = 3$, $N_S = 1080$, salvage factor 0.75 of initial investment cost of 110, expand factor 0.3 of prevailing underlying value with expand investment of 0.2 of initial investment cost).

	Constant risk-free rate: $r_f =$				Stochastic risk-free rate: $\beta/\alpha =$			
S_0	0.03	0.04	0.05	0.06	0.03	0.04	0.05	0.06
50	-27.335	-27.493	-27.500	-27.500	-27.218	-27.450	-27.498	-27.500
60	-24.734	-25.358	-25.867	-26.280	-24.601	-25.277	-25.811	-26.241
70	-19.484	-20.150	-20.726	-21.227	-19.379	-20.078	-20.672	-21.184
80	-12.155	-12.688	-13.143	-13.530	-12.072	-12.627	-13.094	-13.488
90	-3.237	-3.583	-3.859	-4.075	-3.169	-3.529	-3.814	-4.035
100	6.856	6.699	6.600	6.551	6.915	6.748	6.643	6.590
110	17.801	17.809	17.864	17.960	17.853	17.854	17.905	17.997
120	29.357	29.502	29.683	29.894	29.405	29.544	29.721	29.929
130	41.343	41.596	41.875	42.176	41.387	41.635	41.911	42.209
140	53.630	53.967	54.321	54.690	53.670	54.003	54.355	54.721
150	66.131	66.532	66.943	67.361	66.168	66.565	66.974	67.391

Source: ts-3-3 and ts-3-6

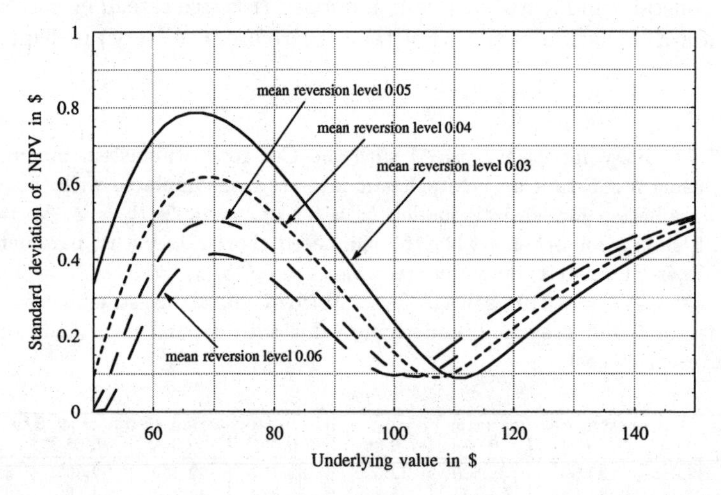

Fig. 5.37. Standard deviation of NPV for case 2 priced with the Cox-Ross-Rubinstein binomial tree method and using a stochastic risk-free rate simulated with the Vasicek model (ts-3-6; corresponding to Tables 5.47 and 5.48).

One finding of case 1 is also evident in case 2: The NPV priced by using a stochastic risk-free rate is not much different from the NPV priced by using the corresponding constant risk-free rate, see Table 5.48. However, this table also displays a remarkable result: In case 2 the NPV increases with an increasing risk-free rate for higher but not for lower values of S_0. This is rather surprising. However, in case 3 this result can be found irrespective of the choice of S_0. An explanation of this phenomenon - already seen in Trigeorgis [132], page 361, Table 11.1 - will be given below when the results of case 3 will be displayed.

The graph of the standard deviation in case 2 is totally different from the corresponding graph in case 1. It is interesting to observe that a higher level of the risk-free rate results in a lower standard deviation for lower values of S_0 but in higher standard deviations for higher values of S_0. For these higher values of S_0 the influence of the option to expand can clearly be seen.

Tables 5.49 and 5.50 display the ROV and the NPV for case 3, respectively, using a constant and a stochastically modelled risk-free rate. The corresponding standard deviation of the NPV (and, therefore, the ROV) is displayed in Figure 5.38. These tables show a very interesting and unexpected result: Flexible investment projects can exhibit a higher NPV for a higher risk-free interest rate as the discount factor. Intuitively, it would be expected to arrive at a lower NPV for a higher discount factor independent of the investment project considered. However, as Tables 5.46 (for case 1), 5.48 (for case 2), and 5.50 (for case 3) show, the real options embedded in the project and the underlying value play a crucial role to determine the NPV.

Table 5.49. ROV for case 3 priced with the Cox-Ross-Rubinstein binomial tree method, using a constant and a stochastic interest rate simulated with the Vasicek model (parameters for Vasicek model: $\alpha = 0.6$, $\sigma_r = 0.02$, $T_r = 4$, $N_r = 7200$, Taylor 1.5 simulation scheme with 100 simulated short-rate paths; parameters for the Cox-Ross-Rubinstein binomial tree: $S_{min} = 50$, $S_{max} = 150$, $M = 50$, $\sigma_S = 0.25$, $T_S = 3$, $N_S = 1080$ for the constant risk-free rate and $N_S = 300$ for the stochastic risk-free rate, salvage factor 0.75 of initial investment cost of 110, expand factor 0.3 of prevailing underlying value with expand investment of 0.2 of initial investment cost).

	Constant risk-free rate: $r_f =$				Stochastic risk-free rate: $\beta/\alpha =$			
S_0	0.03	0.04	0.05	0.06	0.03	0.04	0.05	0.06
50	60.051	60.057	60.063	60.070	60.051	60.056	60.062	60.069
60	50.408	50.441	50.480	50.525	50.408	50.440	50.477	50.520
70	41.688	41.805	41.940	42.088	41.704	41.818	41.947	42.091
80	34.638	34.904	35.201	35.526	34.658	34.928	35.227	35.552
90	29.700	30.174	30.685	31.230	29.751	30.215	30.719	31.260
100	26.848	27.545	28.283	29.060	26.920	27.608	28.340	29.113
110	25.750	26.655	27.603	28.590	25.825	26.723	27.665	28.647
120	25.975	27.067	28.197	29.360	26.027	27.113	28.242	29.403
130	27.113	28.359	29.635	30.935	27.188	28.428	29.699	30.995
140	28.859	30.228	31.618	33.024	28.919	30.281	31.667	33.069
150	31.007	32.473	33.953	35.439	31.075	32.535	34.009	35.490

Source: ts-3-4 and ts-3-7

Table 5.50. NPV for case 3 priced with the Cox-Ross-Rubinstein binomial tree method, using a constant and a stochastic interest rate simulated with the Vasicek model (parameters for Cox-Ingersoll-Ross model: $\alpha = 0.6$, $\sigma_r = 0.02$, $T_r = 4$, $N_r = 7200$, Taylor 1.5 simulation scheme with 100 simulated short-rate paths; parameters for the Cox-Ross-Rubinstein binomial tree: $S_{min} = 50$, $S_{max} = 150$, $M = 50$, $\sigma_S = 0.25$, $T_S = 3$, $N_S = 1080$ for the constant risk-free rate and $N_S = 300$ for the stochastic risk-free rate, salvage factor 0.75 of initial investment cost of 110, expand factor 0.3 of prevailing underlying value with expand investment of 0.2 of initial investment cost).

	Constant risk-free rate: $r_f =$				Stochastic risk-free rate: $\beta/\alpha =$			
S_0	0.03	0.04	0.05	0.06	0.03	0.04	0.05	0.06
50	0.051	0.057	0.063	0.070	0.051	0.056	0.062	0.069
60	0.408	0.441	0.480	0.525	0.408	0.440	0.477	0.520
70	1.688	1.805	1.940	2.088	1.704	1.818	1.947	2.091
80	4.638	4.904	5.201	5.526	4.658	4.928	5.227	5.552
90	9.700	10.174	10.685	11.230	9.751	10.215	10.719	11.260
100	16.848	17.545	18.283	19.060	16.920	17.608	18.340	19.113
110	25.750	26.655	27.603	28.590	25.825	26.723	27.665	28.647
120	35.975	37.067	38.197	39.360	36.027	37.113	38.242	39.403
130	47.113	48.359	49.635	50.935	47.188	48.428	49.699	50.995
140	58.859	60.228	61.618	63.024	58.919	60.281	61.667	63.069
150	71.007	72.473	73.953	75.439	71.075	72.535	74.009	75.490

Source: ts-3-4 and ts-3-7

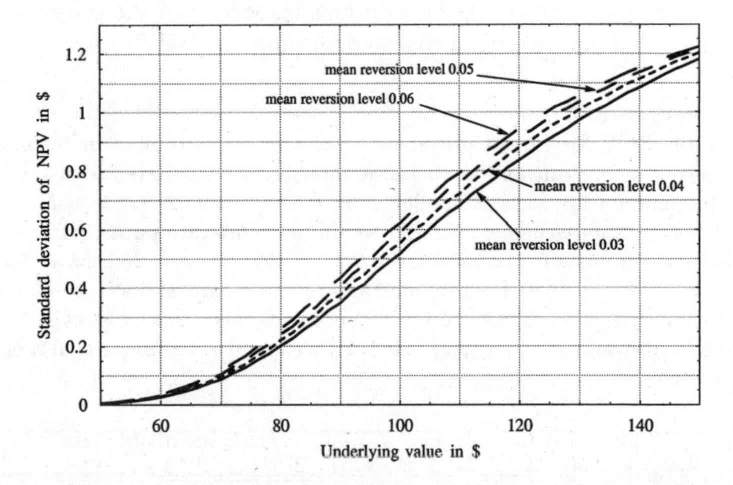

Fig. 5.38. Standard deviation of NPV for case 3 priced with the Cox-Ross-Rubinstein binomial tree method and using a stochastic risk-free rate simulated with the Vasicek model (ts-3-7; corresponding to Tables 5.49 and 5.50).

The reason for that phenomenon lies in the real options types. Obviously, if there are no real options embedded, a higher risk-free rate implies (independent of the underlying value) a lower NPV - a well known fact. If as in case 1 an American put option is included in the investment project, a higher risk-free rate implies (independent of the underlying value) a lower NPV as displayed in Table 5.46 - a well known fact as well. This holds also true for a European style put option and could be seen in test situation 1 of this thesis, when both the Black-Scholes formula and a numerical method were used to calculate the NPV of a project.

However, the situation changes completely for case 2. The option to expand the investment (by 30%) for a constant fraction of 20% of the initial investment cost $I_0 = 110$ makes the difference. The project can only be expanded once at the end of the 3-year construction period. For this real options case, a higher risk-free rate yields a lower NPV only for low values of the underlying (see Table 5.48) but yields a higher NPV for higher values. The reason lies in the option to expand the investment at $T_S = 3$. This option will be exercised for high but not for low underlying values.

A higher risk-free rate implies a higher risk-adjusted probability according to equation (4.3) since it leaves the up- and down-factors u and d of the Cox-Ross-Rubinstein binomial tree unchanged. Assuming a high underlying value, the option to expand will be exercised and the value for $E(i,j)$ at the last node ($j = N_S$) of the Cox-Ross-Rubinstein tree (see Section 4.2.1 for the notation and for a detailed description of this valuation method) will be much higher than without the option to expand (case 1). It is also important that the incremental increase in $E(i,j)$ is higher for the up-branch than for the down-branch at the last step of the binomial tree. Therefore, and because of the higher risk-adjusted probability for the formulas of Figure 4.2, a higher risk-free rate in the presence of an exercised option to expand increases $E(i,j)$ for the project on the pre-last step of the Cox-Ross-Rubinstein tree, which results in a higher overall NPV. This conclusion does not hold if there is no expanding, i.e., for lower values of the underlying.

Finally, for case 3 a higher risk-free rate implies a higher NPV independent of the underlying value as shown in Table 5.50. Here, the effect observed and described in case 2 is strengthened by the option to defer the investment project, which yields positive NPV values for lower underlying values (opposite to case 2).

This phenomenon is not totally new. Trigeorgis also mentioned it when analyzing an investment project using the real options approach[19]. He analyzed a similar yet more complex project than real options case 3 of this thesis but only applied a constant risk-free interest rate. For his project, he used an option to expand the investment as well. As already mentioned, his project of a real-life situation served as a role model for case 3 of this thesis. Table

[19] See Trigeorgis [132], Section 11.5.

11.1 on page 361 of Trigeorgis [132] clearly shows how a higher real risk-free rate increases the combined flexibility value of the project (and, hence, the NPV as well). For example, Trigeorgis calculated a combined flexibility value of 93.7 for a risk-free rate of 1% and a combined flexibility value of 109.7 if the risk-free rate is 4%.

The influence of the short-rate volatility on the investment project. This part analyzes the influence of the short-rate volatility on the ROV, the NPV, and the corresponding standard deviation. The same choice of parameters as above was used as the start scenario. The risk-free interest rate is simulated using the Vasicek model. All parameters are specified in the tables below. In case 3 the defer period is exactly one year.

Tables 5.51 and 5.52 display the ROV and the NPV for case 1, respectively, using a constant and a stochastically modelled risk-free rate. The corresponding standard deviation of the NPV (and, therefore, the ROV) is displayed in Figure 5.39.

Table 5.51. ROV for case 1 priced with the Cox-Ross-Rubinstein binomial tree method, using a constant and a stochastic interest rate simulated with the Vasicek model (parameters for Vasicek model: $\alpha = 0.6$, $\beta = 0.018$, $T_r = 4$, $N_r = 7200$, $r_0 = 0.03$, Taylor 1.5 simulation scheme with 100 simulated short-rate paths; parameters for the Cox-Ross-Rubinstein binomial tree: $S_{min} = 50$, $S_{max} = 150$, $M = 50$, $\sigma_S = 0.25$, $T_S = 3$, $N_S = 1080$, salvage factor 0.75 of initial investment cost of 110).

S_0	Constant $r_f = 0.03$	Stochastic risk-free interest rate: $\sigma_r =$ 0.02	0.015	0.01	0.005
50	32.500	32.549	32.511	32.501	32.500
60	23.558	23.706	23.625	23.581	23.561
70	16.943	17.062	16.995	16.958	16.942
80	12.110	12.194	12.144	12.116	12.106
90	8.620	8.679	8.643	8.622	8.615
100	6.123	6.165	6.138	6.123	6.118
110	4.345	4.377	4.357	4.345	4.341
120	3.089	3.114	3.097	3.088	3.086
130	2.199	2.219	2.207	2.199	2.197
140	1.569	1.585	1.575	1.569	1.568
150	1.126	1.138	1.130	1.126	1.124

Source: ts-3-2 and ts-3-5

Both tables show an interesting result for the ROV and the NPV: The mean of the ROV and the NPV for case 1 remains nearly unchanged regardless of the volatility σ_r of the stochastic term structure model. However, this result makes sense intuitively since the short-rate volatility does not influence the mean reversion level of the short-rate model, but only the variation of the short rates. The effect of the latter is observable in Figure 5.39: the higher

Table 5.52. NPV for case 1 priced with the Cox-Ross-Rubinstein binomial tree method, using a constant and a stochastic interest rate simulated with the Vasicek model (parameters for Vasicek model: $\alpha = 0.6$, $\beta = 0.018$, $T_r = 4$, $N_r = 7200$, $r_0 = 0.03$, Taylor 1.5 simulation scheme with 100 simulated short-rate paths; parameters for the Cox-Ross-Rubinstein binomial tree: $S_{min} = 50$, $S_{max} = 150$, $M = 50$, $\sigma_S = 0.25$, $T_S = 3$, $N_S = 1080$, salvage factor 0.75 of initial investment cost of 110).

S_0	Constant $r_f = 0.03$	Stochastic risk-free interest rate: $\sigma_r =$ 0.02	0.015	0.01	0.005
50	-27.500	-27.451	-27.489	-27.499	-27.500
60	-26.442	-26.294	-26.375	-26.419	-26.439
70	-23.057	-22.938	-23.005	-23.042	-23.058
80	-17.890	-17.806	-17.856	-17.884	-17.894
90	-11.380	-11.321	-11.357	-11.378	-11.385
100	-3.877	-3.835	-3.862	-3.877	-3.882
110	4.345	4.377	4.357	4.345	4.341
120	13.089	13.114	13.097	13.088	13.086
130	22.199	22.219	22.207	22.199	22.197
140	31.569	31.585	31.575	31.569	31.568
150	41.126	41.138	41.130	41.126	41.124

Source: ts-3-2 and ts-3-5

the short-rate volatility σ_r the higher the volatility of the ROV and the NPV. This volatility is very low if the real option of case 1, the option to abandon, is deeply out-of-the-money or deeply in-the-money. The peaks of the graphs in Figure 5.39 occur almost at an identical value of S_0 but the peak value increases approximately proportionally to the increase in σ_r.

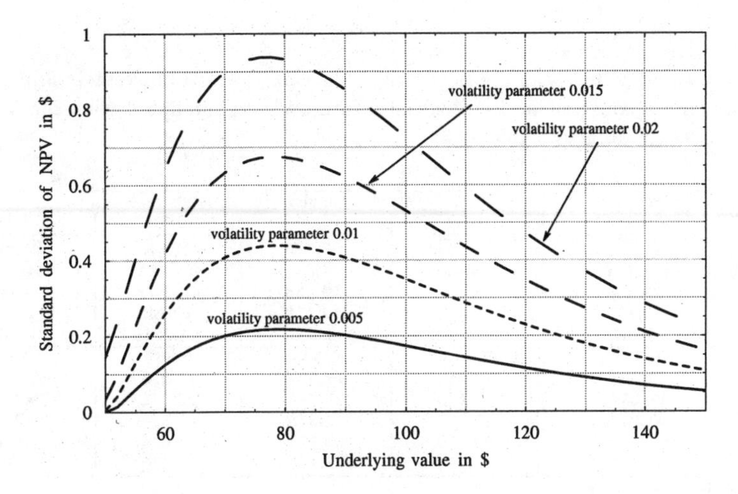

Fig. 5.39. Standard deviation of NPV for case 1 priced with the Cox-Ross-Rubinstein binomial tree method and using a stochastic risk-free rate simulated with the Vasicek model (ts-3-5; corresponding to Tables 5.51 and 5.52).

Tables 5.53 and 5.54 display the ROV and the NPV for case 2, respectively, using a constant and a stochastically modelled risk-free rate. The corresponding standard deviation of the NPV and, consequently, the ROV is displayed in Figure 5.40.

Table 5.53. ROV for case 2 priced with the Cox-Ross-Rubinstein binomial tree method, using a constant and a stochastic interest rate simulated with the Vasicek model (parameters for Vasicek model: $\alpha = 0.6$, $\beta = 0.018$, $T_r = 4$, $N_r = 7200$, $r_0 = 0.03$, Taylor 1.5 simulation scheme with 100 simulated short-rate paths; parameters for the Cox-Ross-Rubinstein binomial tree: $S_{min} = 50$, $S_{max} = 150$, $M = 50$, $\sigma_S = 0.25$, $T_S = 3$, $N_S = 1080$, salvage factor 0.75 of initial investment cost of 110, expand factor 0.3 of prevailing underlying value with expand investment of 0.2 of initial investment cost).

	Constant	Stochastic risk-free interest rate: $\sigma_r =$			
S_0	$r_f = 0.03$	0.02	0.015	0.01	0.005
50	32.665	32.782	32.715	32.684	32.668
60	25.266	25.399	25.325	25.285	25.267
70	20.516	20.621	20.562	20.529	20.515
80	17.845	17.928	17.882	17.855	17.844
90	16.763	16.831	16.795	16.773	16.763
100	16.856	16.915	16.886	16.867	16.858
110	17.801	17.853	17.829	17.813	17.804
120	19.357	19.405	19.385	19.371	19.362
130	21.343	21.387	21.370	21.358	21.349
140	23.630	23.670	23.655	23.644	23.636
150	26.131	26.168	26.156	26.146	26.138

Source: ts-3-3 and ts-3-6

Table 5.54. NPV for case 2 priced with the Cox-Ross-Rubinstein binomial tree method, using a constant and a stochastic interest rate simulated with the Vasicek model (corresponding to Table 5.53).

	Constant	Stochastic risk-free interest rate: $\sigma_r =$			
S_0	$r_f = 0.03$	0.02	0.015	0.01	0.005
50	-27.335	-27.218	-27.285	-27.316	-27.332
60	-24.734	-24.601	-24.675	-24.715	-24.733
70	-19.484	-19.379	-19.438	-19.471	-19.485
80	-12.155	-12.072	-12.118	-12.145	-12.156
90	-3.237	-3.169	-3.205	-3.227	-3.237
100	6.856	6.915	6.886	6.867	6.858
110	17.801	17.853	17.829	17.813	17.804
120	29.357	29.405	29.385	29.371	29.362
130	41.343	41.387	41.370	41.358	41.349
140	53.630	53.670	53.655	53.644	53.636
150	66.131	66.168	66.156	66.146	66.138

Source: ts-3-3 and ts-3-6

Tables 5.53 and 5.54 show that the mean of the ROV and the NPV for case 2 remains nearly unchanged regardless of the volatility parameter σ_r of the stochastic term structure model. This was also found for case 1. The effect of the volatility parameter on the standard deviation of the NPV (and, consequently, the ROV) is observable in Figure 5.40: the higher the short-rate volatility σ_r the higher the volatility of the ROV and the NPV. This was found for case 2 as well.

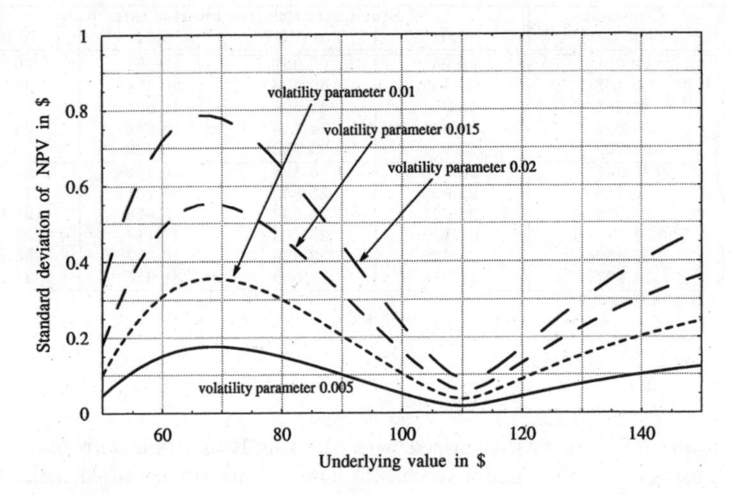

Fig. 5.40. Standard deviation of NPV for case 2 priced with the Cox-Ross-Rubinstein binomial tree method and using a stochastic risk-free rate simulated with the Vasicek model (ts-3-6; corresponding to Tables 5.53 and 5.54).

Tables 5.55 and 5.56 display the ROV and the NPV for case 3, respectively, using a constant and a stochastically modelled risk-free rate. The corresponding standard deviation of the NPV (and, therefore, the ROV) is displayed in Figure 5.41.

As can be seen in Table 5.55, the volatility parameter has an insignificant influence on the ROV. This result is independent of the underlying value S_0. However, the influence of the volatility parameter on the corresponding standard deviation of the NPV (and, therefore, the ROV) is significant as can be seen in Figure 5.41.

Table 5.55. ROV for case 3 priced with the Cox-Ross-Rubinstein binomial tree method, using a constant and a stochastic interest rate simulated with the Vasicek model (parameters for Vasicek model: $\alpha = 0.6$, $\beta = 0.018$, $T_r = 4$, $N_r = 7200$, $r_0 = 0.03$, Taylor 1.5 simulation scheme with 100 simulated short-rate paths; parameters for the Cox-Ross-Rubinstein binomial tree: $S_{min} = 50$, $S_{max} = 150$, $M = 50$, $\sigma_S = 0.25$, $T_S = 3$, $N_S = 1080$ for the constant risk-free rate and $N_S = 300$ for the stochastic risk-free rate, salvage factor 0.75 of initial investment cost of 110, expand factor 0.3 of prevailing underlying value with expand investment of 0.2 of initial investment cost).

	Constant	Stochastic risk-free interest rate: $\sigma_r =$			
S_0	$r_f = 0.03$	0.02	0.015	0.01	0.005
50	60.051	60.051	60.051	60.051	60.050
60	50.408	50.408	50.406	50.404	50.404
70	41.688	41.704	41.698	41.694	41.692
80	34.638	34.658	34.644	34.634	34.628
90	29.700	29.751	29.733	29.720	29.712
100	26.848	26.920	26.898	26.882	26.871
110	25.750	25.825	25.800	25.781	25.768
120	25.975	26.027	25.999	25.978	25.962
130	27.113	27.188	27.163	27.144	27.130
140	28.859	28.919	28.896	28.877	28.862
150	31.007	31.075	31.055	31.037	31.023

Source: ts-3-4 and ts-3-7

Table 5.56. NPV for case 3 priced with the Cox-Ross-Rubinstein binomial tree method, using a constant and a stochastic interest rate simulated with the Vasicek model (parameters for Vasicek model: $\alpha = 0.6$, $\beta = 0.018$, $T_r = 4$, $N_r = 7200$, $r_0 = 0.03$, Taylor 1.5 simulation scheme with 100 simulated short-rate paths; parameters for the Cox-Ross-Rubinstein binomial tree: $S_{min} = 50$, $S_{max} = 150$, $M = 50$, $\sigma_S = 0.25$, $T_S = 3$, $N_S = 1080$ for the constant risk-free rate and $N_S = 300$ for the stochastic risk-free rate, salvage factor 0.75 of initial investment cost of 110, expand factor 0.3 of prevailing underlying value with expand investment of 0.2 of initial investment cost).

	Constant	Stochastic risk-free interest rate: $\sigma_r =$			
S_0	$r_f = 0.03$	0.02	0.015	0.01	0.005
50	0.051	0.051	0.051	0.051	0.050
60	0.408	0.408	0.406	0.404	0.404
70	1.688	1.704	1.698	1.694	1.692
80	4.638	4.658	4.644	4.634	4.628
90	9.700	9.751	9.733	9.720	9.712
100	16.848	16.920	16.898	16.882	16.871
110	25.750	25.825	25.800	25.781	25.768
120	35.975	36.027	35.999	35.978	35.962
130	47.113	47.188	47.163	47.144	47.130
140	58.859	58.919	58.896	58.877	58.862
150	71.007	71.075	71.055	71.037	71.023

Source: ts-3-4 and ts-3-7

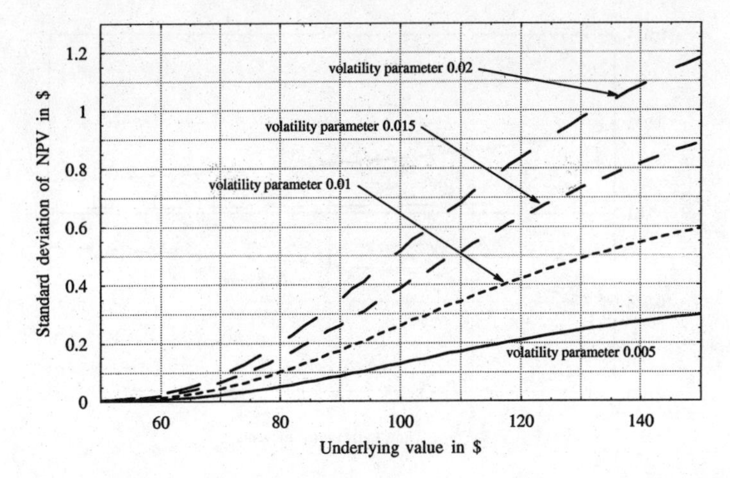

Fig. 5.41. Standard deviation of NPV for case 3 priced with the Cox-Ross-Rubinstein binomial tree method and using a stochastic risk-free rate simulated with the Vasicek model (ts-3-7; corresponding to Tables 5.55 and 5.56).

5.6.3 Analysis for the Cox-Ingersoll-Ross Model

This section analyzes the influence of adding new real options on the investment project when the risk-free interest rate is simulated by using the Cox-Ingersoll-Ross model. This is done for cases 1, 2 and 3. In the following, the analysis will be less extensive than for the Vasicek model for two reasons: First, a particular focus was put on the Vasicek model, which is an equilibrium model like the Cox-Ingersoll-Ross model, as it was the very first stochastic interest rate model historically. Second, and even more importantly, the results of the analysis using the Cox-Ingersoll-Ross model do not differ significantly from the results for the Vasicek model. Therefore, only the project's NPV will be displayed in the following but not the ROV any more.

The influence of real options types on the investment project. In order to investigate the influence of the various real options on the project's NPV the Cox-Ingersoll-Ross model was used with a mean reversion level of 0.03. All parameters are specified in the tables below. In case 3 the defer period is exactly one year.

Figure 5.42 displays an example of the development over time of the term structure of interest rates. Figure 5.43 shows the NPV for all three cases 1, 2 and 3.

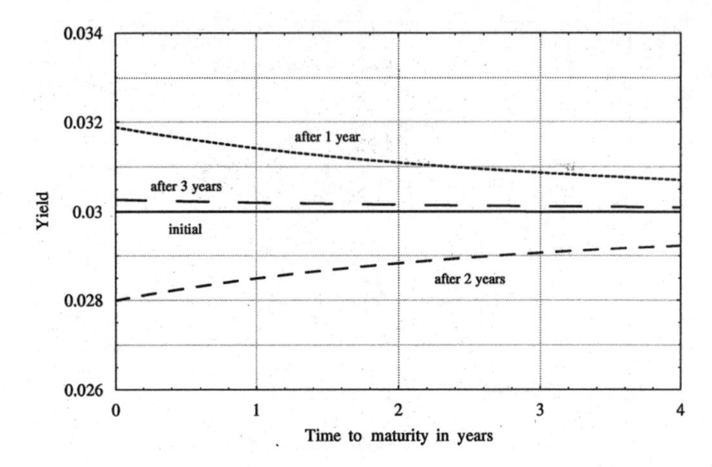

Fig. 5.42. Example of the development of the term structure of interest rates if the Cox-Ingersoll-Ross model is used (parameters for the Cox-Ingersoll-Ross model: $\alpha = 0.6$, $\beta = 0.018$, $T_r = 4$, $N_r = 7200$, $r_0 = 0.03$, $\sigma_r = 0.02$, Taylor 1.5 simulation scheme).

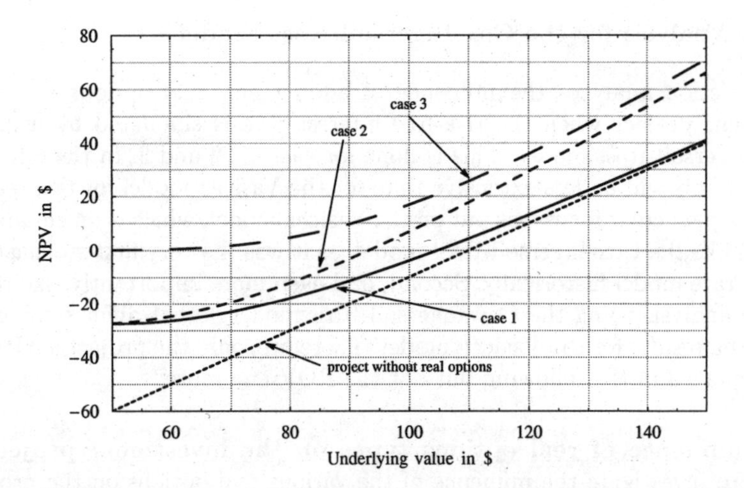

Fig. 5.43. NPV for cases 1, 2 and 3 priced with the Cox-Ross-Rubinstein binomial tree method and using a stochastic risk-free rate simulated with the Cox-Ingersoll-Ross model (ts-3-8, ts-3-9, and ts-3-10; parameters for the Cox-Ingersoll-Ross model: $\alpha = 0.6$, $\beta = 0.018$, $T_r = 4$, $N_r = 7200$, $r_0 = 0.03$, $\sigma_r = 0.02$, Taylor 1.5 simulation scheme for 100 simulated short-rate paths; parameters for the Cox-Ross-Rubinstein binomial tree: $S_{min} = 50$, $S_{max} = 150$, $M = 50$, $\sigma_S = 0.25$, $T_S = 3$, $N_S = 1080$ for case 1 and case 2, $N_S = 300$ for case 3, salvage factor 0.75 of initial investment cost of 110, expand factor 0.3 of prevailing underlying value with expand investment of 0.2 of initial investment cost).

Figure 5.42 shows that for the same choice of parameters as in the Vasicek model the development of the term structure over time "swings" less than in Figure 5.31. This result is consistent with the result of Schulmerich[20] that for a given short-rate time series the estimation of the volatility parameter varies considerably depending on the assumption of a Vasicek model or a Cox-Ingersoll-Ross model. For this reason, the results of the previous section with the Vasicek model cannot be directly compared to the results of this section when the Cox-Ingersoll-Ross model is used and the same choices of parameters apply.

The influence of the short-rate mean reversion level on the investment project. This part analyzes the influence of the mean reversion level on the NPV and the standard deviation of the NPV. The volatility parameter for the Cox-Ingersoll-Ross model remains constant at 0.08 but the mean reversion level varies. All parameters are specified in the tables below. In case 3 the defer period is exactly one year. Table 5.57 displays the NPV for case 1 using a constant and a stochastically modelled risk-free rate. The corresponding standard deviation of the NPV (and, therefore, the ROV) is displayed in Figure 5.44.

Table 5.57. NPV for case 1 priced with the Cox-Ross-Rubinstein binomial tree method, using a constant and a stochastic interest rate simulated with the Cox-Ingersoll-Ross model (parameters for Cox-Ingersoll-Ross model: $\alpha = 0.6$, $\sigma_r = 0.08$, $T_r = 4$, $N_r = 7200$, Taylor 1.5 simulation scheme with 100 simulated short-rate paths; parameters for the Cox-Ross-Rubinstein binomial tree: $S_{min} = 50$, $S_{max} = 150$, $M = 50$, $\sigma_S = 0.25$, $T_S = 3$, $N_S = 1080$, salvage factor 0.75 of initial investment cost of 110).

	Constant risk-free rate: $r_f =$				Stochastic risk-free rate: $\beta/\alpha =$			
S_0	0.03	0.04	0.05	0.06	0.03	0.04	0.05	0.06
50	-27.500	-27.500	-27.500	-27.500	-27.496	-27.500	-27.500	-27.500
60	-26.442	-26.950	-27.283	-27.463	-26.402	-26.910	-27.248	-27.433
70	-23.057	-23.830	-24.494	-25.066	-23.033	-23.801	-24.462	-25.033
80	-17.890	-18.679	-19.385	-20.018	-17.881	-18.664	-19.366	-19.997
90	-11.380	-12.084	-12.717	-13.289	-11.379	-12.077	-12.706	-13.274
100	-3.877	-4.465	-4.991	-5.464	-3.879	-4.462	-4.984	-5.454
110	4.345	3.874	3.454	3.080	4.342	3.875	3.459	3.088
120	13.089	12.718	12.392	12.103	13.086	12.720	12.396	12.110
130	22.199	21.913	21.663	21.444	22.198	21.914	21.667	21.450
140	31.569	31.350	31.161	30.996	31.568	31.352	31.164	31.001
150	41.126	40.959	40.815	40.692	41.125	40.960	40.818	40.697

Source: ts-3-2 and ts-3-8

Table 5.57 shows that - as in the Vasicek model - the mean reversion level has a significant influence on the NPV (and, therefore, the ROV, which is not displayed here) of the investment project. The results obtained by using the

[20] See Schulmerich [117], pages 187-188, Tables H.9 and H.10.

Vasicek model cannot be directly compared with those obtained by using the Cox-Ingersoll-Ross model due to the fact that the short-rate volatility terms in both term structure models are not identical but very different.

The short-rate volatility parameter in the Cox-Ingersoll-Ross model has to be much higher than the short-rate volatility parameter in the Vasicek model to arrive at the same "variability" in the interest rate, because in the Vasicek model the short-rate volatility is the parameter σ_r itself while in the Cox-Ingersoll-Ross model the short-rate volatility is not the volatility parameter alone but adjusted by the square root of the current level of the risk-free rate $\sigma_r\sqrt{r_t}$. Since the risk-free rate is far below 1, the prevailing short-rate volatility in the Cox-Ingersoll-Ross model is well below the volatility parameter itself. Hence the volatility parameter in the Cox-Ingersoll-Ross model needs to be much higher than the volatility parameter in the Vasicek model to get similar variations of the term structure graphs over time. Figures 5.43 (for the Cox-Ingersoll-Ross model) and 5.31 (for the Vasicek model) indicate that relationship since for both figures the parameters are identical.

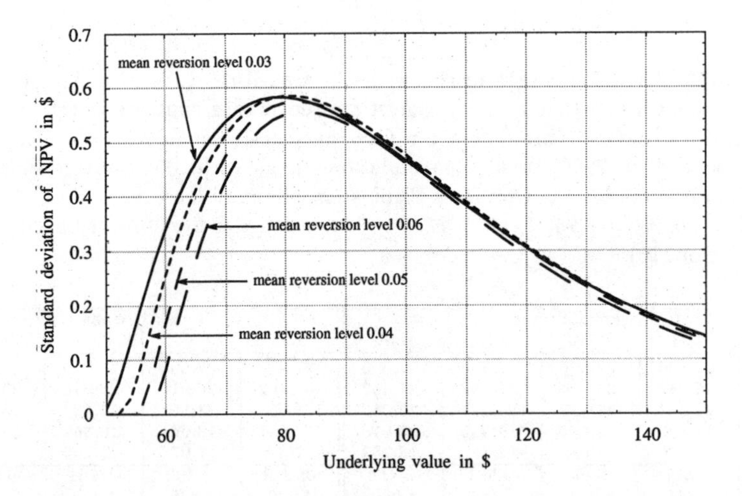

Fig. 5.44. Standard deviation of NPV for case 1 priced with the Cox-Ross-Rubinstein binomial tree method and using a stochastic risk-free rate simulated with the Cox-Ingersoll-Ross model (ts-3-8; corresponding to Table 5.57).

An interesting result can be seen in Figure 5.44 for the NPV's standard deviation in case 1 when using the Cox-Ingersoll-Ross model for modelling the risk-free rate. This figure should be seen in connection with Figure 5.36 when the standard deviation of case 1 was analyzed using the Vasicek model. The resulting graphs in both figures are different: For the Vasicek model the graphs for the standard deviation shift down for a higher mean reversion level. For the Cox-Ingersoll-Ross model the graphs also shift down for a higher mean reversion level but only if the real option is in- and at-the-money, and by a lower degree. If the option is out-of-the-money the graphs are almost the same for different mean reversion levels. This clearly indicates a further area of future research.

As for case 2 the NPV is displayed in Table 5.58 and the corresponding standard deviation in Figure 5.45. Table 5.58 shows that the results for the NPV are significantly different for various mean reversion levels. Figure 5.45 for the Cox-Ingersoll-Ross term structure model and Figure 5.37 for the Vasicek model display a similar shape of the graphs but the movements of the graphs occur at different strengths.

Finally, the NPV in case 3 is given in Table 5.59 with the corresponding standard deviation being displayed in Figure 5.46. Table 5.59 shows, especially when compared with cases 1 and 2, that the mean reversion level has the most significant influence in case 3 when there are 3 different real options. For example, the NPV increases by 13% for $S_0 = 100$ and for a mean reversion level of 0.06 compared with a mean reversion level of only 0.03.

Table 5.58. NPV for case 2 priced with the Cox-Ross-Rubinstein binomial tree method, using a constant and a stochastic interest rate simulated with the Cox-Ingersoll-Ross model (parameters for Cox-Ingersoll-Ross model: $\alpha = 0.6$, $\sigma_r = 0.08$, $T_r = 4$, $N_r = 7200$, Taylor 1.5 simulation scheme with 100 simulated short-rate paths; parameters for the Cox-Ross-Rubinstein binomial tree: $S_{min} = 50$, $S_{max} = 150$, $M = 50$, $\sigma_S = 0.25$, $T_S = 3$, $N_S = 1080$, salvage factor 0.75 of initial investment cost of 110, expand factor 0.3 of prevailing underlying value with expand investment of 0.2 of initial investment cost).

	Constant risk-free rate: $r_f =$				Stochastic risk-free rate: $\beta/\alpha =$			
S_0	0.03	0.04	0.05	0.06	0.03	0.04	0.05	0.06
50	-27.335	-27.493	-27.500	-27.500	-27.302	-27.473	-27.500	-27.500
60	-24.734	-25.358	-25.867	-26.280	-24.701	-25.323	-25.832	-26.248
70	-19.484	-20.150	-20.726	-21.227	-19.461	-20.121	-20.694	-21.193
80	-12.155	-12.688	-13.143	-13.530	-12.137	-12.663	-13.113	-13.495
90	-3.237	-3.583	-3.859	-4.075	-3.219	-3.557	-3.828	-4.040
100	6.856	6.699	6.600	6.551	6.877	6.726	6.633	6.589
110	17.801	17.809	17.864	17.960	17.824	17.839	17.899	17.999
120	29.357	29.502	29.683	29.894	29.384	29.534	29.718	29.933
130	41.343	41.596	41.875	42.176	41.372	41.629	41.912	42.215
140	53.630	53.967	54.321	54.690	53.660	54.000	54.358	54.728
150	66.131	66.532	66.943	67.361	66.162	66.566	66.979	67.399

Source: ts-3-3 and ts-3-9

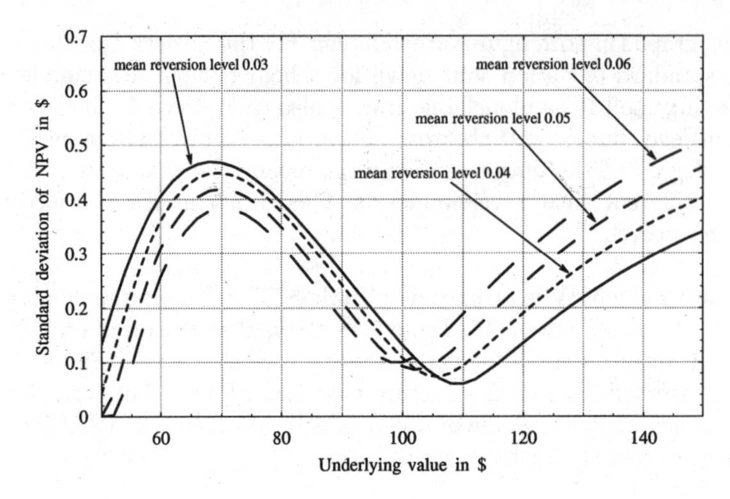

Fig. 5.45. Standard deviation of NPV for case 2 priced with the Cox-Ross-Rubinstein binomial tree method and using a stochastic risk-free rate simulated with the Cox-Ingersoll-Ross model (ts-3-9; corresponding to Table 5.58).

Table 5.59. NPV for case 3 priced with the Cox-Ross-Rubinstein binomial tree method, using a constant and a stochastic interest rate simulated with the Cox-Ingersoll-Ross model (parameters for Cox-Ingersoll-Ross model: $\alpha = 0.6$, $\sigma_r = 0.08$, $T_r = 4$, $N_r = 7200$, Taylor 1.5 simulation scheme with 100 simulated short-rate paths; parameters for the Cox-Ross-Rubinstein binomial tree: $S_{min} = 50$, $S_{max} = 150$, $M = 50$, $\sigma_S = 0.25$, $T_S = 3$, $N_S = 1080$ for the constant and $N_S = 300$ for the stochastic risk-free rate, salvage factor 0.75 of initial investment cost of 110, expand factor 0.3 of prevailing underlying value with expand investment of 0.2 of initial investment cost).

S_0	Constant risk-free rate: $r_f =$				Stochastic risk-free rate: $\beta/\alpha =$			
	0.03	0.04	0.05	0.06	0.03	0.04	0.05	0.06
50	0.051	0.057	0.063	0.070	0.051	0.056	0.062	0.069
60	0.408	0.441	0.480	0.525	0.405	0.438	0.477	0.520
70	1.688	1.805	1.940	2.088	1.697	1.814	1.945	2.091
80	4.638	4.904	5.201	5.520	4.640	4.919	5.224	5.552
90	9.700	10.174	10.685	11.230	9.730	10.203	10.714	11.261
100	16.848	17.545	18.283	19.060	16.895	17.594	18.335	19.115
110	25.750	26.655	27.603	28.590	25.798	26.709	27.661	28.651
120	35.975	37.067	38.197	39.360	35.999	37.100	38.239	39.408
130	47.113	48.359	49.635	50.935	47.166	48.419	49.700	51.003
140	58.859	60.228	61.618	63.024	58.900	60.274	61.669	63.078
150	71.007	72.473	73.953	75.439	71.061	72.532	74.013	75.501

Source: ts-3-4 and ts-3-10

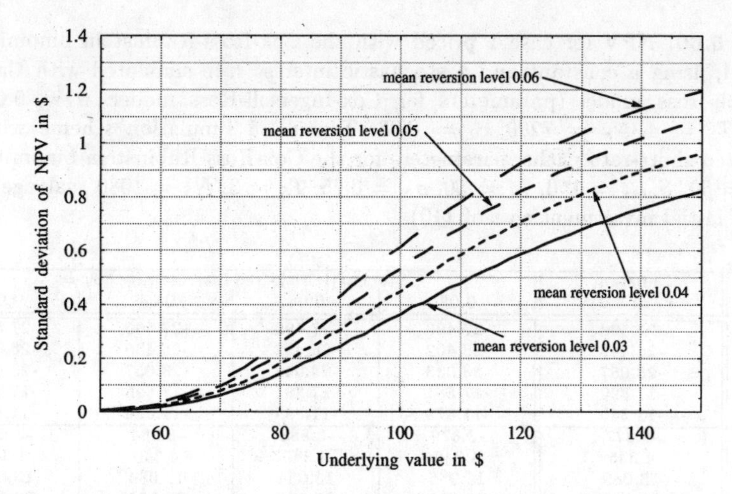

Fig. 5.46. Standard deviation of NPV for case 3 priced with the Cox-Ross-Rubinstein binomial tree method and using a stochastic risk-free rate simulated with the Cox-Ingersoll-Ross model (ts-3-10; corresponding to Table 5.59).

The influence of the short-rate volatility on the investment project. This part analyzes the influence of the short-rate volatility on the NPV and its standard deviation. The mean reversion level of the Cox-Ingersoll-Ross model remains constant at 0.03 while the volatility parameter is changed gradually. The NPV is determined using the Cox-Ross-Rubinstein binomial tree method. In case 3 the defer period is exactly one year. Table 5.60 displays the NPV for case 1 using a constant and a stochastically modelled risk-free rate. The corresponding standard deviation of the NPV is displayed in Figure 5.47.

Table 5.60 shows that the choice of the volatility parameter in the short-rate model has no significant influence on the NPV of the project in case 1. It also shows that when using the Cox-Ingersoll-Ross model the NPV is almost identical to the NPV when the model's mean reversion level is used as a constant risk-free rate. This result is the same when the Vasicek model is used and will also be obtained for cases 2 and 3. Furthermore, the graphs of the NPV's standard deviation in Figure 5.47 for the Cox-Ingersoll-Ross model look similar to those of Figure 5.39 when the Vasicek model is used. Evidently the standard deviation of the project's NPV heavily depends on the short-rate volatility parameter - as opposed to the NPV value itself, which is almost independent of this parameter.

Table 5.61 displays the NPV for case 2 with the corresponding standard deviation being displayed in Figure 5.48. As in case 1, the NPV of the project in case 2 is almost independent of the short-rate volatility and is the same when using the mean reversion level as a constant risk-free rate instead of a

Table 5.60. NPV for case 1 priced with the Cox-Ross-Rubinstein binomial tree method, using a constant and a stochastic interest rate simulated with the Cox-Ingersoll-Ross model (parameters for Cox-Ingersoll-Ross model: $\alpha = 0.6$, $\beta = 0.018$, $T_r = 4$, $N_r = 7200$, $r_0 = 0.03$, Taylor 1.5 simulation scheme with 100 simulated short-rate paths; parameters for the Cox-Ross-Rubinstein binomial tree: $S_{min} = 50$, $S_{max} = 150$, $M = 50$, $\sigma_S = 0.25$, $T_S = 3$, $N_S = 1080$, salvage factor 0.75 of initial investment cost of 110).

S_0	Constant $r_f = 0.03$	Stochastic risk-free interest rate: $\sigma_r =$ 0.08	0.06	0.04	0.02
50	-27.500	-27.496	-27.499	-27.500	-27.500
60	-26.442	-26.402	-26.422	-26.436	-26.442
70	-23.057	-23.033	-23.048	-23.057	-23.061
80	-17.890	-17.881	-17.891	-17.895	-17.895
90	-11.380	-11.379	-11.384	-11.386	-11.385
100	-3.877	-3.879	-3.883	-3.884	-3.882
110	4.345	4.342	4.340	4.340	4.341
120	13.089	13.086	13.084	13.084	13.085
130	22.199	22.198	22.196	22.196	22.197
140	31.569	31.568	31.567	31.567	31.567
150	41.126	41.125	41.124	41.123	41.124

Source: ts-3-2 and ts-3-8

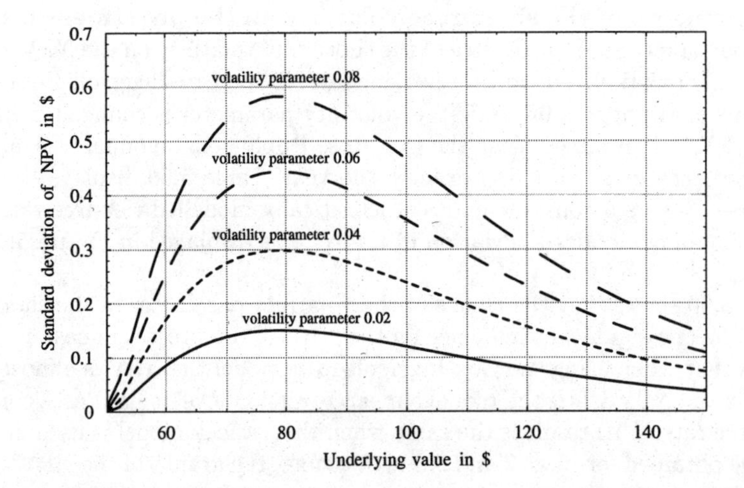

Fig. 5.47. Standard deviation of NPV for case 1 priced with the Cox-Ross-Rubinstein binomial tree method and using a stochastic risk-free rate simulated with the Cox-Ingersoll-Ross model (ts-3-8; corresponding to Table 5.60).

stochastically modelled risk-free interest rate. When comparing Figure 5.48 (Cox-Ingersoll-Ross model) with Figure 5.40 (Vasicek model), the shape of the graphs are the same while the standard deviations are generally larger

Table 5.61. NPV for case 2 priced with the Cox-Ross-Rubinstein binomial tree method, using a constant and a stochastic interest rate simulated with the Cox-Ingersoll-Ross model (parameters for Cox-Ingersoll-Ross model: $\alpha = 0.6$, $\beta = 0.018$, $T_r = 4$, $N_r = 7200$, $r_0 = 0.03$, Taylor 1.5 simulation scheme with 100 simulated short-rate paths; parameters for the Cox-Ross-Rubinstein binomial tree: $S_{min} = 50$, $S_{max} = 150$, $M = 50$, $\sigma_S = 0.25$, $T_S = 3$, $N_S = 1080$, salvage factor 0.75 of initial investment cost of 110, expand factor 0.3 of prevailing underlying value with expand investment of 0.2 of initial investment cost).

S_0	Constant $r_f = 0.03$	Stochastic risk-free interest rate: $\sigma_r =$ 0.08	0.06	0.04	0.02
50	-27.335	-27.302	-27.317	-27.328	-27.334
60	-24.734	-24.701	-24.718	-24.730	-24.735
70	-19.484	-19.461	-19.475	-19.484	-19.487
80	-12.155	-12.137	-12.148	-12.155	-12.158
90	-3.237	-3.219	-3.230	-3.236	-3.239
100	6.856	6.877	6.867	6.860	6.856
110	17.801	17.824	17.814	17.807	17.802
120	29.357	29.384	29.373	29.365	29.360
130	41.343	41.372	41.362	41.353	41.347
140	53.630	53.660	53.649	53.641	53.634
150	66.131	66.162	66.152	66.143	66.136

Source: ts-3-3 and ts-3-9

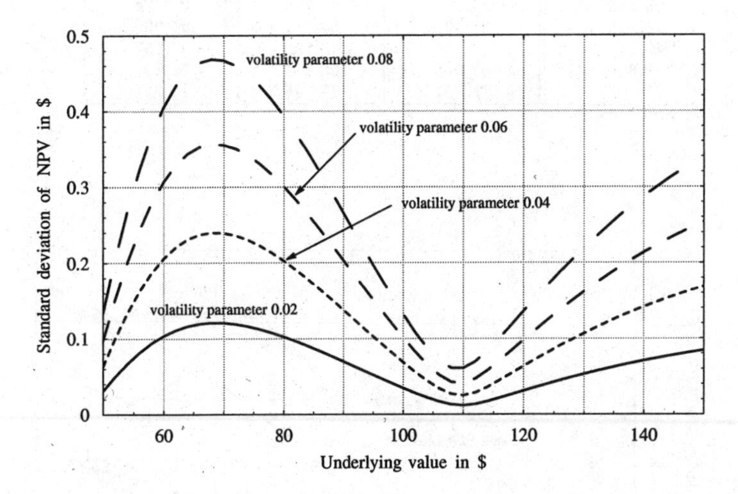

Fig. 5.48. Standard deviation of NPV for case 2 priced with the Cox-Ross-Rubinstein binomial tree method and using a stochastic risk-free rate simulated with the Cox-Ingersoll-Ross model (ts-3-9; corresponding to Table 5.61).

(for any given underlying value S_0) for the Vasicek model than for the Cox-Ingersoll-Ross model. This, again, is due to the different volatility term of the two short-rate models, as elaborated above.

Table 5.62 displays the NPV for case 3 with the corresponding standard deviation being displayed in Figure 5.49.

Table 5.62. NPV for case 3 priced with the Cox-Ross-Rubinstein binomial tree method, using a constant and a stochastic interest rate simulated with the Cox-Ingersoll-Ross model (parameters for Cox-Ingersoll-Ross model: $\alpha = 0.6, \beta = 0.018, T_r = 4, N_r = 7200, r_0 = 0.03$, Taylor 1.5 simulation scheme with 100 simulated short-rate paths; parameters for the Cox-Ross-Rubinstein binomial tree: $S_{min} = 50, S_{max} = 150, M = 50, \sigma_S = 0.25, T_S = 3, N_S = 1080$ for the constant risk-free rate and $N_S = 300$ for the stochastic risk-free rate, salvage factor 0.75 of initial investment cost of 110, expand factor 0.3 of prevailing underlying value with expand investment of 0.2 of initial investment cost).

	Constant	Stochastic risk-free interest rate: $\sigma_r =$			
S_0	$r_f = 0.03$	0.08	0.06	0.04	0.02
50	0.051	0.051	0.051	0.050	0.050
60	0.408	0.405	0.404	0.404	0.403
70	1.688	1.697	1.694	1.692	1.691
80	4.638	4.640	4.635	4.630	4.628
90	9.700	9.730	9.721	9.715	9.710
100	16.848	16.895	16.884	16.875	16.869
110	25.750	25.798	25.785	25.774	25.766
120	35.975	35.999	35.982	35.969	35.959
130	47.113	47.166	47.150	47.137	47.127
140	58.859	58.900	58.884	58.870	58.859
150	71.007	71.061	71.045	71.031	71.020

Source: ts-3-4 and ts-3-10

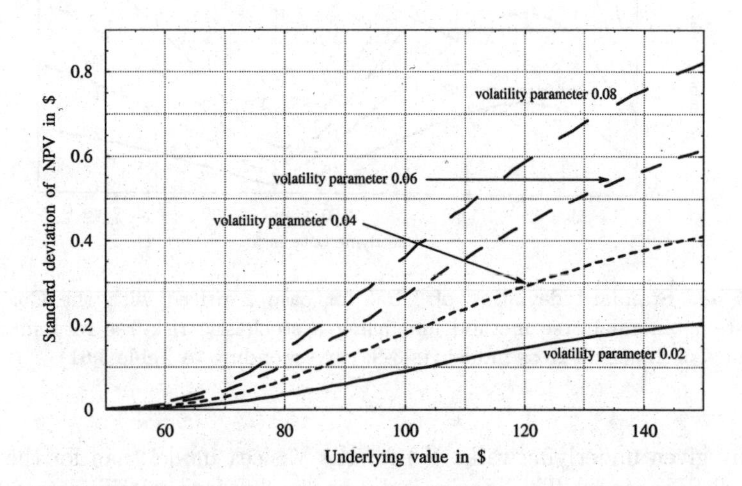

Fig. 5.49. Standard deviation of NPV for case 3 priced with the Cox-Ross-Rubinstein binomial tree method and using a stochastic risk-free rate simulated with the Cox-Ingersoll-Ross model (ts-3-10; corresponding to Table 5.62).

The interpretation of the results in Table 5.62 is identical to the interpretation in cases 1 and 2. The same is true for the graphs in Figure 5.49 when compared with the corresponding graphs in Figure 5.41 for the Vasicek model.

5.6.4 Recapitulation of the Main Results in Test Situation 3

Test situation 3 was the first test situation to analyze the influence of various equilibrium term structure models on real options valuation in three real options cases. Several interesting results were found:

- The chosen salvage factor and the chosen expand factor have a very significant influence on the project's NPV and standard deviation. Therefore, it is imperative to determine these two factors as accurately as possible in the capital budgeting process. Even little deviations can have significant effects on the investment project, independent of the method used to determine the risk-free rate, i.e., this phenomenon occurs for a constant and a non-constant risk-free interest rate as well.
- The level of the interest rate (as the constant risk-free rate in the deterministic case or the mean reversion level for a stochastically modelled risk-free rate) has an important impact on the NPV of a project. The higher the interest rate level, the higher the NPV, if the project contains an option to expand, an option to defer, and an option to abandon (case 3). This effect can also be seen without the option to defer (case 2) but only for higher underlying values. It cannot be observed for case 1, which only contains an option to abandon. These phenomena are independent of the chosen interest rate model and also appear if the risk-free rate is constant.
- The NPV's standard deviation changes dramatically for different interest rate levels. However, it is not possible to establish a general rule that a higher interest rate level increases or decreases the NPV's standard deviation, since this depends on the case and the chosen underlying value.
- The NPV of a project (which is the mean of the NPVs for many simulated short-rate paths) is virtually independent of the chosen volatility parameter in the short-rate model. The NPV's standard deviation, however, is highly dependent on the volatility parameter: the higher the volatility parameter, the higher the NPV's standard deviation. This holds true for both the Vasicek model and the Cox-Ingersoll-Ross model.
- The NPV for a stochastically modelled risk-free rate and the NPV obtained when using the model's mean reversion level are almost identical and independent of the chosen short-rate volatility. This holds true for both equilibrium models.

- Comparing the results of the Cox-Ingersoll-Ross model with the results of the Vasicek model is difficult since for the same choice of parameters, the Vasicek model produces wider swings in the term structure over time than the Cox-Ingersoll-Ross model. This is due to the fact that the Cox-Ingersoll-Ross model includes a volatility term that depends on the prevailing interest rate level. For the same choice of parameters in the two term structure models, the standard deviations for the Cox-Ingersoll-Ross model are therefore lower than the corresponding standard deviations for the Vasicek model. This holds true for all cases analyzed.

5.7 Test Situation 4: Real Options Valuation with a Stochastic Interest Rate Using No-Arbitrage Models

This section analyzes the test results when the risk-free interest rate is simulated using various no-arbitrage models. The obligatory use of market data as input parameters is a common feature of all these models. All models presented in this thesis use the yield curve of U.S. Zero bonds as an input parameter whereby the model calculates the term structure for the first time point $t_0 = 0$ as the initial yield curve. The initial yield curve is given by yields for various times to maturity. The method of the cubic spline interpolation is used to interpolate these data and get a complete yield curve. The appropriate number of discrete times to maturity (TTM) is critical to the application of the cubic spline method. If too many TTMs are chosen, the interpolation becomes rather extrem, which results in an oscillating instantaneous forward rate curve. Therefore, the number of TTMs used for the cubic spline interpolation is limited to the following: 1 mos., 3 mos., 6 mos., 9 mos., 12 mos., 2 yrs., 3 yrs., 4 yrs., 5 yrs., and 6 yrs.

The first part (5.7.1) of this section analyzes the Ho-Lee model. The second part (5.7.2) analyzes the Hull-White one-factor model in comparison with the Hull-White two-factor model. The goal is to investigate if the second risk factor in the term structure model has a significant influence on the real options pricing in all three cases. The third part (5.7.3) compares the Ho-Lee model with the Hull-White one-factor model. The section is concluded with a summary of the most important results (5.7.4).

5.7.1 Ho-Lee Model

Here, the initial yield curve of the model is always the inverted U.S. Zero bond yield curve as of December 1, 2000. Figure 5.50 displays this curve for up to 6 years of TTM.

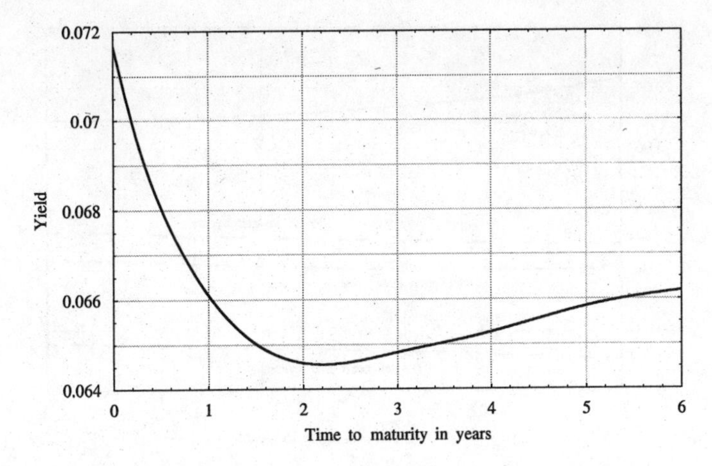

Fig. 5.50. Yield curve of the U.S. Zero bonds on December 1, 2000.

Below, the influence of three parameters on the investment project is subsequently analyzed: the influence of the salvage factor, the expand factor and the short-rate volatility.

The influence of the salvage factor on the investment project. This part uses the following parameters for the Ho-Lee model: $T_r = 4, N_r = 7200, \sigma_r = 0.015$, Taylor 1.5 simulation scheme with the U.S. Zero yield curve from December 1, 2000 as input parameter for each of the 100 simulated short-rate paths. The NPV is determined by using the Cox-Ross-Rubinstein binomial tree method whereby the parameter specification is stated in the corresponding table or figure. Figure 5.51 displays an example of the development of the term structure of interest rates over time when the Ho-Lee model and the parameter specification above is used.

Table 5.63 compares the ROV for various salvage factors in case 1. The standard deviation that corresponds to the ROV of Table 5.63 is graphically displayed in Figure 5.52. Table 5.63 cannot directly be compared with Figure 5.11 which displays the ROV for case 1 when a constant risk-free rate of $r_f = 0.03$ is used since the Ho-Lee model does not have a mean reversion level (see Section 3.3.2). As displayed in Figure 5.52, the standard deviation is skewed to the right for all salvage factors[21]. The higher the salvage factor, the higher the peak of the graphs. For higher salvage factors, the peak is reached for higher values of the underlying S_0.

[21] This can be clearly seen for all salvage factors above 0.15. For the salvage factors 0 and 0.15 the standard deviations are too small and cannot be displayed properly in the diagram any more.

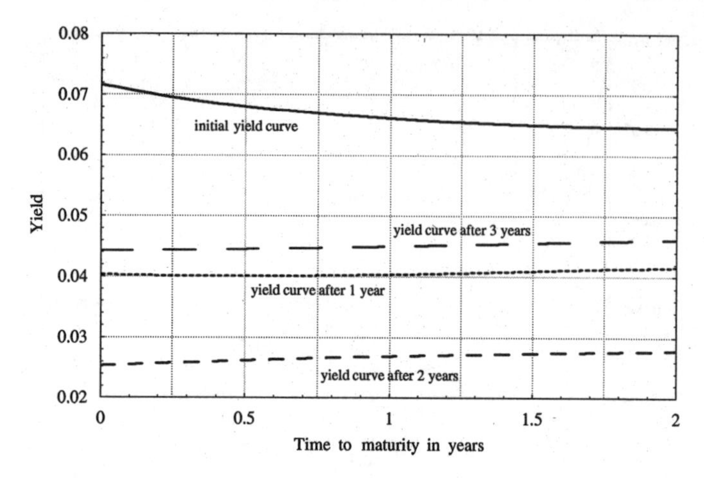

Fig. 5.51. Example of the development of the term structure of interest rates if the Ho-Lee model is used (parameters for the Ho-Lee model: $T_r = 4$, $N_r = 7200, \sigma_r = 0.015$, Taylor 1.5 simulation scheme with the U.S. Zero yield curve from December 1, 2000 as input parameter).

Table 5.63. ROV in case 1 with a variable salvage factor priced with the Cox-Ross-Rubinstein binomial tree method when the risk-free rate is simulated using the Ho-Lee model (parameters for the Ho-Lee model: $T_r = 4$, $N_r = 7200, \sigma_r = 0.015$, Taylor 1.5 simulation scheme with U.S. Zero yield curve from December 1, 2000 as input parameter for each of the 100 simulated short-rate paths; parameters for the Cox-Ross-Rubinstein binomial tree: $S_{min} = 50$, $S_{max} = 150$, $M = 50$, $\sigma_S = 0.25$, $T_S = 3$, $N_S = 1080$).

S_0	Salvage factor 0	0.15	0.3	0.45	0.6	0.75	0.9
50	0.000000	0.004614	0.661	5.077	16.153	32.500	49.000
60	0.000000	0.001052	0.262	2.608	9.520	22.522	39.000
70	0.000000	0.000268	0.107	1.363	5.720	14.656	29.003
80	0.000000	0.000074	0.046	0.725	3.477	9.690	20.249
90	0.000000	0.000022	0.020	0.392	2.135	6.467	14.280
100	0.000000	0.000007	0.009	0.216	1.323	4.346	10.153
110	0.000000	0.000002	0.004	0.121	0.828	2.939	7.258
120	0.000000	0.000001	0.002	0.069	0.523	2.000	5.216
130	0.000000	0.000000	0.001	0.040	0.334	1.369	3.764
140	0.000000	0.000000	0.001	0.023	0.215	0.942	2.726
150	0.000000	0.000000	0.000	0.014	0.140	0.654	1.984

Source: ts-4-1 and ts-4-4

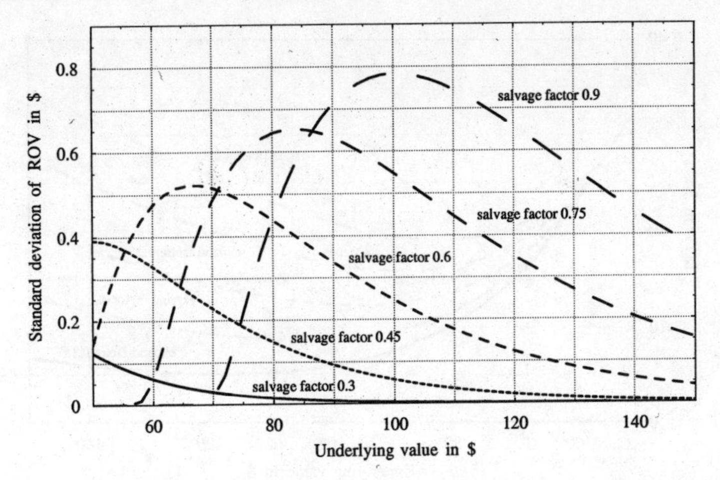

Fig. 5.52. Standard deviation of ROV in case 1 with a variable salvage factor priced with the Cox-Ross-Rubinstein binomial tree method when the risk-free rate is simulated using the Ho-Lee model (ts-4-1 and ts-4-4; corresponding to Table 5.63).

The influence of the expand factor on the investment project. This part analyzes the influence of the expand factor on the ROV and the ROV's standard deviation for cases 2 and 3. As usual, the parameters for the Ho-Lee model and for the Cox-Ross-Rubinstein binomial tree method to price the investment project are given in the corresponding tables and figures.

Case 2 is analyzed first: Figure 5.53 displays the ROV for various expand factors, the corresponding standard deviation is shown in Figure 5.54. Subsequently, case 3 is investigated: Figure 5.55 displays the ROV for various expand factors, the corresponding standard deviation is shown in Figure 5.56.

When comparing the four Figures 5.53, 5.54, 5.55, and 5.56 with similar figures for the Cox-Ingersoll-Ross model (see, e.g., Figures 5.15, 5.16, 5.17, and 5.18), the graphs do not show much difference. However, as already mentioned above, the Ho-Lee model cannot be directly compared to an equilibrium model due to the mean reversion feature. Therefore, a further comparison of these four figures was not considered.

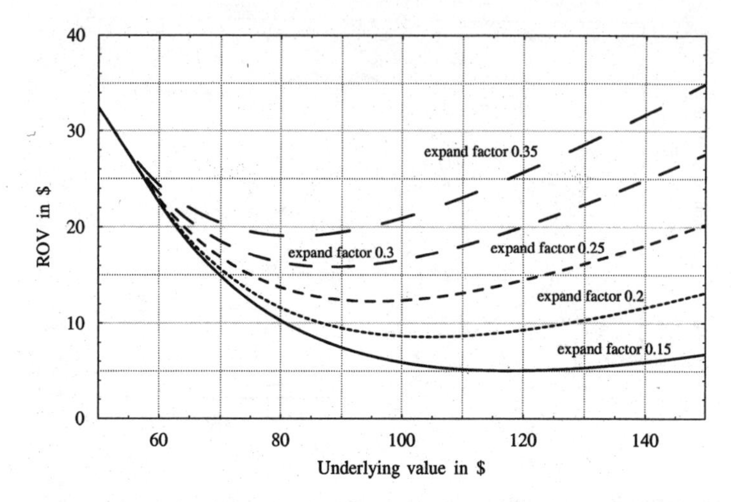

Fig. 5.53. ROV in case 2 with a variable expand factor priced when the risk-free rate is simulated using the Ho-Lee model (ts-4-2 and ts-4-5; parameters for the Ho-Lee model: $T_r = 4$, $N_r = 7200, \sigma_r = 0.015$, Taylor 1.5 simulation scheme with U.S. Zero yield curve from December 1, 2000 as input parameter for each of the 100 simulated short-rate paths; parameters for the Cox-Ross-Rubinstein binomial tree: $S_{min} = 50$, $S_{max} = 150$, $M = 50$, $\sigma_S = 0.25$, $T_S = 3$, $N_S = 1080$, salvage factor 0.75 of initial investment cost of 110, expand investment of 0.2 of initial investment cost).

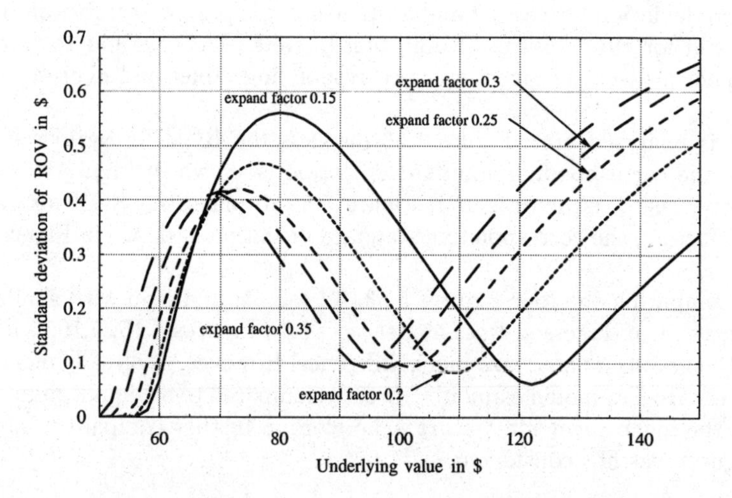

Fig. 5.54. Standard deviation of ROV in case 2 with a variable expand factor priced when the risk-free rate is simulated using the Ho-Lee model (ts-4-2 and ts-4-5; corresponding to Figure 5.53).

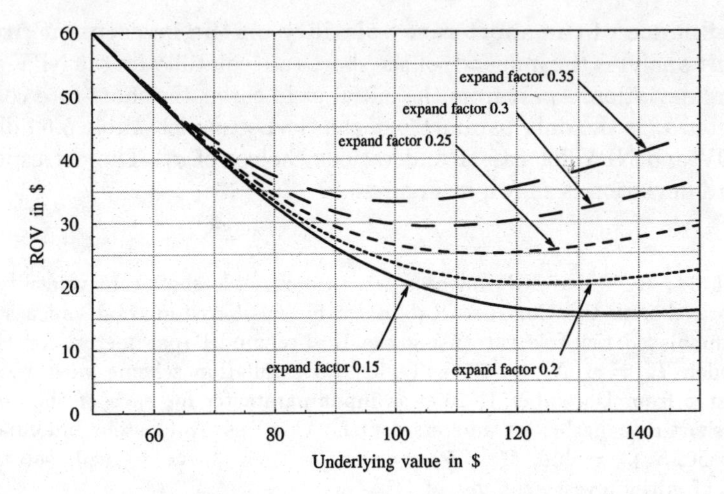

Fig. 5.55. ROV in case 3 with a variable expand factor priced when the risk-free rate is simulated using the Ho-Lee model (ts-4-3 and ts-4-6; parameters for the Ho-Lee model: $T_r = 4$, $N_r = 7200$, $\sigma_r = 0.015$, Taylor 1.5 simulation scheme with U.S. Zero yield curve from December 1, 2000 as input parameter for each of the 100 simulated short-rate paths; parameters for the Cox-Ross-Rubinstein binomial tree: $S_{min} = 50$, $S_{max} = 150$, $M = 50$, $\sigma_S = 0.25$, $T_S = 3$, $N_S = 300$, salvage factor 0.75 of initial investment cost of 110, expand investment of 0.2 of initial investment cost).

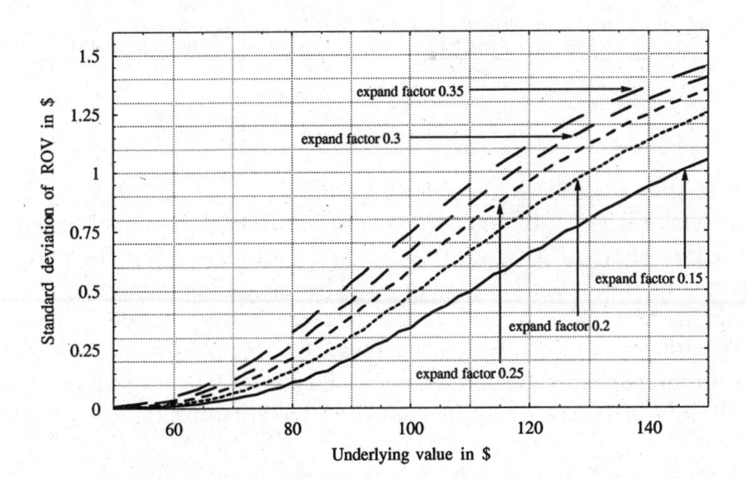

Fig. 5.56. Standard deviation of ROV in case 3 with a variable expand factor priced when the risk-free rate is simulated using the Ho-Lee model (ts-4-3 and ts-4-6; corresponding to Figure 5.55).

The influence of the short-rate volatility on the investment project. This part analyzes the influence of the short-rate volatility on the NPV and its standard deviation. Apart from the initial yield curve, the short-rate volatility parameter σ_r is the only parameter of the Ho-Lee model. Table 5.64 displays the ROV and NPV for case 1 and various choices of σ_r. The corresponding standard deviation is shown in Figure 5.57.

Table 5.64. ROV and NPV in case 1 with a variable short-rate volatility parameter priced with the Cox-Ross-Rubinstein binomial tree method and a stochastically simulated risk-free rate using the Ho-Lee model (parameters for the Ho-Lee model: $T_r = 4$, $N_r = 7200$, Taylor 1.5 simulation scheme with U.S. Zero yield curve from December 1, 2000 as input parameter for each of the 100 simulated short-rate paths; parameters for the Cox-Ross-Rubinstein binomial tree: $S_{min} = 50$, $S_{max} = 150$, $M = 50$, $\sigma_S = 0.25$, $T_S = 3$, $N_S = 1080$, salvage factor 0.75 of initial investment cost of 110).

	ROV for $\sigma_r =$				NPV for $\sigma_r =$			
S_0	0.005	0.01	0.015	0.02	0.005	0.01	0.015	0.02
50	32.500	32.500	32.500	32.500	-27.500	-27.500	-27.500	-27.500
60	22.501	22.507	22.522	22.555	-27.499	-27.493	-27.478	-27.445
70	14.600	14.619	14.656	14.721	-25.400	-25.381	-25.344	-25.279
80	9.636	9.654	9.690	9.750	-20.364	-20.346	-20.310	-20.250
90	6.419	6.436	6.467	6.518	-13.581	-13.564	-13.533	-13.482
100	4.305	4.320	4.346	4.389	-5.695	-5.680	-5.654	-5.611
110	2.904	2.916	2.939	2.973	2.904	2.916	2.939	2.973
120	1.971	1.981	2.000	2.028	11.971	11.981	12.000	12.028
130	1.346	1.354	1.369	1.391	21.346	21.354	21.369	21.391
140	0.924	0.931	0.942	0.960	30.924	30.931	30.942	30.960
150	0.639	0.645	0.654	0.668	40.639	40.645	40.654	40.668

Source: ts-4-4

Table 5.64 shows that there is almost no difference for the ROV (and, therefore, the NPV) when a different short-rate volatility parameter is used. However, a higher short-rate volatility results in a slightly higher ROV and NPV irrespective of the underlying value S_0. This is the tendency in case 1 (and also later in cases 2 and 3), but the increase is very small. For example, $\text{ROV}(S_0 = 90, \sigma_r = 0.005) = 6.419$ versus $\text{ROV}(S_0 = 90, \sigma_r = 0.02) = 6.518$, i.e., for an increase in the short-rate volatility from 0.005 by factor 4 to 0.02 the ROV only increases by $6.518 - 6.419 = 0.099$, an insignificant amount.

On the other hand, Figure 5.57 is similar to the corresponding figures for the Vasicek model (see Figure 5.39) and the Cox-Ingersoll-Ross model (see Figure 5.47). However, as already elaborated, a direct comparison of the Ho-Lee model with these two equilibrium models is not possible.

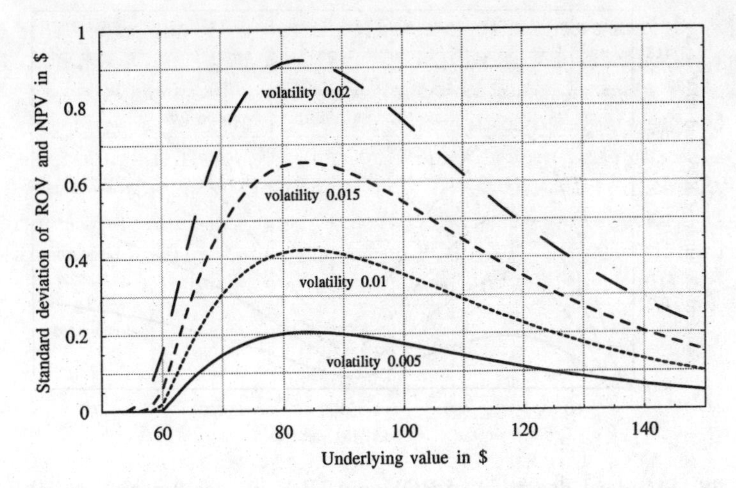

Fig. 5.57. Standard deviation of ROV and NPV in case 1 with a variable short-rate volatility parameter priced when the risk-free rate is simulated using the Ho-Lee model (ts-4-4; corresponding to Table 5.64).

The results in case 2 are displayed in Table 5.65 for the ROV and NPV; the corresponding standard deviation is shown in Figure 5.58.

Table 5.65. ROV and NPV in case 2 with a variable short-rate volatility parameter priced with the Cox-Ross-Rubinstein binomial tree method and a risk-free rate simulated with the Ho-Lee model (parameters for the Ho-Lee model: $T_r = 4$, $N_r = 7200$, Taylor 1.5 simulation scheme with U.S. Zero yield curve from December 1, 2000 as input parameter for each of the 100 simulated short-rate paths; parameters for the Cox-Ross-Rubinstein binomial tree: $S_{min} = 50$, $S_{max} = 150$, $M = 50$, $\sigma_S = 0.25$, $T_S = 3$, $N_S = 1080$, salvage factor 0.75 of initial investment cost of 110, expand factor 0.3 of prevailing underlying value with expand investment of 0.2 of initial investment cost).

	ROV for $\sigma_r =$				NPV for $\sigma_r =$			
S_0	0.005	0.01	0.015	0.02	0.005	0.01	0.015	0.02
50	32.500	32.500	32.500	32.500	-27.500	-27.500	-27.500	-27.500
60	23.463	23.476	23.503	23.554	-26.537	-26.524	-26.497	-26.446
70	18.482	18.499	18.534	18.595	-21.518	-21.501	-21.466	-21.405
80	16.267	16.284	16.318	16.376	-13.733	-13.716	-13.682	-13.624
90	15.823	15.839	15.870	15.921	-4.177	-4.161	-4.130	-4.079
100	16.540	16.554	16.580	16.623	6.540	6.554	6.580	6.623
110	18.021	18.032	18.054	18.088	18.021	18.032	18.054	18.088
120	20.011	20.019	20.036	20.063	30.011	30.019	30.036	30.063
130	22.335	22.340	22.353	22.373	42.335	42.340	42.353	42.373
140	24.879	24.883	24.892	24.906	54.879	54.883	54.892	54.906
150	27.574	27.575	27.581	27.591	67.574	67.575	67.581	67.591

Source: ts-4-5

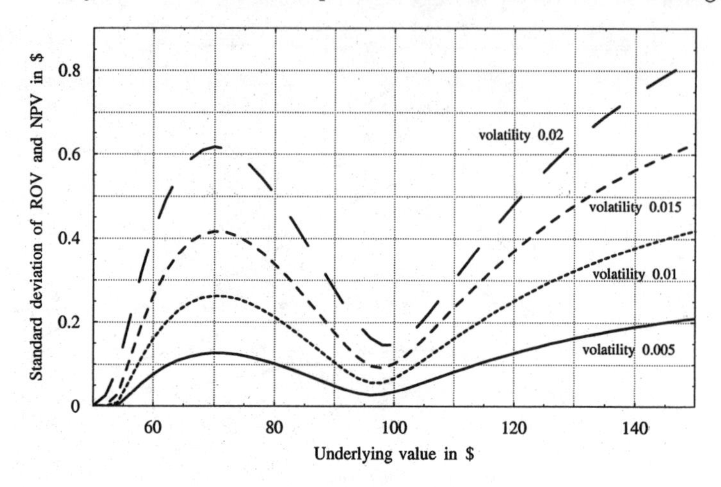

Fig. 5.58. Standard deviation of ROV and NPV in case 2 with a variable short-rate volatility parameter priced when the risk-free rate is simulated using the Ho-Lee model (ts-4-5; corresponding to Table 5.65).

The findings in case 2 are similar to the findings in case 1. A higher short-rate volatility leads to an increase of ROV and NPV. For example, $\mathrm{ROV}(S_0 = 90, \sigma_r = 0.005) = 15.823$ versus $\mathrm{ROV}(S_0 = 90, \sigma_r = 0.02) = 15.921$, i.e, for an increase in the short-rate volatility from 0.005 by factor 4 to 0.02 the ROV only increases by $15.921 - 15.823 = 0.098$, an insignificant amount. On the other hand, Figure 5.58 is similar to the corresponding figures for the Vasicek model (see Figure 5.40) and the Cox-Ingersoll-Ross model (see Figure 5.48).

The results in case 3 are displayed in Table 5.66 and in Figure 5.59 with similar findings as in cases 1 and 2. However, the increase in the ROV (and NPV) for a higher short-rate volatility is slightly stronger than in cases 1 and 2 but still insignificant, e.g., $\mathrm{ROV}(S_0 = 90, \sigma_r = 0.005) = 31.642$ versus $\mathrm{ROV}(S_0, \sigma_r = 0.02) = 31.765$, which gives an increase of $31.765 - 31.642 = 0.123$.

5.7.2 Hull-White One-Factor Model versus Hull-White Two-Factor Model

In this section various investment projects are compared by applying risk-free rates stochastically simulated according to the Hull-White one-factor and two-factor models. The investment projects are cases 1, 2 and 3. A specific choice of parameters is selected for this comparison, which is stated in the description of the tables and figures. The central question that has to be answered is whether the additional risk factor in the stochastic term structure model is of any influence on the real options pricing in the three cases considered.

Table 5.66. ROV and NPV in case 3 with a variable short-rate volatility parameter priced with the Cox-Ross-Rubinstein binomial tree method and a risk-free rate simulated with the Ho-Lee model (parameters for the Ho-Lee model: $T_r = 4$, $N_r = 7200$, Taylor 1.5 simulation scheme with U.S. Zero yield curve from December 1, 2000 as input parameter for each of the 100 simulated short-rate paths; parameters for the Cox-Ross-Rubinstein binomial tree: $S_{min} = 50$, $S_{max} = 150$, $M = 50$, $\sigma_S = 0.25$, $T_S = 3$, $N_S = 300$, salvage factor 0.75 of initial investment cost of 110, expand factor 0.3 of prevailing underlying value with expand investment of 0.2 of initial investment cost).

S_0	ROV for $\sigma_r =$ 0.005	0.01	0.015	0.02	NPV for $\sigma_r =$ 0.005	0.01	0.015	0.02
50	60.075	60.075	60.076	60.077	0.075	0.075	0.076	0.077
60	50.553	50.556	50.562	50.569	0.553	0.556	0.562	0.569
70	42.198	42.208	42.222	42.241	2.198	2.208	2.222	2.241
80	35.785	35.802	35.827	35.861	5.785	5.802	5.827	5.861
90	31.642	31.671	31.712	31.765	11.642	11.671	11.712	11.765
100	29.649	29.685	29.735	29.799	19.649	19.685	19.735	19.799
110	29.317	29.358	29.413	29.485	29.317	29.358	29.413	29.485
120	30.184	30.225	30.280	30.351	40.184	40.225	40.280	40.351
130	31.856	31.898	31.953	32.021	51.856	51.898	51.953	52.021
140	33.991	34.032	34.084	34.149	63.991	64.032	64.084	64.149
150	36.457	36.496	36.544	36.603	76.457	76.496	76.544	76.603

Source: ts-4-6

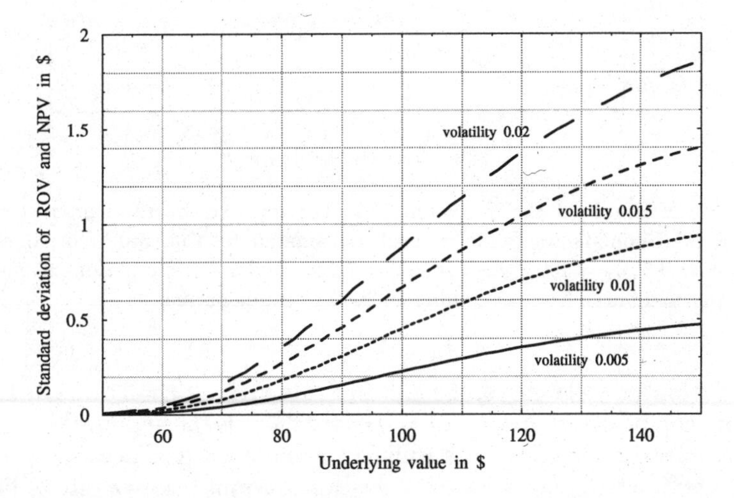

Fig. 5.59. Standard deviation of the ROV and NPV in case 3 with a variable short-rate volatility parameter priced when the risk-free rate is simulated using the Ho-Lee model (ts-4-6; corresponding to Table 5.66).

Both Hull-White models contain a mean reversion feature whereby the mean reversion force is determined by parameter α in the short-rate model. Function θ, which is derived using the initial yield curve (see Section 3.3.2), plays an important role for the mean reversion level, which is not constant, but time-dependent. For the two-factor model the mean reversion level is further influenced by a second stochastic process, an Ornstein-Uhlenbeck process. The Hull-White two-factor model contains two volatility parameters. The first one is the short-rate volatility parameter in the short-rate process itself ($\sigma_{r,1}$), the second one is the volatility parameter in the Ornstein-Uhlenbeck process ($\sigma_{r,2}$).

Figures 5.60 and 5.61 display examples of the development of the term structure of interest rates over time when the Hull-White one-factor model and the Hull-White two-factor model are used, respectively.

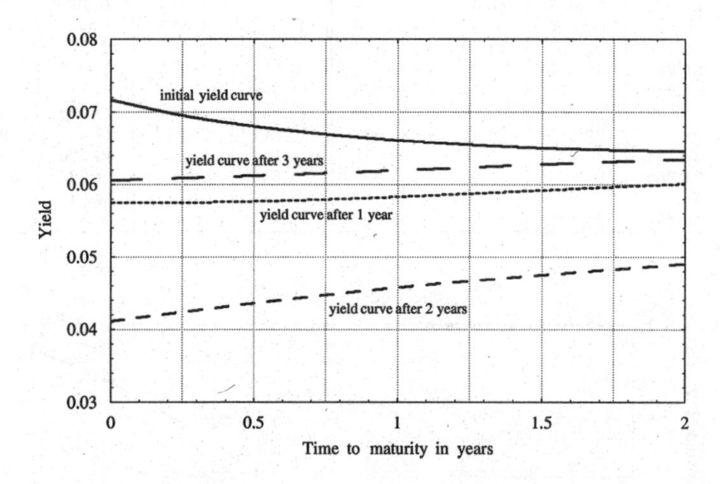

Fig. 5.60. Example of the development of the term structure of interest rates if the Hull-White one-factor model is used (parameters for the Hull White one-factor model: $\alpha = 0.3$, $\sigma_r = 0.02$, $T_r = 4$, $N_r = 7200$, Taylor 1.5 simulation scheme with U.S. Zero yield curve from December 1, 2000 as input parameter).

Given the initial yield curve, the risk-free rate r_f for $T_S = 3$ years as time to maturity can be determined as $r_f = 0.06482$. This is the appropriate constant risk-free rate for the Cox-Ross-Rubinstein binomial tree according to Hull and White[22], which allows the NPV for this constant risk-free rate to be used as a benchmark for both stochastic term structure models. In addition, the influence of the short-rate volatility parameters on the investment project will be investigated as well. For the two Hull-White models it does not make sense to analyze the results for "different mean reversion levels" as done in Section 5.6 for the equilibrium models. It is more advisable to analyze the simulation results for various choices of α, the mean reversion force.

[22] See Hull & White [62], page 344.

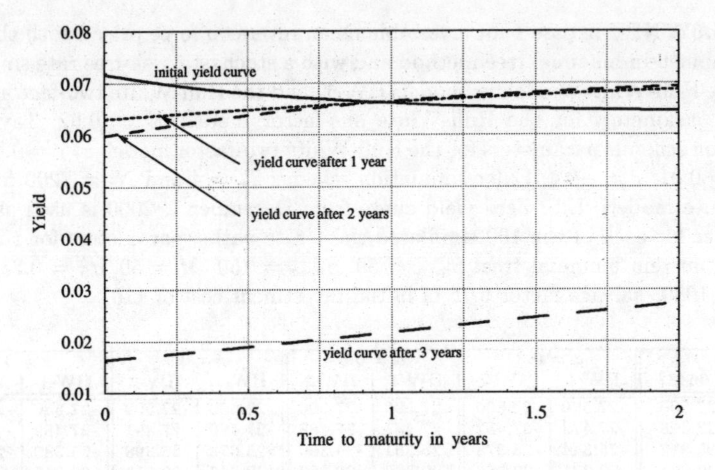

Fig. 5.61. Example of the development of the term structure of interest rates if the Hull-White two-factor model is used (parameters for the Hull White two-factor model: $\alpha = 0.3$, $\sigma_{r,1} = 0.02$, $\xi = 1$, $\sigma_{r,2} = 0.01$, $\rho = -0.4$, $T_r = 4$, $N_r = 7200$, Euler simulation scheme and U.S. Zero yield curve from December 1, 2000 as input parameter).

The influence of parameter α on the investment project. This section analyzes how the mean reversion force α influences the investment project and how the Hull-White one-factor and two-factor models differ in their influence on the investment project. This analysis is done case by case whereby only the NPV will be displayed. For case 1 the salvage factor is always 0.75 of the initial investment cost of $I_0 = 110$. The NPV for various values of α and the two Hull-White models are displayed in Table 5.67. The corresponding standard deviation of the NPV is displayed in Table 5.68.

Table 5.67 shows that there is almost no difference between the results of the Hull-White one-factor model and those of the Hull-White two-factor model for any given α. The same holds true when the NPVs for different values of α are considered. When comparing the NPVs derived by applying either the Hull-White one- or two-factor model with the NPVs derived by using the appropriate risk-free rate (3-year spot rate as of December 1, 2000), there is no significant difference either. However, the corresponding standard deviations shown in Table 5.68 display a significant difference. For a given underlying value a higher value of α implies a lower standard deviation. Economically, this makes sense since parameter α as the mean reversion force determines how "quick" the interest rate reverts back to its mean. A quick mean reversion leads to less volatility in the term structure of interest rates, which results in a lower standard deviation of the NPV.

Table 5.67. NPV in case 1 for a variable mean reversion force priced with the Cox-Ross-Rubinstein binomial tree method and with a stochastic risk-free rate simulated using the Hull-White one-factor model (HW-1) and the Hull-White two-factor model (HW-2; parameters for the Hull-White one-factor model: $\sigma_r = 0.02$, Taylor 1.5 simulation scheme; parameters for the Hull-White two-factor model: $\sigma_{r,1} = 0.02$, $\xi = 1$, $\sigma_{r,2} = 0.01$, $\rho = -0.4$, Euler simulation scheme; $T_r = 4$ and $N_r = 7200$ for both Hull-White models, U.S. Zero yield curve from December 1, 2000 is used as input parameter for each of the 100 simulated short-rate paths; parameters for the Cox-Ross-Rubinstein binomial tree: $S_{min} = 50$, $S_{max} = 150$, $M = 50$, $\sigma_S = 0.25$, $T_S = 3$, $N_S = 1080$, salvage factor 0.75 of initial investment cost of 110).

	$r_f =$	$\alpha = 0.3$		$\alpha = 0.5$		$\alpha = 0.7$		$\alpha = 0.9$	
S_0	0.06482	HW-1	HW-2	HW-1	HW-2	HW-1	HW-2	HW-1	HW-2
50	-27.500	-27.500	-27.500	-27.500	-27.500	-27.500	-27.500	-27.500	-27.500
60	-27.498	-27.473	-27.482	-27.481	-27.488	-27.486	-27.491	-27.489	-27.493
70	-25.312	-25.344	-25.378	-25.361	-25.390	-25.372	-25.398	-25.380	-25.403
80	-20.301	-20.321	-20.358	-20.336	-20.369	-20.346	-20.375	-20.353	-20.379
90	-13.545	-13.549	-13.581	-13.563	-13.590	-13.571	-13.595	-13.576	-13.598
100	-5.675	-5.672	-5.697	-5.682	-5.704	-5.689	-5.708	-5.693	-5.710
110	2.914	2.922	2.902	2.913	2.896	2.908	2.893	2.904	2.892
120	11.976	11.985	11.971	11.978	11.966	11.974	11.963	11.971	11.962
130	21.349	21.358	21.347	21.352	21.343	21.348	21.341	21.346	21.339
140	30.925	30.933	30.925	30.929	30.922	30.926	30.920	30.924	30.919
150	40.640	40.647	40.641	40.643	40.638	40.641	40.637	40.640	40.636

Source: ts-4-7, ts-4-10, and ts-4-15

Table 5.68. Standard deviation of NPV in case 1 for a variable mean reversion force priced with the Cox-Ross-Rubinstein binomial tree method and with a stochastic risk-free rate simulated using the Hull-White one-factor and two-factor models, corresponding to Table 5.67.

	$\alpha = 0.3$		$\alpha = 0.5$		$\alpha = 0.7$		$\alpha = 0.9$	
S_0	HW-1	HW-2	HW-1	HW-2	HW-1	HW-2	HW-1	HW-2
50	0.000	0.000	0.000	0.000	0.000	0.000	0.000	0.000
60	0.060	0.057	0.043	0.038	0.032	0.027	0.025	0.020
70	0.518	0.456	0.448	0.392	0.394	0.345	0.352	0.307
80	0.683	0.608	0.592	0.525	0.521	0.462	0.465	0.411
90	0.667	0.596	0.576	0.514	0.505	0.450	0.449	0.399
100	0.576	0.516	0.495	0.443	0.433	0.387	0.383	0.342
110	0.467	0.419	0.400	0.359	0.349	0.312	0.308	0.276
120	0.365	0.328	0.312	0.280	0.271	0.244	0.239	0.214
130	0.280	0.252	0.238	0.214	0.207	0.186	0.182	0.163
140	0.211	0.190	0.180	0.162	0.155	0.140	0.136	0.123
150	0.159	0.143	0.135	0.121	0.116	0.104	0.102	0.092

Source: ts-4-7 and ts-4-10

For case 2 the salvage factor is always 0.75 of the initial investment cost of $I_0 = 110$. The expand factor is 0.3 of the then prevailing underlying value and the expand investment is 0.2 of I_0. The NPV for various values of α and the two Hull-White models is displayed in Table 5.69. The corresponding standard deviations of the NPV are displayed in Table 5.70.

Table 5.69. NPV in case 2 for a variable mean reversion force priced with the Cox-Ross-Rubinstein binomial tree method and with a stochastic risk-free rate simulated using the Hull-White one-factor and two-factor models (parameters for the Hull-White one-factor model: $\sigma_r = 0.02$, Taylor 1.5 simulation scheme; parameters for the Hull-White two-factor model: $\sigma_{r,1} = 0.02$, $\xi = 1$, $\sigma_{r,2} = 0.01$, $\rho = -0.4$, Euler simulation scheme; $T_r = 4$ and $N_r = 7200$ for both Hull-White models, U.S. Zero yield curve from December 1, 2000 is used as input parameter for each of the 100 simulated short-rate paths; parameters for the Cox-Ross-Rubinstein binomial tree: $S_{min} = 50$, $S_{max} = 150$, $M = 50$, $\sigma_S = 0.25$, $T_S = 3$, $N_S = 1080$, salvage factor 0.75 of initial investment cost of 110, expand factor 0.3 of prevailing underlying value with expand investment of 0.2 of initial investment cost).

	$r_f =$	$\alpha = 0.3$		$\alpha = 0.5$		$\alpha = 0.7$		$\alpha = 0.9$	
S_0	0.06482	HW-1	HW-2	HW-1	HW-2	HW-1	HW-2	HW-1	HW-2
50	-27.500	-27.500	-27.500	-27.500	-27.500	-27.500	-27.500	-27.500	-27.500
60	-26.450	-26.487	-26.513	-26.501	-26.523	-26.510	-26.529	-26.516	-26.532
70	-21.445	-21.457	-21.494	-21.476	-21.508	-21.488	-21.517	-21.496	-21.517
80	-13.694	-13.672	-13.706	-13.691	-13.721	-13.704	-13.730	-13.712	-13.731
90	-4.160	-4.116	-4.142	-4.135	-4.157	-4.147	-4.166	-4.155	-4.169
100	6.544	6.600	6.582	6.583	6.568	6.571	6.558	6.563	6.554
110	18.019	18.081	18.069	18.065	18.055	18.054	18.046	18.047	18.040
120	30.005	30.069	30.062	30.055	30.049	30.045	30.041	30.037	30.033
130	42.327	42.391	42.388	42.378	42.376	42.369	42.368	42.362	42.359
140	54.871	54.934	54.934	54.922	54.923	54.914	54.915	54.908	54.905
150	67.565	67.627	67.629	67.616	67.618	67.608	67.611	67.602	67.601

Source: ts-4-8, ts-4-11, and ts-4-15

Table 5.70. Standard deviation of NPV in case 2 for a variable mean reversion force priced with the Cox-Ross-Rubinstein binomial tree method and with a stochastic risk-free rate simulated using the Hull-White one-factor and two-factor models, corresponding to Table 5.69.

	$\alpha = 0.3$		$\alpha = 0.5$		$\alpha = 0.7$		$\alpha = 0.9$	
S_0	HW-1	HW-2	HW-1	HW-2	HW-1	HW-2	HW-1	HW-2
50	0.000	0.000	0.000	0.000	0.000	0.000	0.000	0.000
60	0.291	0.253	0.254	0.220	0.226	0.195	0.205	0.173
70	0.455	0.401	0.403	0.354	0.362	0.317	0.329	0.283
80	0.369	0.326	0.328	0.289	0.296	0.261	0.270	0.234
90	0.198	0.174	0.179	0.157	0.164	0.144	0.151	0.129
100	0.129	0.117	0.107	0.097	0.093	0.084	0.082	0.078
110	0.260	0.238	0.216	0.196	0.182	0.165	0.156	0.152
120	0.400	0.363	0.334	0.303	0.285	0.258	0.246	0.235
130	0.513	0.465	0.431	0.390	0.368	0.333	0.319	0.303
140	0.600	0.543	0.505	0.457	0.433	0.391	0.376	0.355
150	0.665	0.602	0.561	0.507	0.481	0.434	0.418	0.394

Source: ts-4-8 and ts-4-11

Table 5.69 shows that for case 2 as well the influence of parameter α is insignificant. The comparison of the NPVs for a constant risk-free rate versus a stochastically modelled risk-free rate offers an interesting result in case 2. In general the NPV is slightly lower for a constant risk-free rate than for a stochastically simulated rate using one of the Hull-White models if S_0 is not too low. The difference is higher for higher values of the underlying. This phenomenon could not be observed in case 1.

For case 3 the salvage factor is always 0.75 of the initial investment cost of $I_0 = 110$. The expand factor is 0.3 of the then prevailing underlying value, and the expand investment is 0.2 of I_0. The defer period is exactly one year. The NPV for various values of α and the two Hull-White models are displayed in Table 5.71. The corresponding standard deviation of the NPV is displayed in Table 5.72.

Table 5.71. NPV in case 3 for a variable mean reversion force priced with the Cox-Ross-Rubinstein binomial tree method and with a stochastic risk-free rate simulated using the Hull-White one-factor and two-factor models (parameters for the Hull-White one-factor model: $\sigma_r = 0.02$, Taylor 1.5 simulation scheme; parameters for the Hull-White two-factor model: $\sigma_{r,1} = 0.02$, $\xi = 1$, $\sigma_{r,2} = 0.01$, $\rho = -0.4$, Euler simulation scheme; $T_r = 4$ and $N_r = 7200$ for both Hull-White models, U.S. Zero yield curve from December 1, 2000 is used as input parameter for each of the 100 simulated short-rate paths; parameters for the Cox-Ross-Rubinstein binomial tree: $S_{min} = 50$, $S_{max} = 150$, $M = 50$, $\sigma_S = 0.25$, $T_S = 3$, $N_S = 300$, salvage factor 0.75 of initial investment cost of 110, expand factor 0.3 of prevailing underlying value with expand investment of 0.2 of initial investment cost).

	$r_f =$	$\alpha = 0.3$		$\alpha = 0.5$		$\alpha = 0.7$		$\alpha = 0.9$	
S_0	0.06482	HW-1	HW-2	HW-1	HW-2	HW-1	HW-2	HW-1	HW-2
50	0.072	0.076	0.076	0.076	0.076	0.075	0.076	0.075	0.076
60	0.538	0.559	0.560	0.557	0.558	0.555	0.557	0.554	0.556
70	2.153	2.214	2.218	2.208	2.213	2.204	2.210	2.201	2.208
80	5.694	5.813	5.824	5.802	5.815	5.795	5.810	5.791	5.806
90	11.496	11.689	11.706	11.671	11.692	11.659	11.682	11.652	11.676
100	19.453	19.706	19.732	19.684	19.714	19.670	19.702	19.661	19.694
110	29.082	29.383	29.414	29.358	29.394	29.342	29.381	29.331	29.372
120	39.921	40.250	40.288	40.225	40.268	40.209	40.255	40.198	40.245
130	51.575	51.924	51.966	51.899	51.946	51.883	51.932	51.871	51.922
140	63.698	64.059	64.104	64.034	64.084	64.018	64.071	64.007	64.061
150	76.157	76.521	76.569	76.498	76.550	76.483	76.537	76.472	76.528

Source: ts-4-9, ts-4-12, and ts-4-15

The analysis of Table 5.71 shows that the influence of parameter α on the NPV is insignificant and that the NPVs do not differ significantly between the two Hull-White models. Table 5.71 also shows that for high values of the underlying the NPV is significantly less for a constant risk-free rate than for any of the two Hull-White models. This already has occured in case 2 but is even more obvious in case 3. The interpretation of the standard deviation for case 3 (see Table 5.72) is analogous to the interpretation in cases 1 and 2: A higher α yields a lower standard deviation independent of the underlying value. The data of Table 5.72 are graphically displayed in Figures 5.62 and 5.63.

Table 5.72. Standard deviation of NPV in case 3 for a variable mean reversion force priced with the Cox-Ross-Rubinstein binomial tree method and with a stochastic risk-free rate simulated using the Hull-White one-factor and two-factor models, corresponding to Table 5.71.

	$\alpha = 0.3$		$\alpha = 0.5$		$\alpha = 0.7$		$\alpha = 0.9$	
S_0	HW-1	HW-2	HW-1	HW-2	HW-1	HW-2	HW-1	HW-2
50	0.007	0.006	0.007	0.006	0.006	0.005	0.006	0.005
60	0.043	0.038	0.040	0.035	0.037	0.032	0.035	0.030
70	0.141	0.124	0.129	0.113	0.120	0.104	0.112	0.097
80	0.315	0.277	0.286	0.250	0.264	0.228	0.246	0.211
90	0.522	0.460	0.472	0.412	0.433	0.375	0.402	0.346
100	0.749	0.659	0.671	0.585	0.612	0.530	0.565	0.486
110	0.958	0.842	0.852	0.743	0.772	0.669	0.710	0.611
120	1.143	1.005	1.011	0.882	0.911	0.789	0.832	0.717
130	1.287	1.130	1.130	0.986	1.013	0.879	0.923	0.795
140	1.404	1.234	1.228	1.072	1.096	0.951	0.995	0.858
150	1.494	1.313	1.301	1.136	1.158	1.005	1.047	0.904

Source: ts-4-9 and ts-4-12

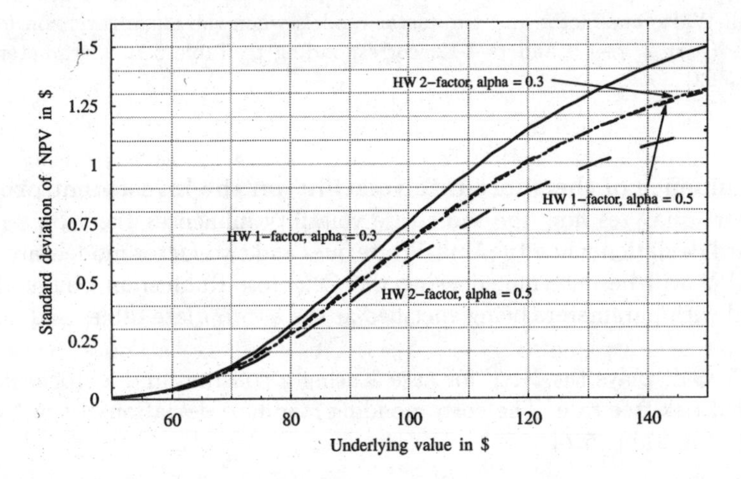

Fig. 5.62. Standard deviation of NPV in case 3 priced with the Cox-Ross-Rubinstein binomial tree method using a stochastic risk-free rate simulated with the Hull-White one-factor and two-factor models when the mean reversion force is variable (part 1; ts-4-9 and ts-4-12; corresponding to Table 5.72; parameters for the Hull-White one-factor model: $\sigma_r = 0.02$, Taylor 1.5 simulation scheme; parameters for the Hull-White two-factor model: $\sigma_{r,1} = 0.02$, $\xi = 1$, $\sigma_{r,2} = 0.01$, $\rho = -0.4$, Euler simulation scheme; $T_r = 4$ and $N_r = 7200$ for both Hull-White models, U.S. Zero yield curve from December 1, 2000 is used as input parameter for each of the 100 simulated short-rate paths; parameters for the Cox-Ross-Rubinstein binomial tree: $S_{min} = 50$, $S_{max} = 150$, $M = 50$, $\sigma_S = 0.25$, $T_S = 3$, $N_S = 300$, salvage factor 0.75 of initial investment cost of 110, expand factor 0.3 of prevailing underlying value with expand investment of 0.2 of initial investment cost).

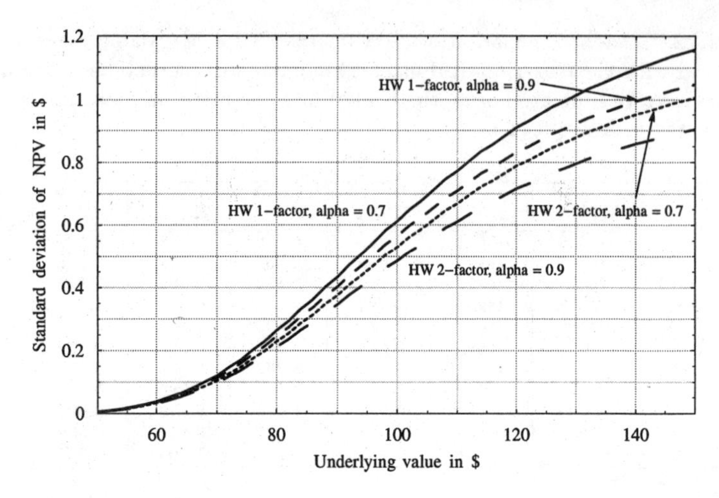

Fig. 5.63. Standard deviation of NPV in case 3 priced with the Cox-Ross-Rubinstein binomial tree method using a stochastic risk-free rate simulated with the Hull-White one-factor and two-factor models when the mean reversion force is variable (part 2; ts-4-9 and ts-4-12; corresponding to Table 5.72; parameters: see Figure 5.62).

The influence of the short-rate volatility on the investment project. This part analyzes how the short-rate volatility influences the NPV and its standard deviation when the Hull-White one- and two-factor models are used. The NPV will be determined using the Cox-Ross-Rubinstein binomial tree method with parameters being specified at the appropriate tables and figures.

Table 5.73 displays the NPV for case 1 using a constant and a stochastically modelled risk-free rate. The corresponding standard deviation of the NPV is displayed in Table 5.74.

For a given value of $\sigma_{r,1}$ the NPVs for both Hull-White models are similar. For both the Hull-White one-factor and two-factor models a higher value of $\sigma_{r,1}$ yields a slightly higher NPV independent of the chosen underlying value. According to Table 5.74, the standard deviations are highly dependent on the volatility parameter since the standard deviation of both Hull-White models gets larger for a higher value of $\sigma_{r,1}$, irrespective of the underlying value.

Table 5.75 displays the NPV for case 2 using a constant and a stochastically modelled risk-free rate. The corresponding standard deviations of the NPV are displayed in Table 5.76.

Table 5.73. NPV in case 1 for a variable short-rate volatility priced with the Cox-Ross-Rubinstein binomial tree method and with a stochastic risk-free rate simulated using the Hull-White one-factor and two-factor models (parameters for the Hull-White one-factor model: $\alpha = 0.3$, $\sigma_r = \sigma_{r,1}$, Taylor 1.5 simulation scheme; parameters for the Hull-White two-factor model: $\alpha = 0.3$, $\xi = 1$, $\sigma_{r,2} = 0.01$, $\rho = -0.4$, Euler simulation scheme; $T_r = 4$ and $N_r = 7200$ for both Hull-White models, U.S. Zero yield curve from December 1, 2000 is used as input parameter for each of the 100 simulated short-rate paths; parameters for the Cox-Ross-Rubinstein binomial tree: $S_{min} = 50$, $S_{max} = 150$, $M = 50$, $\sigma_S = 0.25$, $T_S = 3$, $N_S = 1080$, salvage factor 0.75 of initial investment cost of 110).

S_0	$r_f =$ 0.06482	$\sigma_{r,1} = 0.005$ HW-1	HW-2	$\sigma_{r,1} = 0.01$ HW-1	HW-2	$\sigma_{r,1} = 0.015$ HW-1	HW-2	$\sigma_{r,1} = 0.02$ HW-1	HW-2
50	-27.500	-27.500	-27.500	-27.500	-27.500	-27.500	-27.500	-27.500	-27.500
60	-27.498	-27.500	-27.500	-27.496	-27.497	-27.487	-27.492	-27.473	-27.482
70	-25.312	-25.405	-25.405	-25.396	-25.407	-25.376	-25.398	-25.344	-25.378
80	-20.301	-20.371	-20.371	-20.364	-20.377	-20.348	-20.373	-20.321	-20.358
90	-13.545	-13.589	-13.588	-13.584	-13.595	-13.572	-13.593	-13.549	-13.581
100	-5.675	-5.703	-5.702	-5.700	-5.708	-5.689	-5.707	-5.672	-5.697
110	2.914	2.897	2.898	2.899	2.893	2.907	2.894	2.922	2.902
120	11.976	11.965	11.966	11.967	11.963	11.974	11.964	11.985	11.971
130	21.349	21.341	21.342	21.343	21.339	21.348	21.341	21.358	21.347
140	30.925	30.920	30.921	30.921	30.919	30.926	30.921	30.933	30.925
150	40.640	40.636	40.637	40.637	40.636	40.641	40.637	40.647	40.641

Source: ts-4-7, ts-4-10, and ts-4-15

Table 5.74. Standard deviation of NPV in case 1 for a variable short-rate volatility priced with the Cox-Ross-Rubinstein binomial tree method and with a stochastic risk-free rate simulated using the Hull-White one-factor and two-factor models, corresponding to Table 5.73.

S_0	$\sigma_{r,1} = 0.005$ HW-1	HW-2	$\sigma_{r,1} = 0.01$ HW-1	HW-2	$\sigma_{r,1} = 0.015$ HW-1	HW-2	$\sigma_{r,1} = 0.02$ HW-1	HW-2
50	0.000	0.000	0.000	0.000	0.000	0.000	0.000	0.000
60	0.001	0.002	0.010	0.008	0.028	0.024	0.060	0.057
70	0.120	0.127	0.244	0.214	0.375	0.327	0.518	0.456
80	0.164	0.177	0.330	0.294	0.502	0.443	0.683	0.608
90	0.161	0.178	0.325	0.292	0.492	0.436	0.667	0.596
100	0.140	0.157	0.281	0.254	0.426	0.379	0.576	0.516
110	0.113	0.129	0.228	0.207	0.345	0.308	0.467	0.419
120	0.089	0.102	0.178	0.163	0.270	0.241	0.365	0.328
130	0.068	0.079	0.137	0.125	0.207	0.185	0.280	0.252
140	0.051	0.060	0.103	0.094	0.156	0.140	0.211	0.190
150	0.038	0.045	0.077	0.071	0.117	0.105	0.159	0.143

Source: ts-4-7 and ts-4-10

Table 5.75. NPV in case 2 for a variable short-rate volatility priced with the Cox-Ross-Rubinstein binomial tree method and with a stochastic risk-free rate simulated using the Hull-White one-factor and two-factor models (parameters for the Hull-White one-factor model: $\alpha = 0.3$, $\sigma_r = \sigma_{r,1}$, Taylor 1.5 simulation scheme; parameters for the Hull-White two-factor models: $\alpha = 0.3$, $\xi = 1$, $\sigma_{r,2} = 0.01$, $\rho = -0.4$, Euler simulation scheme; $T_r = 4$ and $N_r = 7200$ for both Hull-White models, U.S. Zero yield curve from December 1, 2000 is used as input parameter for each of the 100 simulated short-rate paths; parameters for the Cox-Ross-Rubinstein binomial tree: $S_{min} = 50$, $S_{max} = 150$, $M = 50$, $\sigma_S = 0.25$, $T_S = 3$, $N_S = 1080$, salvage factor 0.75 of initial investment cost of 110, expand factor 0.3 of prevailing underlying value with expand investment of 0.2 of initial investment cost).

	$r_f =$	$\sigma_{r,1} = 0.005$		$\sigma_{r,1} = 0.01$		$\sigma_{r,1} = 0.015$		$\sigma_{r,1} = 0.02$	
S_0	0.06482	HW-1	HW-2	HW-1	HW-2	HW-1	HW-2	HW-1	HW-2
50	-27.500	-27.500	-27.500	-27.500	-27.500	-27.500	-27.500	-27.500	-27.500
60	-26.450	-26.538	-26.539	-26.530	-26.538	-26.513	-26.530	-26.487	-26.513
70	-21.445	-21.519	-21.521	-21.510	-21.523	-21.489	-21.514	-21.457	-21.494
80	-13.694	-13.735	-13.735	-13.724	-13.736	-13.704	-13.726	-13.672	-13.706
90	-4.160	-4.177	-4.178	-4.166	-4.174	-4.146	-4.163	-4.116	-4.142
100	6.544	6.542	6.541	6.554	6.548	6.573	6.561	6.600	6.582
110	18.019	18.026	18.025	18.038	18.035	18.057	18.049	18.081	18.069
120	30.005	30.018	30.017	30.030	30.028	30.047	30.043	30.069	30.062
130	42.327	42.344	42.342	42.356	42.355	42.372	42.370	42.391	42.388
140	54.871	54.890	54.888	54.902	54.902	54.917	54.917	54.934	54.934
150	67.565	67.585	67.583	67.597	67.598	67.611	67.613	67.627	67.629

Source: ts-4-8, ts-4-11, and ts-4-15

Table 5.76. Standard deviation of NPV in case 2 for a variable short-rate volatility priced with the Cox-Ross-Rubinstein binomial tree method and with a stochastic risk-free rate simulated using the Hull-White one-factor and two-factor models, corresponding to Table 5.75.

	$\sigma_{r,1} = 0.005$		$\sigma_{r,1} = 0.01$		$\sigma_{r,1} = 0.015$		$\sigma_{r,1} = 0.02$	
S_0	HW-1	HW-2	HW-1	HW-2	HW-1	HW-2	HW-1	HW-2
50	0.000	0.000	0.000	0.000	0.000	0.000	0.000	0.000
60	0.064	0.062	0.131	0.113	0.205	0.176	0.291	0.253
70	0.106	0.104	0.215	0.186	0.330	0.286	0.455	0.401
80	0.086	0.083	0.174	0.150	0.267	0.232	0.369	0.326
90	0.044	0.040	0.089	0.076	0.140	0.121	0.198	0.174
100	0.030	0.042	0.061	0.060	0.094	0.086	0.129	0.117
110	0.066	0.087	0.132	0.126	0.197	0.180	0.260	0.238
120	0.101	0.128	0.202	0.190	0.302	0.274	0.400	0.363
130	0.130	0.161	0.259	0.242	0.387	0.350	0.513	0.465
140	0.151	0.186	0.302	0.282	0.452	0.408	0.600	0.543
150	0.167	0.205	0.334	0.311	0.500	0.452	0.665	0.602

Source: ts-4-8 and ts-4-11

Table 5.75 shows that the NPVs for any given underlying value are very similar regardless of the choice of a constant risk-free rate or of one of the two Hull-White models for simulating the risk-free interest rate. It is evident that for case 2 the influence of the volatility parameter $\sigma_{r,1}$ is insignificant. However, a higher volatility parameter results in a higher standard deviation for any given underlying value, see Table 5.76.

Table 5.77. NPV in case 3 for a variable short-rate volatility priced with the Cox-Ross-Rubinstein binomial tree method and with a stochastic risk-free rate simulated using the Hull-White one-factor and two-factor models (parameters for the Hull-White one-factor model: $\alpha = 0.3$, $\sigma_r = \sigma_{r,1}$, Taylor 1.5 simulation scheme; parameters for the Hull-White two-factor model: $\alpha = 0.3$, $\xi = 1$, $\sigma_{r,2} = 0.01$, $\rho = -0.4$, Euler simulation scheme; $T_r = 4$ and $N_r = 7200$ for both Hull-White models, U.S. Zero yield curve from December 1, 2000 is used as input parameter for each of the 100 simulated short-rate paths; parameters for the Cox-Ross-Rubinstein binomial tree: $S_{min} = 50$, $S_{max} = 150$, $M = 50$, $\sigma_S = 0.25$, $T_S = 3$, $N_S = 300$, salvage factor 0.75 of initial investment cost of 110, expand factor 0.3 of prevailing underlying value with expand investment of 0.2 of initial investment cost).

	$r_f =$	$\sigma_{r,1} = 0.005$		$\sigma_{r,1} = 0.01$		$\sigma_{r,1} = 0.015$		$\sigma_{r,1} = 0.02$	
S_0	0.06482	HW-1	HW-2	HW-1	HW-2	HW-1	HW-2	HW-1	HW-2
50	0.072	0.075	0.075	0.075	0.075	0.075	0.076	0.076	0.076
60	0.538	0.551	0.552	0.553	0.554	0.556	0.557	0.559	0.560
70	2.153	2.194	2.196	2.198	2.202	2.205	2.209	2.214	2.218
80	5.694	5.778	5.783	5.786	5.793	5.797	5.807	5.813	5.824
90	11.496	11.631	11.639	11.644	11.657	11.663	11.679	11.689	11.706
100	19.453	19.635	19.645	19.652	19.669	19.676	19.698	19.706	19.732
110	29.082	29.302	29.314	29.322	29.342	29.349	29.376	29.383	29.414
120	39.921	40.169	40.181	40.189	40.213	40.216	40.248	40.250	40.288
130	51.575	51.842	51.854	51.863	51.888	51.891	51.926	51.924	51.966
140	63.698	63.978	63.990	63.999	64.026	64.026	64.064	64.059	64.104
150	76.157	76.444	76.456	76.465	76.492	76.491	76.530	76.521	76.569

Source: ts-4-9, ts-4-12, and ts-4-15

Table 5.78. Standard deviation of NPV in case 3 for a variable short-rate volatility priced with the Cox-Ross-Rubinstein binomial tree method and with a stochastic risk-free rate simulated using the Hull-White one-factor and two-factor models, corresponding to Table 5.77.

	$\sigma_{r,1} = 0.005$		$\sigma_{r,1} = 0.01$		$\sigma_{r,1} = 0.015$		$\sigma_{r,1} = 0.02$	
S_0	HW-1	HW-2	HW-1	HW-2	HW-1	HW-2	HW-1	HW-2
50	0.002	0.002	0.004	0.003	0.005	0.005	0.007	0.006
60	0.011	0.010	0.021	0.018	0.032	0.028	0.043	0.038
70	0.035	0.033	0.070	0.061	0.105	0.092	0.141	0.124
80	0.079	0.076	0.158	0.136	0.237	0.206	0.315	0.277
90	0.131	0.129	0.262	0.227	0.393	0.342	0.522	0.460
100	0.189	0.189	0.377	0.328	0.564	0.490	0.749	0.659
110	0.241	0.247	0.482	0.421	0.721	0.627	0.958	0.842
120	0.288	0.300	0.576	0.505	0.862	0.750	1.143	1.005
130	0.324	0.343	0.647	0.569	0.968	0.844	1.287	1.130
140	0.354	0.379	0.706	0.623	1.057	0.921	1.404	1.234
150	0.376	0.406	0.750	0.663	1.123	0.980	1.494	1.313

Source: ts-4-9 and ts-4-12

Table 5.77 displays the NPV for case 3 using a constant and a stochastically modelled risk-free rate. The corresponding standard deviation of the NPV is shown in Table 5.78 and graphically displayed in Figures 5.64 and 5.65.

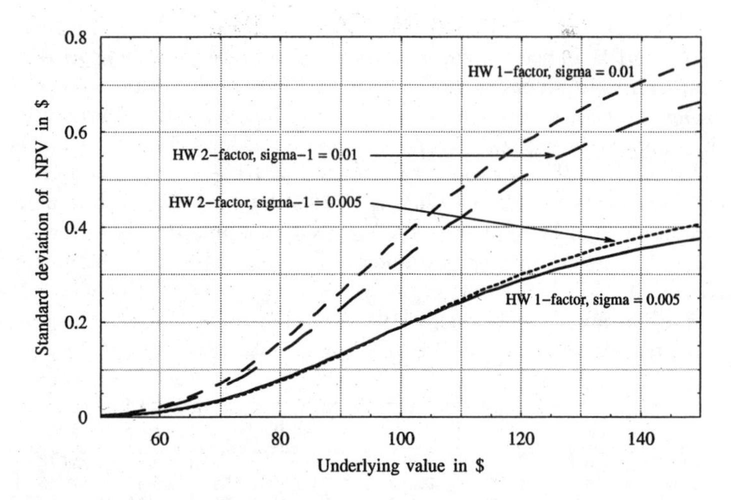

Fig. 5.64. Standard deviation of NPV in case 3 priced with the Cox-Ross-Rubinstein binomial tree method using a stochastic risk-free rate simulated with the Hull-White one-factor and two-factor models with variable short-rate volatility (part 1; ts-4-9 and ts-4-12; corresponding to Table 5.78; parameters: see Table 5.77).

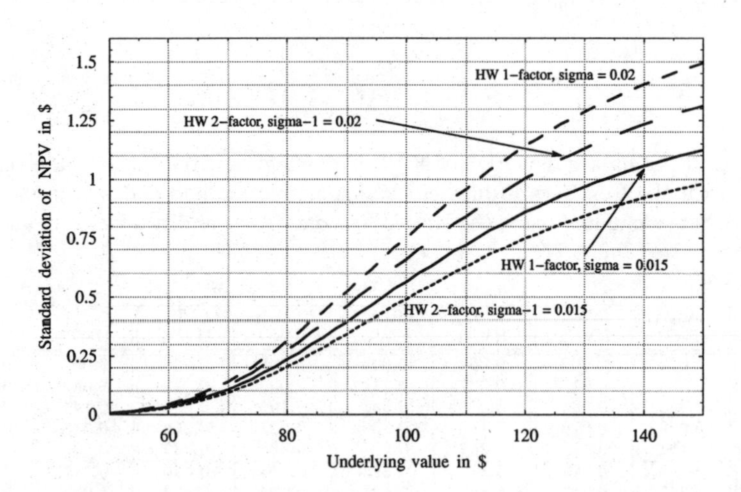

Fig. 5.65. Standard deviation of NPV in case 3 priced with the Cox-Ross-Rubinstein binomial tree method using a stochastic risk-free rate simulated with the Hull-White one-factor and two-factor models with variable short-rate volatility (part 2; ts-4-9 and ts-4-12; corresponding to Table 5.78; parameters: see Table 5.77).

Table 5.77 shows that the NPV is lower for a constant risk-free rate than for a stochastically simulated risk-free rate. This holds especially true for high values of the underlying.

5.7.3 Ho-Lee Model versus Hull-White One-Factor Model

In this final section of test situation 4, the results when using the Ho-Lee model versus using the Hull-White one-factor model will be compared. As already seen in the previous section, the results of the NPV (and, therefore, the ROV) of an investment project are almost identical when using the Hull-White one-factor and the Hull-White two-factor models to simulate the risk-free interest rate.

In this section, only the NPV will be compared but not the standard deviations. As seen in the previous section, parameter α of the Hull-White one-factor model has a significant influence on the standard deviation of any case, whereas the Ho-Lee model does not include this parameter. A comparison of the standard deviations for the Ho-Lee model versus the Hull-White one-factor model is therefore ruled out. Table 5.79 displays the NPV in case 1 for both the Ho-Lee model and the Hull-White one-factor model. The NPVs for cases 2 and 3 are displayed in Tables 5.80 and 5.81, respectively.

Table 5.79. NPV in case 1 priced with the Cox-Ross-Rubinstein binomial tree method and with a stochastic risk-free rate simulated using the Ho-Lee model and the Hull-White one-factor model (parameters for both term structure models: $T_r = 4$ and $N_r = 7200$, Taylor 1.5 simulation scheme, U.S. Zero yield curve from December 1, 2000 is used as input parameter for each of the 100 simulated short-rate paths; parameters for the Cox-Ross-Rubinstein binomial tree: $S_{min} = 50$, $S_{max} = 150$, $M = 50$, $\sigma_S = 0.25$, $T_S = 3$, $N_S = 1080$, salvage factor 0.75 of initial investment cost of 110).

	$\sigma_r = 0.005$			$\sigma_r = 0.015$		
	Ho-Lee	Hull-White one-factor model		Ho-Lee	Hull-White one-factor model	
S_0		$\alpha = 0.3$	$\alpha = 0.9$		$\alpha = 0.3$	$\alpha = 0.9$
50	-27.500	-27.500	-27.500	-27.500	-27.500	-27.500
60	-27.499	-27.500	-27.500	-27.478	-27.487	-27.495
70	-25.400	-25.405	-25.405	-25.344	-25.376	-25.393
80	-20.364	-20.371	-20.371	-20.310	-20.348	-20.363
90	-13.581	-13.589	-13.588	-13.533	-13.572	-13.584
100	-5.695	-5.703	-5.702	-5.654	-5.689	-5.699
110	2.904	2.897	2.897	2.939	2.907	2.899
120	11.971	11.965	11.965	12.000	11.974	11.967
130	21.346	21.341	21.341	21.369	21.348	21.343
140	30.924	30.920	30.920	30.942	30.926	30.921
150	40.639	40.636	40.636	40.654	40.641	40.637

Source: ts-4-4 and ts-4-7

Table 5.80. NPV in case 2 priced with the Cox-Ross-Rubinstein binomial tree method and with a stochastic risk-free rate simulated using the Ho-Lee model and the Hull-White one-factor model (parameters for both term structure models: $T_r = 4$ and $N_r = 7200$, Taylor 1.5 simulation scheme, U.S. Zero yield curve from December 1, 2000 is used as input parameter for each of the 100 simulated short-rate paths; parameters for the Cox-Ross-Rubinstein binomial tree: $S_{min} = 50$, $S_{max} = 150$, $M = 50$, $\sigma_S = 0.25$, $T_S = 3$, $N_S = 1080$, salvage factor 0.75 of initial investment cost of 110, expand factor 0.3 of prevailing underlying value with expand investment of 0.2 of initial investment cost).

	$\sigma_r = 0.005$			$\sigma_r = 0.015$		
	Ho-Lee	Hull-White one-factor model		Ho-Lee	Hull-White one-factor model	
S_0		$\alpha = 0.3$	$\alpha = 0.9$		$\alpha = 0.3$	$\alpha = 0.9$
50	-27.500	-27.500	-27.500	-27.500	-27.500	-27.500
60	-26.537	-26.538	-26.539	-26.497	-26.513	-26.528
70	-21.518	-21.519	-21.521	-21.466	-21.489	-21.509
80	-13.733	-13.735	-13.736	-13.682	-13.704	-13.725
90	-4.177	-4.177	-4.179	-4.130	-4.146	-4.167
100	6.540	6.542	6.539	6.580	6.573	6.552
110	18.021	18.026	18.023	18.054	18.057	18.036
120	30.011	30.018	30.014	30.036	30.047	30.028
130	42.335	42.344	42.340	42.353	42.372	42.353
140	54.879	54.890	54.886	54.892	54.917	54.899
150	67.574	67.585	67.581	67.581	67.611	67.594

Source: ts-4-5 and ts-4-8

Table 5.81. NPV in case 3 priced with the Cox-Ross-Rubinstein binomial tree method and with a stochastic risk-free rate simulated using the Ho-Lee model and the Hull-White one-factor model (parameters for both term structure models: $T_r = 4$ and $N_r = 7200$, Taylor 1.5 simulation scheme, U.S. Zero yield curve from December 1, 2000 is used as input parameter for each of the 100 simulated short-rate paths; parameters for the Cox-Ross-Rubinstein binomial tree: $S_{min} = 50$, $S_{max} = 150$, $M = 50$, $\sigma_S = 0.25$, $T_S = 3$, $N_S = 300$, salvage factor 0.75 of initial investment cost of 110, expand factor 0.3 of prevailing underlying value with expand investment of 0.2 of initial investment cost).

	$\sigma_r = 0.005$			$\sigma_r = 0.015$		
	Ho-Lee	Hull-White one-factor model		Ho-Lee	Hull-White one-factor model	
S_0		$\alpha = 0.3$	$\alpha = 0.9$		$\alpha = 0.3$	$\alpha = 0.9$
50	0.075	0.075	0.075	0.076	0.075	0.075
60	0.553	0.551	0.551	0.562	0.556	0.553
70	2.198	2.194	2.192	2.222	2.205	2.197
80	5.785	5.778	5.775	5.827	5.797	5.784
90	11.642	11.631	11.627	11.712	11.663	11.641
100	19.649	19.635	19.630	19.735	19.676	19.648
110	29.317	29.302	29.296	29.413	29.349	29.317
120	40.184	40.169	40.162	40.280	40.216	40.183
130	51.856	51.842	51.834	51.953	51.891	51.857
140	63.991	63.978	63.970	64.084	64.026	63.993
150	76.457	76.444	76.436	76.544	76.491	76.458

Source: ts-4-6 and ts-4-9

For parameter α, the values 0.3 and 0.9 were chosen. For the short-rate volatility σ_r, the values 0.005 and 0.015 were chosen[23]. According to Table 5.79, for various choices of parameter α the NPV for case 1 is almost identical for $\sigma_r = 0.005$ and only insignificantly different for $\sigma_r = 0.015$ when using the Ho-Lee model or the Hull-White one-factor model to simulate the risk-free interest rate. When comparing the NPV for case 2, see Table 5.80, the differences in the NPVs for the two stochastic term structure models are similar to the results of case 1. In case 3 these differences are slightly higher for 0.015 as the short-rate volatility parameter, yet still insignificant.

5.7.4 Recapitulation of the Main Results in Test Situation 4

In test situation 4, the influence of several no-arbitrage stochastic interest rate models on an investment project was investigated. In detail, the Ho-Lee model and the Hull-White one-factor and two-factor models were analyzed for all three real options cases. The findings are summarized in the following:

- The influence of the salvage factor on the investment project is similar for the Ho-Lee model and for the equilibrium models considered in test situation 3. The same is true for the influence of the expand factor on the investment project.
- For the Ho-Lee model the only parameter needed apart from the initial yield curve is the short-rate volatility parameter. For all cases 1, 2 and 3 the NPV is almost independent of the short-rate volatility parameter, i.e., the volatility parameter has an insignificant influence on the NPV. However, the influence of the volatility parameter on the NPV's standard deviation is significant: the higher the short-rate volatility parameter, the higher the NPV's standard deviation.
- The Hull-White one-factor model contains a mean reversion force α unlike the Ho-Lee model which is not mean reverting. In addition to parameter α, the Hull-White two-factor model contains a variable term modelled by an Ornstein-Uhlenbeck process. For both Hull-White models the influence of parameter α on the NPV for all cases 1, 2 and 3 is insignificant.
- For both the Hull-White one-factor model and the two-factor model the NPV is almost independent of the volatility parameter(s).

[23] The reason for this choice of σ_r is that for historical backtesting (test situation 5) a value of 0.01 is chosen as the short-rate volatility parameter based on capital markets data. Therefore, the values 0.005 and 0.015 for the short-rate volatility parameter are reasonable extreme cases in practice.

- The NPV for the Ho-Lee model and the Hull-White one-factor model is similar in cases 1, 2 and 3, independent of the mean reversion force and the volatility.
- When the three no-arbitrage models of this test situation are used to simulate the risk-free interest rate, the analysis of the NPV reveals that it is almost independent of the chosen term structure model. As already mentioned above, the influence of parameter α and the volatility parameter(s) is insignificant as well. Therefore, for these no-arbitrage models, the main influence on the NPV can only stem from the initial yield curve. This result will be heavily drawn on in the following test situation 5 with the consequence that historical backtesting is carried out only for one of the no-arbitrage models presented. Since the Ho-Lee model only needs one additional input parameter apart from the initial yield curve, only the Ho-Lee model will be used for historical backtesting.

5.8 Test Situation 5: Real Options Valuation in Historical Backtesting

The common feature of the previous test situations was the real options valuation done for hypothetical scenarios but not for real scenarios. This final test situation refers to reality when historical backtesting is used as the benchmark, the only benchmark that can be valid. In the following, several methods to derive the risk-free interest rate for the Cox-Ross-Rubinstein binomial tree will be compared with this benchmark for various capital markets scenarios (inverted, flat, and normal initial yield curve) and all three real options cases. The risk-free interest rate will be either constant, or implied in the current yield curve (Hull's approach), or stochastically simulated (Ho-Lee model), or the real risk-free rates that prevailed during the historical backtesting period will be applied. This means is detail:

1. **Constant risk-free interest rate:**
   This method uses a constant risk-free rate that is determined on the day the investment project was started. If the maturity is 3 years, the risk-free rate in the Cox-Ross-Rubinstein tree is the yield of a risk-free Zero bond with maturity 3 years on that day[24]. Here, the risk-free U.S. Zero yield is used.
2. **Implied forward rates:**
   The implied forward rates approach explained in Section 4.4 is based on Hull [62], page 357. From the spot rate curve on the day the project was started the implied forward rates are determined and used as the

[24] See Hull [62], footnote on page 344.

future spot rates for the Cox-Ross-Rubinstein tree based on the unbiased expectations theory. The risk-free rate in the tree is updated every three months.

3. **Historical interest rates:**
   This method uses the risk-free rates that actually prevailed during the historical backtesting period. This is the central backtesting approach which provides a benchmark for all other methods to derive the risk-free interest rate. According to Section 4.4.2 the risk-free rate is updated every three months.
4. **Ho-Lee term structure model:**
   Here, the future term structure of interest rates is stochastically simulated with the Ho-Lee model. This model needs the term structure on the day the project was started as an input parameter besides the volatility σ_r of the short-rate process.

For all test situations the Cox-Ross-Rubinstein binomial tree method is used, applying the following parameters: $S_{min} = 50$, $S_{max} = 150$, $M = 50$, $\sigma_S = 0.25$, $T_S = 3$ with an initial investment cost I_0 of 110. Parameter N_S that determines the time step $\Delta t = \frac{T_S}{N_S}$ is 1080 for cases 1, 2 and 3 except for the Ho-Lee model in case 3, where it is 300. As usual, the number of simulated short-rate paths for the stochastic term structure model is 100. For case 1 the salvage factor is 0.75 of the initial investment cost. For cases 2 and 3 the expand factor is 0.3 with an expand investment of 0.2 of the initial investment cost. In case 3 the defer period is exactly one year, i.e., the start of the investment project is either immediately or in exactly one year from today if the NPV is positive then.

As case 3 includes a 1-year defer period as an option to defer, term structure data for four years in total are required here. Data providers such as *Datastream* provide daily data for U.S. Zeros, risk-less bonds, from April 1997 onwards. Especially, the nine months after April 1, 1998 provided a rich pattern of term structure shapes: During the 9 months period from July 1998 until March 1999 the term structure changed tremendously and took a normal, a flat, and an inverted shape on different days. It needs to be mentioned that the inverted shape of the yield curve could only be spotted at the short end but not over the complete maturity spectrum in the capital markets within the last 6 years.

The method of cubic spline interpolation (see Section 3.2.2) is applied to calculate all yield curves needed in this test situation. For example, in order to determine the complete initial yield curve based on the given maturity buckets (input parameter for Ho-Lee model) and to calculate forward rates based on this initial yield curve (Hull's implied forward rates approach), the following TTMs are used: 1 mos., 3 mos., 6 mos., 9 mos., 1 yr., 2 yrs., 3 yrs., 4 yrs., 5 yrs., and 6 yrs. A detailed description of how the risk-free rates are simulated and what risk-free interest rates over which historical backtesting period have to be used is given in Table 5.82.

For calibrating the Ho-Lee model a simple method based on European put options on Zeros was presented in Section 5.2.2. However, finding the appropriate data from far in the past for historical backtesting is hardly possible, in particular for long-term real options if the start of the option was in mid-1990's. Even data providers like *Bloomberg* and *Datastream* do not provide options data for a time period as far in the past. Moreover, the simple distinction whether an option is American or European style cannot be found in the historical database of these data providers even when using SQL languages to operate on these databases. Therefore, this calibration procedure for the Ho-Lee model cannot be used for the historical backtesting part of this thesis (but, of course, it could be used for option pricing with current market data).

Another calibration method which is based on caplets results in a small systematic error as mentioned in Section 5.2.2 as well. Since caplets are quoted for specific Libor rates, using these caplets for calibrating results in parameters that, combined with a Libor curve as the initial yield curve for a no-arbitrage model, are perfectly suited to value swaps and swaptions. However, in this thesis, U.S. Zero yields are used. Therefore, parameter calibration using caplets results in a systematic error for the volatility estimate in the Ho-Lee model.

However, this is not an important issue due to the following reason: As seen in Section 5.7.1 the volatility parameter has virtually no influence on the NPV of the investment project, only on the NPV's standard deviation which is not important for the historical backtesting. Therefore, it is possible to use an approximation for the short-rate volatility parameter in the Ho-Lee model based on a calibration on caplets done by Clewlow and Strickland[25]. They calculated $\sigma_r = 0.0095$ as the Ho-Lee short-rate volatility parameter on January 21, 1995. Therefore, 0.01 is an appropriate approximation to use for σ_r in this thesis, knowing that the specific choice of this parameter has virtually no influence on the outcome of the NPV. Therefore, all following comparisons between the Ho-Lee method and the three other methods are allowed.

Figures 5.66, 5.67, and 5.68 display the three initial yield curves for the three historical backtesting periods. Figure 5.66 shows the inverted yield curve as of October 1, 1998; Figure 5.67 displays the flat yield curve as of July 1, 1998; Figure 5.68 displays the normal yield curve as of March 1, 1999. Finally, Figure 5.69 shows the development of the yield curves between July 1, 1998 and March 1, 1999. The specifications of the parameters needed to determine the risk-free interest rates are displayed in Table 5.82.

[25] See Clewlow & Strickland [31], pages 212-215.

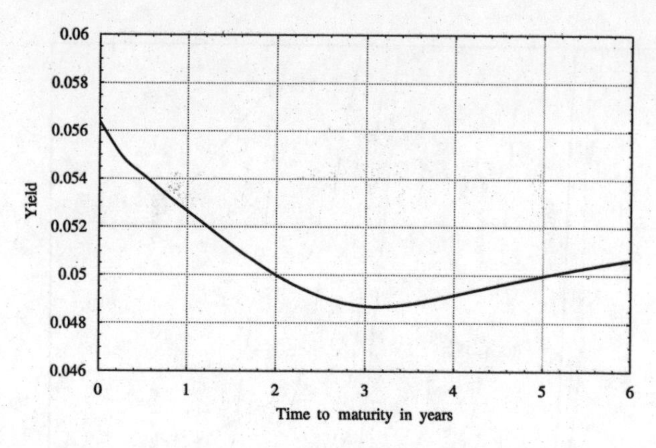

Fig. 5.66. Inverted yield curve for U.S. Zero bonds as of October 1, 1998.

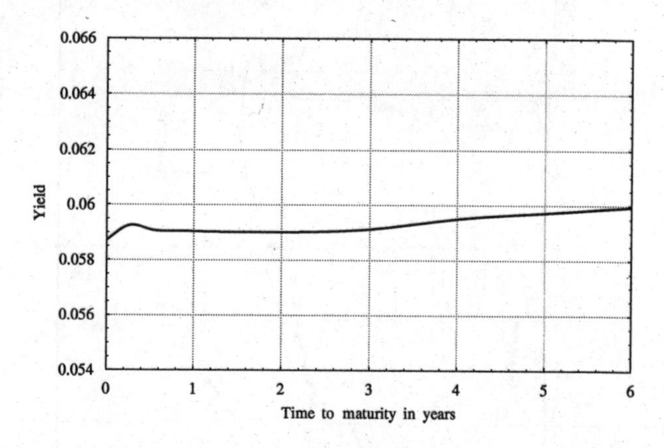

Fig. 5.67. Flat yield curve for U.S. Zero bonds as of July 1, 1998.

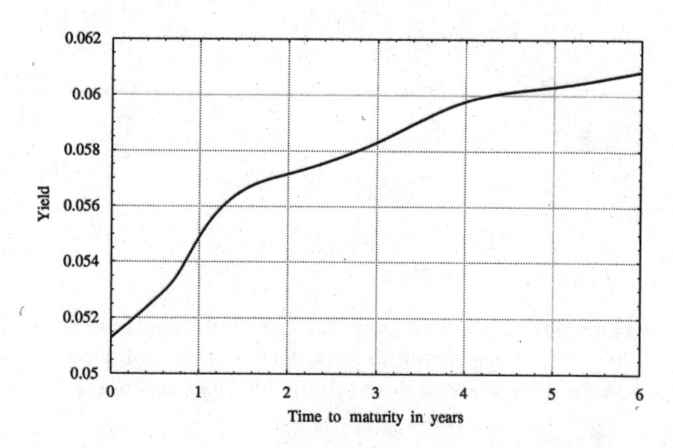

Fig. 5.68. Normal yield curve for U.S. Zero bonds as of March 1, 1999.

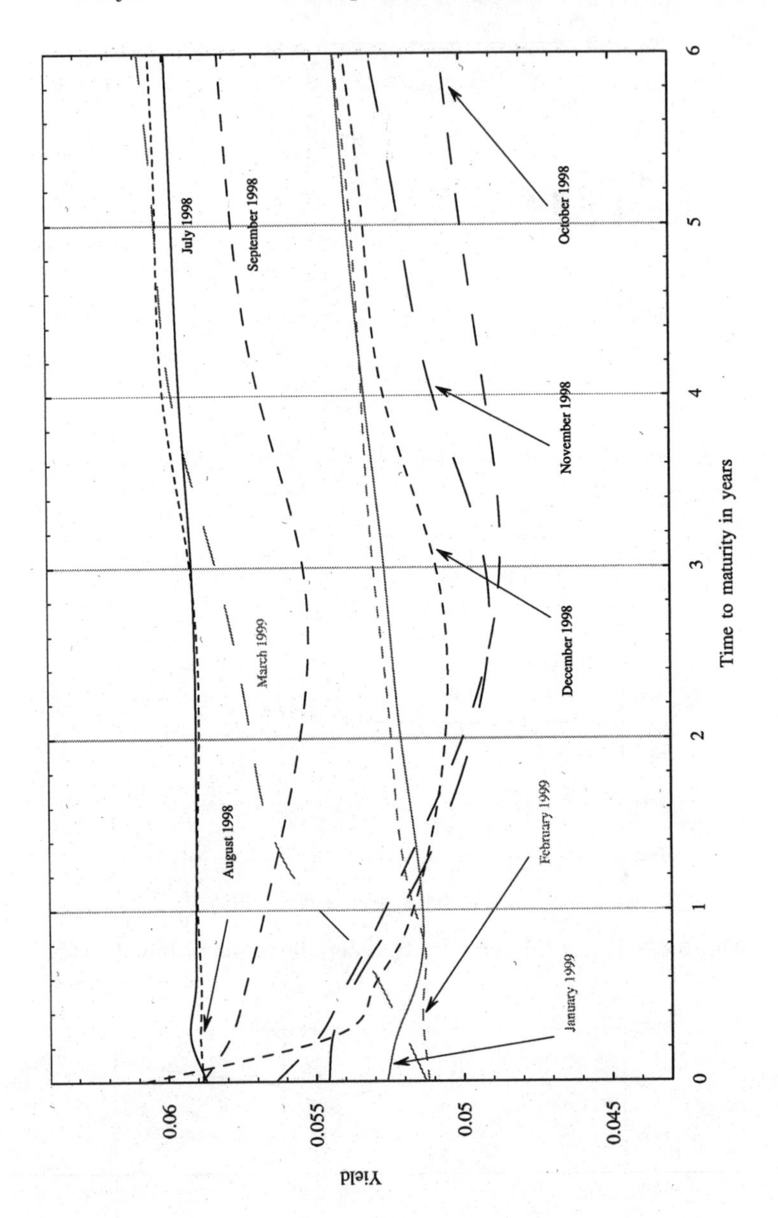

Fig. 5.69. Development of the yield curve for U.S. Zero bonds between July 1, 1998 and March 1, 1999 (beginning-of-month data; yield curves and descriptions for 1998 are in black while yield curves and descriptions for 1999 are in gray).

Table 5.82. Specifications of the parameters needed to determine the risk-free interest rates for test situation 5.

	Shape of the initial term structure and start date of the backtesting period		
Interest rate	**Inverted October 1, 1998**	**Flat July 1, 1998**	**Normal March 1, 1999**
constant r_f	$r_f = r_f(t=3) = 0.048706$ for all three cases	$r_f = r_f(t=3) = 0.059123$ for all three cases	$r_f = r_f(t=3) = 0.058305$ for all three cases
forward rates	$r_f = r_f(t) = f_{t,t+\Delta t}$ This gives: $r_f(0.00) = 0.056359$ $r_f(0.25) = 0.053820$ $r_f(0.50) = 0.052715$ $r_f(0.75) = 0.051068$ $r_f(1.00) = 0.049843$ $r_f(1.25) = 0.048400$ $r_f(1.50) = 0.047134$ $r_f(1.75) = 0.046166$ $r_f(2.00) = 0.045617$ $r_f(2.25) = 0.045515$ $r_f(2.50) = 0.045832$ $r_f(2.75) = 0.046612$ $r_f(3.00) = 0.047895$ $r_f(3.25) = 0.049414$ $r_f(3.50) = 0.050785$ $r_f(3.75) = 0.051864$ with the last four values only for case 3 with the defer period.	$r_f = r_f(t) = f_{t,t+\Delta t}$ This gives: $r_f(0.00) = 0.058717$ $r_f(0.25) = 0.059439$ $r_f(0.50) = 0.058710$ $r_f(0.75) = 0.059069$ $r_f(1.00) = 0.058942$ $r_f(1.25) = 0.058951$ $r_f(1.50) = 0.058977$ $r_f(1.75) = 0.059012$ $r_f(2.00) = 0.059047$ $r_f(2.25) = 0.059116$ $r_f(2.50) = 0.059275$ $r_f(2.75) = 0.059550$ $r_f(3.00) = 0.059969$ $r_f(3.25) = 0.060423$ $r_f(3.50) = 0.060752$ $r_f(3.75) = 0.060903$ with the last four values only for case 3 with the defer period.	$r_f = r_f(t) = f_{t,t+\Delta t}$ This gives: $r_f(0.00) = 0.051295$ $r_f(0.25) = 0.052643$ $r_f(0.50) = 0.054057$ $r_f(0.75) = 0.057007$ $r_f(1.00) = 0.060239$ $r_f(1.25) = 0.060074$ $r_f(1.50) = 0.059380$ $r_f(1.75) = 0.058778$ $r_f(2.00) = 0.058886$ $r_f(2.25) = 0.059616$ $r_f(2.50) = 0.060497$ $r_f(2.75) = 0.061552$ $r_f(3.00) = 0.062801$ $r_f(3.25) = 0.063962$ $r_f(3.50) = 0.064637$ $r_f(3.75) = 0.064667$ with the last four values only for case 3 with the defer period.
historical rates	Input of the following beginning-of-month term structures (maturities 1 mos., 2 mos., ..., 12 mos.; $T_r = 1$, $N_r = 1800$): 10/1998, 01/1999, 04/1999, 07/1999, 10/1999, 01/2000, 04/2000, 07/2000, 10/2000, 01/2001, 04/2001, 07/2001, 10/2001, 01/2002, 04/2002, 07/2002 with the last four quarters only for case 3 with the defer period.	Input of the following beginning-of-month term structures (maturities 1 mos., 2 mos., ..., 12 mos.; $T_r = 1$, $N_r = 1800$): 07/1998, 10/1998, 01/1999, 04/1999, 07/1999, 10/1999, 01/2000, 04/2000, 07/2000, 10/2000, 01/2001, 04/2001, 07/2001, 10/2001, 01/2002, 04/2002 with the last four quarters only for case 3 with the defer period.	Input of the following beginning-of-month term structures (maturities 1 mos., 2 mos., ..., 12 mos.; $T_r = 1$, $N_r = 1800$): 03/1999, 06/1999, 09/1999, 12/1999, 03/2000, 06/2000, 09/2000, 12/2000, 03/2001, 06/2001, 09/2001, 12/2001, 03/2002, 06/2002, 09/2002, 12/2002 with the last four quarters only for case 3 with the defer period.
Ho-Lee	Initial term structure as of October 1, 1998 as input parameter and: $\sigma_r = 0.01$, $T_r = 4, N_r = 7200$.	Initial term structure as of July 1, 1998 as input parameter and: $\sigma_r = 0.01$, $T_r = 4, N_r = 7200$.	Initial term structure as of March 1, 1999 as input parameter and: $\sigma_r = 0.01$, $T_r = 4, N_r = 7200$.

Source: own

5.8.1 Analysis for Case 1

Case 1 will be analyzed by applying the four methods to generate the risk-free rate mentioned above for an inverted, flat and normal initial yield curve, respectively.

Inverted initial yield curve as of October 1, 1998: ROV and NPV are displayed in Tables 5.83 and 5.84, respectively, for an inverted initial yield curve. The historical backtesting period October 1, 1998 to September 30, 2001 is the construction period of the investment project.

Table 5.83. ROV in case 1 for a risk-free rate that is modelled using various methods; initial yield curve as of October 1, 1998 is inverted.

S_0	constant interest rate	implied forward rates	historical interest rates	Ho-Lee term structure model NPV	Std.
50	32.500	32.500	32.500	32.500	0.000
60	22.750	22.652	22.586	22.683	0.136
70	15.586	15.407	15.124	15.444	0.428
80	10.702	10.553	10.164	10.588	0.522
90	7.362	7.254	6.850	7.286	0.503
100	5.074	5.000	4.632	5.027	0.436
110	3.506	3.455	3.145	3.479	0.357
120	2.432	2.398	2.148	2.417	0.283
130	1.693	1.670	1.475	1.686	0.220
140	1.184	1.168	1.018	1.180	0.169
150	0.833	0.822	0.708	0.832	0.128

Source: ts-5-1

Table 5.84. NPV in case 1 for a risk-free rate that is modelled using various methods; initial yield curve as of October 1, 1998 is inverted.

S_0	constant interest rate	implied forward rates	historical interest rates	Ho-Lee term structure model NPV	Std.
50	-27.500	-27.500	-27.500	-27.500	0.000
60	-27.250	-27.348	-27.414	-27.317	0.136
70	-24.414	-24.593	-24.876	-24.556	0.428
80	-19.298	-19.447	-19.836	-19.412	0.522
90	-12.638	-12.746	-13.150	-12.714	0.503
100	-4.926	-5.000	-5.368	-4.973	0.436
110	3.506	3.455	3.145	3.479	0.357
120	12.432	12.398	12.148	12.417	0.283
130	21.693	21.670	21.475	21.686	0.220
140	31.184	31.168	31.018	31.180	0.169
150	40.833	40.822	40.708	40.832	0.128

Source: ts-5-1

Table 5.84 shows the differences in the NPV depending on the risk-free rate that is used. When using the forward rates implied by the initial yield curve, the NPVs are very similar to those when using the Ho-Lee model. However, when using the historical risk-free rate prevailing during the backtesting period from October 1998 until September 2001, the NPV is significantly less than when using any of the other methods. The difference is the highest if the option to abandon is at-the-money.

Figure 5.70 shows the absolute difference between the NPVs and the benchmark for the various approaches. Using the data from Table 5.84 the absolute difference between the constant rate approach and the historical rates benchmark for, e.g., $S_0 = 90$ is calculated as $0.512\ (= |-12.638-(-13.150)|)$. When comparing the Ho-Lee model approach with the historical rates benchmark it is $0.436\ (= |-12.714-(-13.150)|)$ for $S_0 = 90$. Finally, when comparing the implied forward rates approach with the historical rates benchmark it is $0.404\ (= |-12.746-(-13.150)|)$ for $S_0 = 90$.

For every underlying value the difference is smaller when using the Ho-Lee model than when using a constant risk-free rate, but the absolute difference when using the implied forward rates is even lower. Therefore, for this scenario the application of the implied forward rates result in more realistic NPVs than when using the Ho-Lee model and these NPVs are especially better compared to the NPVs calculated when using a constant risk-free rate.

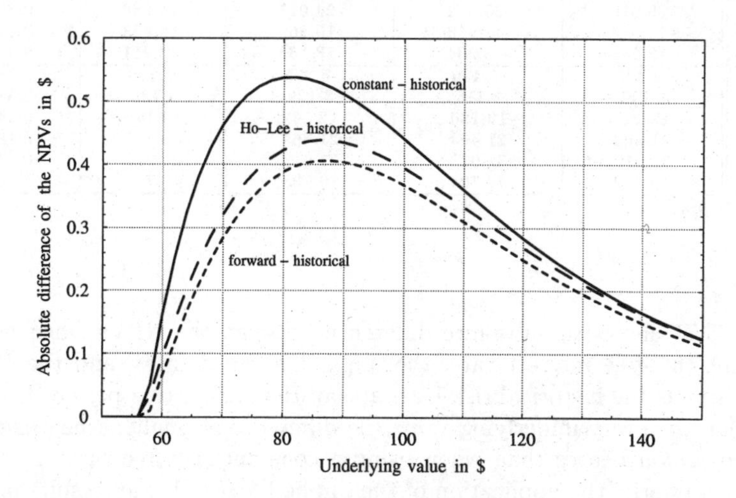

Fig. 5.70. Absolute difference between the NPVs in case 1 for an inverted initial yield curve when using a constant risk-free rate, the implied forward rates, and the Ho-Lee model, respectively, versus using the historical risk-free rates of the backtesting period (ts-5-1, corresponding to Table 5.84).

Flat initial yield curve as of July 1, 1998: The NPV of case 1 is displayed in Table 5.85 for a flat initial yield curve. In the following, only the NPV will be displayed in tables but not the ROV any more. The historical backtesting period is now July 1, 1998 to June 30, 2001, which is the construction period of the investment project.

Table 5.85 shows that the NPV results for stochastically modelled risk-free rates are closer to the historical NPVs than for implied forward rates or for a constant risk-free rate if the underlying value is not too large. If the underlying value is large, the NPV for the implied forward rates is closer to the historical NPV. Moreover, the NPV values for the implied forward rates and the constant risk-free rate are very similar. In theory, the implied forward rates are the same as the constant risk-free interest rate for a flat yield curve. In practice, a yield curve with exactly the same yield for each TTM does not exist, and the implied forward rates are only similar to the "flat" risk-free rate.

Table 5.85. NPV in case 1 for a risk-free rate that is modelled using various methods; initial yield curve as of July 1, 1998 is flat.

S_0	constant interest rate	implied forward rates	historical interest rates	Ho-Lee term structure model NPV	Std.
50	-27.500	-27.500	-27.500	-27.500	0.000
60	-27.452	-27.450	-27.431	-27.431	0.070
70	-25.019	-25.012	-24.917	-24.986	0.344
80	-19.966	-19.958	-19.860	-19.933	0.445
90	-13.241	-13.234	-13.163	-13.211	0.437
100	-5.425	-5.420	-5.375	-5.399	0.380
110	3.111	3.115	3.139	3.133	0.311
120	12.127	12.130	12.142	12.146	0.245
130	21.462	21.464	21.469	21.477	0.189
140	31.010	31.011	31.013	31.022	0.144
150	40.703	40.704	40.703	40.712	0.109

Source: ts-5-2

Figure 5.71 shows the absolute difference between the NPVs when using a constant risk-free interest rate, the implied forward rates, and the Ho-Lee model versus the historical risk-free rates of the backtesting period. It can be seen that for every underlying value the difference is smaller when using the implied forward rates than when using a constant risk-free rate. Therefore, for this scenario, the application of the implied forward rates results in more realistic NPVs than when using a constant risk-free rate. However, the best results are obtained when using the Ho-Lee model as can be seen in Figure 5.71.

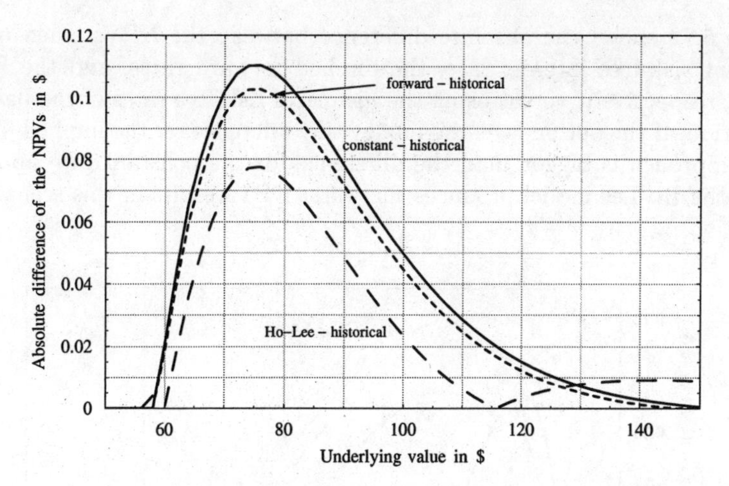

Fig. 5.71. Absolute difference between the NPVs in case 1 for a flat initial yield curve when using a constant risk-free rate, the implied forward rates, and the Ho-Lee model, respectively, versus using the historical risk-free rates of the backtesting period (ts-5-2, corresponding to Table 5.85).

Normal initial yield curve as of March 1, 1999: Table 5.86 displays the NPV in case 1 for a normal initial yield curve. The historical backtesting period is now March 1, 1999 to February 28, 2002, which is the construction period of the investment project.

Table 5.86 shows that the NPV for the Ho-Lee model is much higher than when using the implied forward rates or a constant risk-free rate which delivers the best results here.

Table 5.86. NPV in case 1 for a risk-free rate that is modelled using various methods; initial yield curve as of March 1, 1999 is normal.

S_0	constant interest rate	implied forward rates	historical interest rates	Ho-Lee term structure model NPV	Std.
50	-27.500	-27.500	-27.500	-27.500	0.000
60	-27.442	-27.379	-27.397	-27.360	0.097
70	-24.975	-24.813	-24.888	-24.788	0.352
80	-19.916	-19.778	-19.873	-19.754	0.449
90	-13.196	-13.097	-13.179	-13.074	0.440
100	-5.388	-5.320	-5.379	-5.298	0.384
110	3.140	3.187	3.149	3.205	0.315
120	12.150	12.182	12.160	12.197	0.249
130	21.479	21.501	21.491	21.514	0.193
140	31.022	31.038	31.034	31.048	0.147
150	40.712	40.723	40.723	40.731	0.111

Source: ts-5-3

Figure 5.72 shows the absolute difference between the NPVs when using a constant risk-free interest rate, the implied forward rates, and the Ho-Lee model, respectively, versus using the historical risk-free rates of the backtesting period. If the option is at-the-money the difference for the implied forward rates approach is bigger than the difference for the constant rate approach. Using the Ho-Lee model produces the worst NPV results in this scenario.

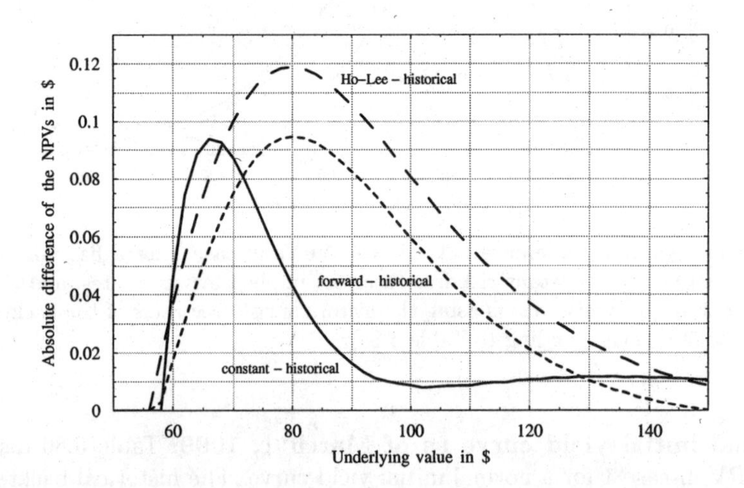

Fig. 5.72. Absolute difference between the NPVs in case 1 for a normal initial yield curve when using a constant risk-free rate, the implied forward rates, and the Ho-Lee model, respectively, versus using the historical risk-free rates of the backtesting period (ts-5-3, corresponding to Table 5.86).

5.8.2 Analysis for Case 2

Analogous to case 1, case 2 will be analyzed by applying the four methods to generate the risk-free rate mentioned above for an inverted, flat and normal yield curve, respectively. The corresponding backtesting periods are identical to case 1, and thus, all interest rate parameters are identical. The only difference to case 1 is that case 2 contains an additional option, an option to expand the project at the end of the 3-year construction period.

Inverted initial yield curve as of October 1, 1998: Table 5.87 displays the NPV for an inverted initial yield curve. As usual, the historical backtesting period, now October 1, 1998 to September 30, 2001, is the construction period of the investment project.

Table 5.87. NPV in case 2 for a risk-free rate that is modelled using various methods; initial yield curve as of October 1, 1998 is inverted.

S_0	constant interest rate	implied forward rates	historical interest rates	Ho-Lee term structure model NPV	Std.
50	-27.500	-27.500	-27.500	-27.500	0.000
60	-25.807	-25.983	-26.127	-25.970	0.233
70	-20.656	-20.822	-21.058	-20.835	0.322
80	-13.088	-13.206	-13.421	-13.223	0.264
90	-3.827	-3.902	-4.030	-3.890	0.148
100	6.609	6.565	6.549	6.625	0.055
110	17.855	17.832	17.925	17.943	0.130
120	29.657	29.649	29.837	29.806	0.227
130	41.838	41.839	42.103	42.034	0.306
140	54.274	54.282	54.606	54.508	0.368
150	66.889	66.901	67.271	67.150	0.415

Source: ts-5-4

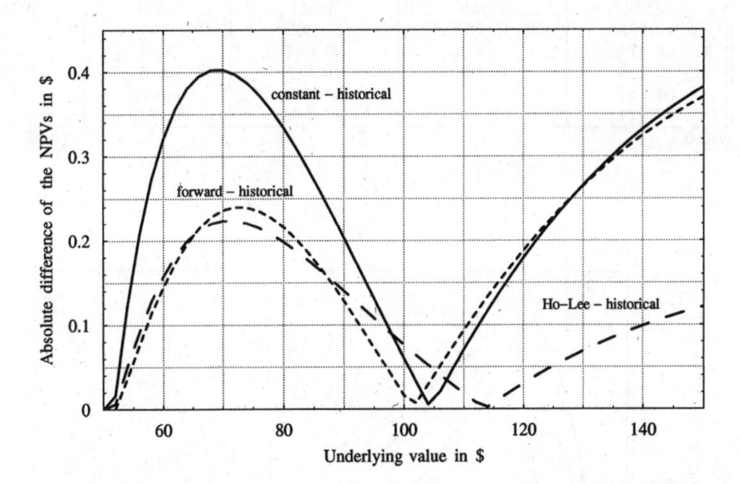

Fig. 5.73. Absolute difference between the NPVs in case 2 for an inverted initial yield curve when using a constant risk-free rate, the implied forward rates, and the Ho-Lee model, respectively, versus using the historical risk-free rates of the backtesting period (ts-5-4, corresponding to Table 5.87).

For this case scenario, the Ho-Lee model is by far the best for higher values of the underlying since its NPV values are closest to the NPV derived from using the historical risk-free interest rates. For these higher values of the underlying, the NPV is almost identical when using the implied forward rates or a constant risk-free rate. However, the situation is totally different for lower underlying values. The NPV is almost identical when using the Ho-Lee model or the implied forward rates but it shows a difference when a constant risk-free rate is used. This is graphically displayed in Figure 5.73 that shows the absolute difference between the NPVs when using a constant risk-free interest rate, the implied forward rates, and the Ho-Lee model, respectively, versus using the historical risk-free rates of the backtesting period.

Flat initial yield curve as of July 1, 1998: The NPV for a flat initial yield curve is displayed in Table 5.88. The historical backtesting period July 1, 1998 to June 30, 2001 is the construction period of the investment project.

Table 5.88. NPV in case 2 for a risk-free rate that is modelled using various methods; initial yield curve as of July 1, 1998 is flat.

S_0	constant interest rate	implied forward rates	historical interest rates	Ho-Lee term structure model NPV	Std.
50	-27.500	-27.500	-27.500	-27.500	0.000
60	-26.248	-26.241	-26.157	-26.223	0.195
70	-21.186	-21.179	-21.059	-21.157	0.294
80	-13.498	-13.492	-13.393	-13.470	0.242
90	-4.059	-4.055	-3.984	-4.033	0.129
100	6.554	6.556	6.605	6.574	0.056
110	17.950	17.951	17.988	17.966	0.143
120	29.874	29.874	29.904	29.884	0.237
130	42.149	42.148	42.175	42.154	0.313
140	54.657	54.655	54.682	54.658	0.373
150	67.324	67.322	67.349	67.323	0.417

Source: ts-5-5

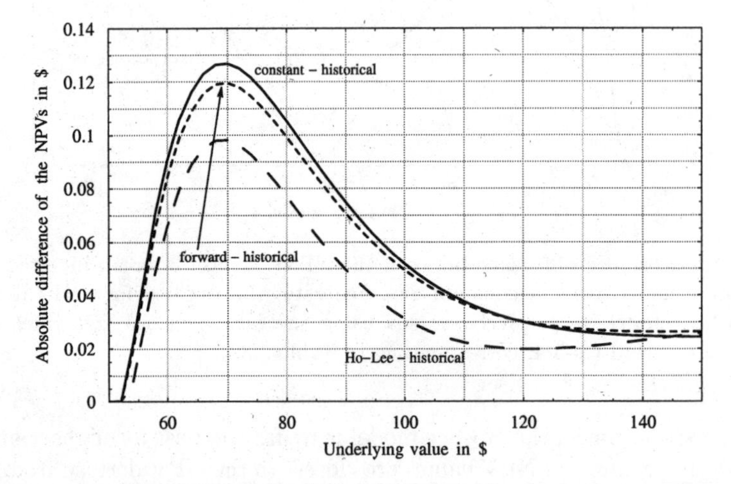

Fig. 5.74. Absolute difference between the NPVs in case 2 for a flat initial yield curve when using a constant risk-free rate, the implied forward rates, and the Ho-Lee model, respectively, versus using the historical risk-free rates of the backtesting period (ts-5-5, corresponding to Table 5.88).

Table 5.88 shows a similar pattern to case 1 with a flat initial yield curve. Again, it is interesting that when using the Ho-Lee model the NPV is much closer to the NPV derived from using the historical risk-free rates of the backtesting period than when using any other method to derive the risk-free rate. The stochastic feature and the initial yield curve as an input parameter better assess the future development of the market, which results in a more accurate NPV. Figure 5.74 shows the absolute difference between the NPVs when using a constant risk-free interest rate, the implied forward rates, and the Ho-Lee model, respectively, versus using the historical risk-free rates of the backtesting period. The graphs for the implied forward rates and a constant risk-free rate are almost identical, a phenomenon already observed and explained in case 1 with a flat initial yield curve.

Normal initial yield curve as of March 1, 1999: The NPV for a normal initial yield curve is displayed in Table 5.89. The historical backtesting period is March 1, 1999 to February 28, 2002, which is the construction period of the investment project.

Table 5.89. NPV in case 2 for a risk-free rate that is modelled using various methods; initial yield curve as of March 1, 1999 is normal.

S_0	constant interest rate	implied forward rates	historical interest rates	Ho-Lee term structure model NPV	Std.
50	-27.500	-27.500	-27.500	-27.500	0.000
60	-26.217	-26.063	-26.150	-26.045	0.207
70	-21.147	-20.997	-21.156	-20.975	0.298
80	-13.468	-13.364	-13.545	-13.342	0.246
90	-4.043	-3.978	-4.154	-3.957	0.134
100	6.556	6.592	6.427	6.610	0.056
110	17.941	17.958	17.800	17.972	0.140
120	29.856	29.860	29.706	29.870	0.234
130	42.124	42.118	41.966	42.125	0.311
140	54.626	54.615	54.461	54.618	0.371
150	67.290	67.274	67.119	67.275	0.416

Source: ts-5-6

Figure 5.75 shows the absolute difference between the NPVs when using a constant risk-free interest rate, the implied forward rates, and the Ho-Lee model, respectively, versus using the historical risk-free rates of the backtesting period. Evidently, the outcome for a normal initial yield curve is totally different from the previous scenario. Now, the Ho-Lee model yields the worst NPV results while a constant risk-free rate results by and large in the best NPVs. Figure 5.75 is similar to Figure 5.72 for case 1 that showed the application of a constant risk-free rate to be the best choice. However, further research is needed to determine the exact influence of the shape of the initial yield curve on the NPV.

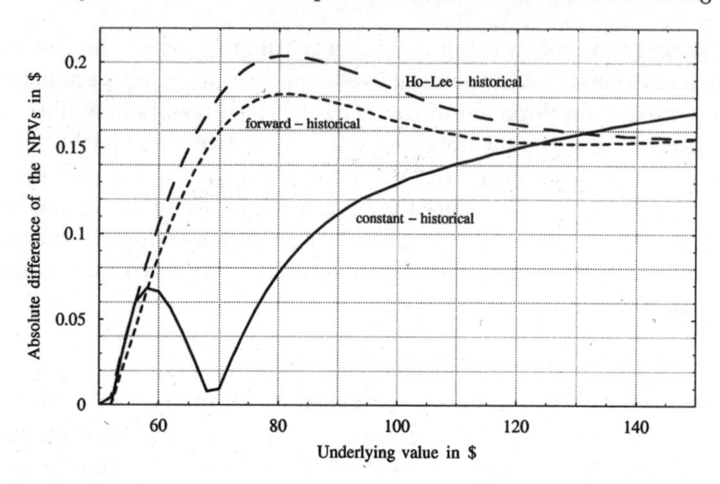

Fig. 5.75. Absolute difference between the NPVs in case 2 for a normal initial yield curve when using a constant risk-free rate, the implied forward rates, and the Ho-Lee model, respectively, versus using the historical risk-free rates of the backtesting period (ts-5-6, corresponding to Table 5.89).

5.8.3 Analysis for Case 3

Case 3 will be analyzed for the four methods to generate the risk-free rates mentioned in Table 5.82 for an inverted, flat or normal initial yield curve, respectively.

Inverted initial yield curve as of October 1, 1998: The NPV of case 3 with an inverted initial yield curve is displayed in Table 5.90. The historical backtesting period is from October 1, 1998 to September 30, 2002, including a 3-year construction period of the investment project.

Since case 3 includes an option to defer the start of the investment project by one year while the construction period of the investment project is 3 years, interest rates over a 4-year time period are necessary. Table 5.90 shows that using the implied forward rates yields the most accurate NPV results independent of the underlying value. Figure 5.76 shows the absolute difference between the NPVs when using a constant risk-free interest rate, the implied forward rates, and the Ho-Lee model, respectively, versus using the historical risk-free rates of the backtesting period. It is interesting to note that for higher values of the underlying, the absolute difference increases when using a constant risk-free rate but decreases when using the implied forward rates or the Ho-Lee model. This, however, is unique to this scenario with an inverted initial yield curve.

Table 5.90. NPV in case 3 for a risk-free rate that is modelled using various methods; initial yield curve as of October 1, 1998 is inverted.

S_0	constant interest rate	implied forward rates	historical interest rates	Ho-Lee term structure model NPV	Std.
50	0.062	0.065	0.063	0.065	0.003
60	0.475	0.497	0.486	0.495	0.017
70	1.921	1.990	1.960	2.002	0.062
80	5.161	5.306	5.247	5.337	0.152
90	10.617	10.854	10.759	10.900	0.259
100	18.185	18.509	18.394	18.581	0.393
110	27.478	27.872	27.749	27.952	0.525
120	38.049	38.495	38.372	38.556	0.655
130	49.468	49.949	49.833	50.032	0.759
140	61.437	61.941	61.834	62.008	0.849
150	73.761	74.278	74.182	74.353	0.924

Source: ts-5-7

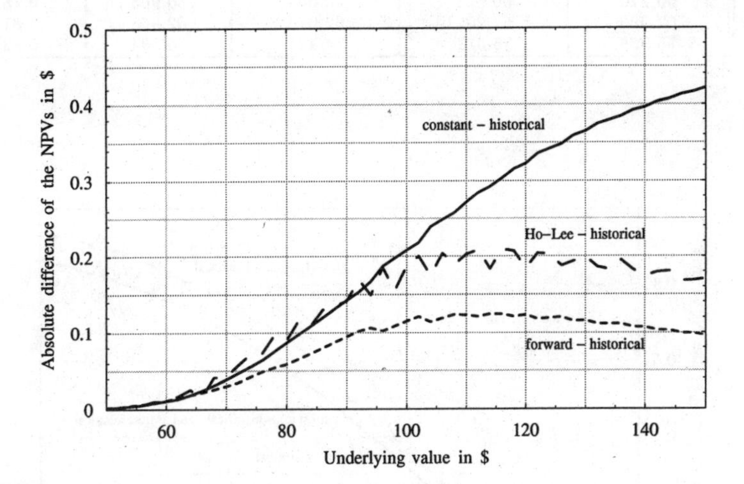

Fig. 5.76. Absolute difference between the NPVs in case 3 for an inverted initial yield curve when using a constant risk-free rate, the implied forward rates, and the Ho-Lee model, respectively, versus using the historical risk-free rates of the backtesting period (ts-5-7, corresponding to Table 5.90).

Flat initial yield curve as of July 1, 1998: The NPV of case 3 is displayed in Table 5.91 for a flat initial yield curve. The historical backtesting period is from July 1, 1998 to June 30, 2002, including a 3-year construction period of the investment project.

Table 5.91 shows that the results using the implied forward rates and a constant risk-free rate yields, as expected, almost identical results for case 3 with a flat initial yield curve. The NPVs derived from using the Ho-Lee model are slightly worse than when using a constant rate or the implied forward rates.

This finding is opposite to case 2 with a flat initial yield curve where the Ho-Lee model yields the best NPV results.

Table 5.91. NPV in case 3 for a risk-free rate that is modelled using various methods; initial yield curve as of July 1, 1998 is flat.

S_0	constant interest rate	implied forward rates	historical interest rates	Ho-Lee term structure model NPV	Std.
50	0.069	0.069	0.064	0.069	0.003
60	0.521	0.520	0.492	0.517	0.021
70	2.074	2.074	1.982	2.081	0.072
80	5.497	5.496	5.298	5.527	0.171
90	11.181	11.180	10.845	11.221	0.289
100	18.991	18.990	18.519	19.058	0.428
110	28.501	28.501	27.911	28.577	0.562
120	39.256	39.257	38.566	39.320	0.688
130	50.820	50.821	50.053	50.903	0.787
140	62.900	62.902	62.076	62.968	0.871
150	75.308	75.311	74.441	75.384	0.937

Source: ts-5-8

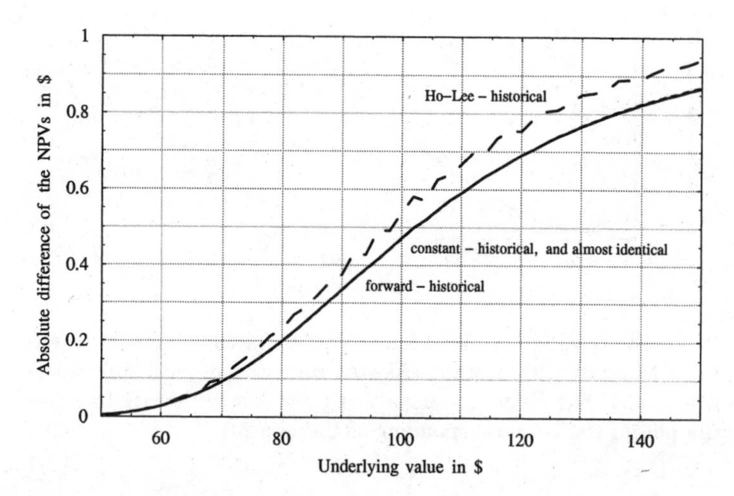

Fig. 5.77. Absolute difference between the NPVs in case 3 for a flat initial yield curve when using a constant risk-free rate, the implied forward rates, and the Ho-Lee model, respectively, versus using the historical risk-free rates of the backtesting period (ts-5-8, corresponding to Table 5.91).

Figure 5.77 displays the absolute difference between the NPVs when using a constant risk-free interest rate and the implied forward rates, respectively, versus using the historical risk-free rates of the backtesting period. The scenario of case 3 shows that when using the implied forward rates or the constant risk-free rate the difference is less than when using the Ho-Lee model. As already seen in cases 1 and 2, the graphs for the constant risk-free rate and the implied forward rates are very similar if the initial yield curve is flat.

Normal initial yield curve as of March 1, 1999: The NPV of case 3 is displayed in Table 5.92 for a normal initial yield curve. The historical backtesting period is from March 1, 1999 to February 28, 2003, including a 3-year construction period of the investment project.

Table 5.92. NPV in case 3 for a risk-free rate that is modelled using various methods; initial yield curve as of March 1, 1999 is normal.

S_0	constant interest rate	implied forward rates	historical interest rates	Ho-Lee term structure model NPV	Std.
50	0.069	0.065	0.066	0.065	0.003
60	0.517	0.498	0.503	0.495	0.020
70	2.062	2.002	2.012	2.009	0.071
80	5.469	5.345	5.352	5.377	0.168
90	11.136	10.936	10.922	10.978	0.287
100	18.926	18.658	18.602	18.727	0.427
110	28.419	28.097	27.984	28.176	0.561
120	39.161	38.802	38.619	38.867	0.688
130	50.713	50.331	50.079	50.416	0.787
140	62.785	62.390	62.072	62.459	0.872
150	75.186	74.785	74.409	74.862	0.938

Source: ts-5-9

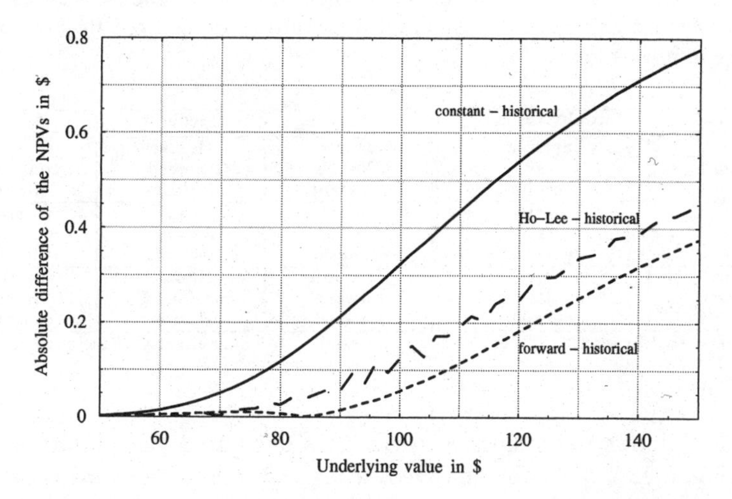

Fig. 5.78. Absolute difference between the NPVs in case 3 for a normal initial yield curve when using a constant risk-free rate, the implied forward rates, and the Ho-Lee model, respectively, versus using the historical risk-free rates of the backtesting period (ts-5-9, corresponding to Table 5.92).

Table 5.92 shows again that the NPV for the implied forward rates is closest to the NPVs derived from using the historical risk-free rates. Using a constant risk-free rate results in the worst NPV graph. Figure 5.78 shows the absolute difference between the NPVs when using a constant risk-free interest rate, the implied forward rates, and the Ho-Lee model, respectively, versus using the historical risk-free rates of the backtesting period. This result is totally different from the result in case 2 with a normal initial yield curve (see Figure 5.75) since in case 3 an option to defer the project start is embedded as opposed to case 2.

5.8.4 Recapitulation of the Main Results in Test Situation 5

The final test situation 5 with the comparison of various methods to determine the risk-free interest rate for the pricing of cases 1, 2 and 3 is the core of the thesis. This test situation provided many interesting results and implications for Corporate Finance practice. An evaluation based on the NPV graphs which show the absolute difference between the results of the various methods is given below[26]. This evaluation is not quantitatively exact but based on the "first look" to determine which of the graphs that display the absolute difference is by and large better for which test scenario. These results are displayed in Table 5.93.

Table 5.93. Overall evaluation of the absolute difference graphs in test situation 5 for a constant risk-free rate, the implied forward rates, and the Ho-Lee modelled risk-free interest rates.

Real option	Initial yield curve	Best	Medium	Worst
case 1	inverted	forward	Ho-Lee	constant
case 1	flat	Ho-Lee	forward	constant*
case 1	normal	constant	forward	Ho-Lee
case 2	inverted	Ho-Lee	forward	constant
case 2	flat	Ho-Lee	forward	constant**
case 2	normal	constant	forward	Ho-Lee
case 3	inverted	forward	Ho-Lee	constant
case 3	flat	forward	constant***	Ho-Lee
case 3	normal	forward	Ho-Lee	constant

* very close to "forward" ** very close to "forward" *** only slightly worse than "forward"

Table 5.93 shows an interesting result for a flat initial yield curve. Independent if case 1, 2 or 3 was analyzed, using the implied forward rates yields (almost) the same NPV results (regardless of the underlying value) as using the constant risk-free rate. This result could have been expected since a flat

[26] See the nine Figures 5.70, 5.71, 5.72, 5.73, 5.74, 5.75, 5.76, 5.77, and 5.78 of test situation 5.

yield curve implies forward rates identical (in theory) or similar (in practice since the yield curve is not completely "flat") to the constant risk-free rate. However, for cases 1 and 2 the application of the Ho-Lee model delivers better results (i.e., NPVs closer to the NPVs derived from using historical risk-free rates); but it leads to worse results for case 3.

From the backtesting results summarized in Table 5.93 it is not possible to conclude that for a certain initial yield curve one specific method is always better suited to derive the risk-free rate than the other ones. However, Table 5.93 shows that for an inverted initial yield curve the implied forward rates or the Ho-Lee model always yield better results than a constant risk-free rate. It also shows that for case 3 it is always optimal to use the implied forward rates. This does not hold true for cases 1 and 2. By and large the Ho-Lee model is the best choice in three scenarios and the implied forward rates are the best one in four scenarios whereas the constant risk-free rate is the best one in only two scenarios. Thus, the application of either the implied forward rates or of stochastically simulated risk-free rates using the Ho-Lee model is better suited to assess the investment project than using a constant risk-free rate in seven out of the nine scenarios.

The implications for Corporate Finance practice are clear. The implementation of a non-constant risk-free interest rate in the Cox-Ross-Rubinstein binomial tree to value an investment project as done in the computer simulation program written by the author for this thesis is relatively easy for Hull's implied forward rates approach. All information about the risk-free interest rates needed for the real options valuation tree can easily be derived from the current yield curve. On the other hand, the implementation of a stochastic term structure model to derive the appropriate risk-free rate for the real options valuation tree is technically difficult and requires detailed knowledge about stochastic calculus and stochastic simulation in discrete time. Accordingly, from the stand point of implementational convenience, Hull's forward rates approach is better suited in Corporate Finance practice than a stochastic term structure model like the Ho-Lee model.

As already mentioned, the implied forward rates approach yields by and large the best NPV results in historical backtesting in four out of nine test scenarios. In addition, this approach yields the second best (medium) NPV results in five out of nine test scenarios according to Table 5.93. This means that in all nine test scenarios analyzed the implied forward rates approach used for real options valuation never yields the worst NPV results whereas the currently prevailing practice to use a constant risk-free rate leads to the worst NPV results in two-thirds (six out of nine) of the test scenarios. Consequently, Hull's approach of using the forward rates that are implied in the current yield curve has to be preferred in general over using a constant risk-free rate for real options valuation. This is the central result of this thesis.

The differences between the various methods to derive the risk-free rate are small on an absolute basis but obvious. However, these differences are much larger on a relative basis, i.e., when considering these differences relative to the NPV for a given underlying value. This is especially true for an NPV close to zero, i.e., if a small improvement in accuracy to determine the NPV has a significant impact and can influence the manager's decision to carry out a project or not.

It could not be determined in test situation 5 whether the results of Table 5.93 mainly depend on the chosen case, the chosen initial yield curve, or the chosen backtesting period. Further analysis needs to be conducted in this respect, e.g., using various backtesting periods for the same case with an inverted, flat and normal initial yield curve, respectively, in order to determine whether the results of Table 5.93 still hold.

5.9 Summary

In this chapter, the theory introduced in Chapters 1-4 was applied to real options in practice. After the introduction in Section 5.1, the aspect of calibration was investigated in Section 5.2. Section 5.2.1 shortly introduced a computational tool developed by Schulmerich to estimate the parameters of various mean reversion short-rate models (especially, the parameters of the Vasicek and the Cox-Ingersoll-Ross models). Calibration methods for the Ho-Lee model were explained in Section 5.2.2. After having presented the prerequisites for an in-depth analysis, the exact research strategy was elaborated in Section 5.3. This strategy comprised five consecutive test situations with the final fifth step, the historical backtesting section, as the central part of Chapter 5 and the thesis as a whole:

1. Section 5.4: *Test situation 1: the Schwartz-Moon model with a deferred project start*

   Variable interest rates in the case of the Schwartz-Moon model with a deferred project start.

2. Section 5.5: *Test situation 2: preliminary tests for real options valuation*

   Preliminary tests to investigate the parameters of the real options valuation tools and to compare the valuation methods.

3. Section 5.6: *Test situation 3: real options valuation with a stochastic interest rate using equilibrium models*

   Influence of the salvage and expand factors in cases 1, 2 and 3 on the real options (Section 5.6.1); analysis of all cases for equilibrium models

(Vasicek model in Section 5.6.2 and Cox-Ingersoll-Ross model in Section 5.6.3).

4. Section 5.7: *Test situation 4: real options valuation with a stochastic interest rate using no-arbitrage models*
   Analysis of cases 1, 2 and 3 for the Ho-Lee model (Section 5.7.1), comparison of the Hull-White one-factor model with the Hull-White two-factor model (Section 5.7.2), and comparison of the Ho-Lee model with the Hull-White one-factor model (Section 5.7.3).

5. Section 5.8: *Test situation 5: real options valuation in historical backtesting*
   Historical backtesting for cases 1, 2 and 3 through a comparison of the real options pricing when using a stochastically modelled risk-free interest rate (Ho-Lee model), a constant rate, interest rates that equal the currently implied forward rates (Hull's approach), and the historical risk-free rates of the backtesting period.

In test situation 2 two modified methods to price real options when the risk-free rate is non-constant were analyzed: the Cox-Ross-Rubinstein binomial tree method and the Trigeorgis log-transformed binomial tree method. The analysis showed that the Trigeorgis method is not practicable at all for non-constant risk-free rates due to the bushy tree problem. Consequently, the Cox-Ross-Rubinstein binomial tree method is the method of choice when it comes to non-constant risk-free interest rates.

Every test situation (with the exception of the first one) was concluded by a recapitulation of the test situation's main results. Among the many different results which were found for each of these test situations, the results from historical backtesting are the most important ones with regard to practical application in Corporate Finance. Therefore, only these are summarized here with the reservation that the results are based on the chosen backtesting period and real options cases analyzed:

> When pricing real options, no matter if single or complex, the implied forward rates (Hull's approach) or the Ho-Lee modelled interest rates as the variable risk-free rates for the real options pricing algorithm should be preferred to a constant risk-free interest rate. In two-thirds of the analyzed scenarios, these two methods yield better NPV results than using a constant risk-free interest rate.
>
> In all nine test scenarios analyzed the implied forward rates approach used for real options valuation never yields the worst NPV results whereas the currently prevailing practice to use a constant risk-free rate leads to the worst NPV results in two-thirds of the test scenarios.

From the stand point of implementational convenience, Hull's implied forward rates approach is better suited in Corporate Finance practice than a stochastic term structure model like the Ho-Lee model. Consequently, Hull's approach of using the forward rates that are implied in the current yield curve has to be preferred in general over using a constant risk-free rate for real options valuation.

6

Summary and Outlook

6.1 Purpose of the Research

This thesis analyzed real options valuation for non-constant risk-free interest rates, especially for stochastically modelled risk-free interest rates, in simulation and historical backtesting. Several real options cases were investigated and combined with various pricing tools and stochastic term structure models. Additionally, preliminary tests were conducted to see how various choices of model inherent parameters (number of simulated short-rate paths, discretization parameters for the time and the state axis of pricing tools, etc.) influence the outcome of the respective pricing tool.

For the preliminary tests, the main real option investigated was an option to abandon the investment project at a specific time for a pre-specified salvage value. Since this is a European put option the Black-Scholes formula can be used as a benchmark. All other tests within this thesis were conducted for three real options (simple and complex), called real options cases:

1. **Case 1:** Option to abandon the project at any time during the construction period for a salvage value. Since such a real option is an American put option, the salvage value is the strike price of this option. The initial investment cost is assumed to be constant.
2. **Case 2:** Option to abandon the project at any time during the construction period for a salvage value (case 1) and option to expand the project once by an expand factor (e.g., expand project by 30%) for an expand investment at the end of the construction period. The expand investment is assumed to be a fraction of the initial investment cost.
3. **Case 3:** Complex real option in case 2 combined with an option to defer the project start by exactly one year. Thus, the project can start today or in exactly one year from today if the investment in one year has a positive NPV. The initial investment cost to start the project in one year is assumed to be the same as if the project is started immediately. If, in one year from now, the NPV will be negative, the project will not be started at all.

The purpose of these three cases was to analyze how the addition of new real options to an investment project affects the value of the project. In all cases various methods to derive the risk-free interest rate for the real options pricing tool were investigated. Besides using a constant risk-free rate, which is nowadays the usual practice in Corporate Finance, the risk-free rates were also simulated by using equilibrium models (Vasicek model and Cox-Ingersoll-Ross model) as well as no-arbitrage term structure models (Ho-Lee model and Hull-White one-factor and two-factor models). Furthermore, Hull's approach of using the forward rates implied in the current spot rate curve was also presented in detail.

However, using non-constant interest rates instead of a constant interest rate required modifications of existing real options pricing methods like the Cox-Ross-Rubinstein binomial tree method and the Trigeorgis log-transformed binomial tree method. Although Hull already suggested to use the Cox-Ross-Rubinstein binomial tree method with non-constant interest rates, a systematic analysis of Hull's approach to use the implied forward rates in comparison with methods that simulate the non-constant interest rates stochastically for various real options cases in simulations and historical backtesting has not yet been done. Moreover, Trigeorgis' model only accommodated a constant risk-free rate but not non-constant interest rates. The application of the historical risk-free rates results in the historical NPV for the corresponding backtesting period and served as a benchmark for all the other methods investigated.

The overall goal of all these real options cases and the analysis conducted was to answer the following three questions posed in Section 5.3:

- How do the various real options influence the total value of the project, i.e., the project value including all real options?
- In the case of stochastically modelled risk-free interest rates, how does the choice of the term structure model and its parameters influence the real options value and accordingly the net present value of the project?
- Does a stochastically modelled risk-free rate better capture the interest rate volatility that could be observed especially between 1999 and 2002 in the capital markets compared with models with a constant risk-free rate (historical backtesting)?

To answer these questions the tests were framed in five test situations. Each test situation focused on a different aspect of the research topic (see Section 1.2 and Section 5.9):

1. Section 5.4: *Test situation 1: the Schwartz-Moon model with a deferred project start*

   Variable interest rates in the case of the Schwartz-Moon model with a deferred project start.

2. Section 5.5: *Test situation 2: preliminary tests for real options valuation*
   Preliminary tests to investigate the parameters of the real options valuation tools and to compare the valuation methods.

3. Section 5.6: *Test situation 3: real options valuation with a stochastic interest rate using equilibrium models*
   Influence of the salvage and expand factors in cases 1, 2 and 3 on the real options (Section 5.6.1); analysis of all cases for equilibrium models (Vasicek model in Section 5.6.2 and Cox-Ingersoll-Ross model in Section 5.6.3).

4. Section 5.7: *Test situation 4: real options valuation with a stochastic interest rate using no-arbitrage models*
   Analysis of cases 1, 2 and 3 for the Ho-Lee model (Section 5.7.1), comparison of the Hull-White one-factor model with the Hull-White two-factor model (Section 5.7.2), and comparison of the Ho-Lee model with the Hull-White one-factor model (Section 5.7.3).

5. Section 5.8: *Test situation 5: real options valuation in historical backtesting*
   Historical backtesting for cases 1, 2 and 3 through a comparison of the real options pricing when using a stochastically modelled risk-free interest rate (Ho-Lee model), a constant rate, interest rates that equal the currently implied forward rates (Hull's approach), and the historical risk-free rates of the backtesting period.

As a final result, rules for the application of interest rate models (either stochastic term structure models or deterministic models like Hull's implied forward rates approach) within the real options approach for capital budgeting were found that improve the existing methods which usually only apply a constant risk-free rate. These results are summarized below.

6.2 Summary of the Research Results

This section gives an account of the main research results and conclusions of test situations 2 - 5. It describes the results of the conducted analysis, but does not contain any mathematical formula or parameter specification. Since the only purpose of test situation 1 (Schwartz-Moon model) was to introduce the idea of a non-constant risk-free interest rate and to show that a decreasing flat term structure can increase the project's NPV if an option to defer is included into the investment project, the first test situation will not be further referred to.

The main results for test situation 2 (preliminary tests for real options valuation) are displayed in Table 6.1. They were subsequently used to conduct the analysis of the following test situations. The most important results for test situation 3 (real options valuation with a stochastic interest rate using equilibrium models) are displayed in Table 6.2 while the main results for test situation 4 (real options valuation with a stochastic interest rate using no-arbitrage models) are given in Table 6.3. After these results are presented, the final test situation 5 (real options valuation in historical backtesting) is summarized in Tables 6.4 and 6.5. A comparison of the various methods to derive the risk-free interest rate and their effect on the NPV calculation for the three real options cases is provided in Table 6.4.

Table 6.1. Summary of the main results for test situation 2: preliminary tests for real options valuation.

Area of analysis	Main results
Log-transformed explicit versus implicit finite difference method	Valuation results for log-transformed explicit and implicit finite difference methods are almost identical; since stability and consistency restrictions are more strict for explicit than for implicit method the latter is preferred for practical application.
Log-transformed finite difference methods versus Black-Scholes formula	Log-transformed finite difference methods yield inaccurate results for longer times to maturity (e.g., above one year) compared to values calculated with the Black-Scholes formula; since real options are usually long-term options, the log-transformed finite difference methods are not well suited for real options pricing.
Cox-Ross-Rubinstein versus Trigeorgis log-transformed binomial tree method	Real options values derived from Cox-Ross-Rubinstein method and Trigeorgis log-transformed binomial tree method are almost identical for both a constant and a stochastic risk-free interest rate; however, computational time for the Trigeorgis method with non-constant interest rates is extremely long compared with the Cox-Ross-Rubinstein method due to the bushy tree problem inherent in the Trigeorgis method for non-constant risk-free interest rates even for shorter times to maturity; since the Trigeorgis method is not practicable at all for longer times to maturity, the Cox-Ross-Rubinstein binomial tree method is the best method to use for non-constant risk-free interest rates.
Number of simulated short-rate paths for the Cox-Ross-Rubinstein binomial tree method	Accuracy of valuation stabilizes for the Cox-Ross-Rubinstein method combined with a stochastically modelled risk-free rate, if the number of simulated future short-rate paths is approximately 100 or higher.

Source: Section 5.5.2

Table 6.2. Summary of the main results for test situation 3: real options valuation with a stochastic interest rate using equilibrium models.

Area of analysis	Main results
Influence of salvage and expand factors on investment project	The influence of the salvage factor and the expand factor on a project is very significant when the interest rate is simulated with equilibrium term structure models and when the risk-free rate is constant; these parameters affect both the project's NPV and the corresponding standard deviation; therefore, an accurate estimate of both parameters is very important in the capital budgeting process.
Influence of level of interest rates on NPV	The level of the interest rate (as the constant risk-free interest rate in the deterministic case or the mean reversion level for a stochastically modelled risk-free rate) has a significant impact on the NPV; the NPV increases with higher interest rate levels if the project contains many real options (especially, an option to defer); this phenomenon is independent of the chosen interest rate model and also appears if the risk-free rate is constant.
Influence of level of interest rates on NPV's standard deviation	The NPV's standard deviation changes dramatically for different interest rate levels; however, it is not possible to establish a general rule since a higher interest rate level increases or decreases the NPV's standard deviation depending on the chosen real options case and even on the chosen underlying value.
Influence of short-rate volatility parameter on NPV	The NPV (i.e., the mean of the NPVs for many simulated short-rate paths) is independent of the chosen volatility parameter in the short-rate model; the NPV for a stochastically modelled risk-free interest rate and the NPV derived from using the model's mean reversion level are almost identical.
Influence of short-rate volatility parameter on NPV's standard deviation	The NPV's standard deviation is highly dependent on the short-rate volatility parameter if the risk-free rate is stochastically modelled: the higher the volatility, the higher the NPV's standard deviation.
Vasicek versus Cox-Ingersoll-Ross term structure model	Comparing the results for using the Cox-Ingersoll-Ross model with the results for using the Vasicek model is difficult since for the same choice of parameters the Vasicek model produces wider swings in the term structure over time than the Cox-Ingersoll-Ross model due to the different volatility term in each model.

Source: Section 5.6.4

Table 6.3. Summary of the main results for test situation 4: real options valuation with a stochastic interest rate using no-arbitrage models.

Area of analysis	Main results
Influence of salvage and expand factors on investment project	The influence of the salvage factor on the investment project is the same for the Ho-Lee model as for equilibrium models; the same is true for the influence of the expand factor on the investment project's NPV.
Influence of short-rate volatility parameter in the Ho-Lee model on NPV	For the Ho-Lee model the only parameter besides the initial yield curve is the short-rate volatility parameter. For each case 1, 2 and 3 the NPV is almost independent of the short-rate volatility parameter, i.e., the volatility parameter has an insignificant influence on the NPV.
Influence of short-rate volatility parameter in the Ho-Lee model on NPV's standard deviation	The influence of the short-rate volatility parameter of the Ho-Lee model on the NPV's standard deviation is significant: the higher the short-rate volatility parameter, the higher the NPV's standard deviation.
Comparison of Ho-Lee model versus Hull-White one- and two-factor models	The Hull-White one-factor model contains a mean reversion force unlike the Ho-Lee model which is not mean reverting; for both Hull-White models the influence of the mean reversion force on the NPV for cases 1, 2 and 3 is insignificant. The NPV for all three models is similar in cases 1, 2 and 3, independent of the mean reversion force and the volatility. However, the influence of the mean reversion force on the NPV's standard deviation is significant.
Influence of short-rate volatility parameters of Hull-White one- and two-factor models on NPV	For both the Hull-White one-factor and two-factor term structure models the NPV is virtually independent of the volatility parameters.
Influence of the chosen no-arbitrage term structure model on NPV	A comparison of the NPV for risk-free interest rates simulated by using any of the three no-arbitrage models of this test situation reveals that the NPV is almost independent of the term structure model chosen; the mean reversion force of the Hull-White one- and two-factor models as well as the short-rate volatility parameter(s) have an insignificant influence on the NPV as well; the main influence on the NPV can only stem from the initial yield curve whose effect on the NPV is independent of the chosen no-arbitrage term structure model.

Source: Section 5.7.4

In the following, the results from test situation 5 (real options valuation in historical backtesting) are displayed. This test situation is the heart of the whole thesis. It evaluates how the various methods to derive the risk-free interest rate result by comparing the NPV calculations according to these methods with the NPV that was derived from using the historical risk-free interest rates of the historical backtesting period. Which interest rate calculation methods produced the most accurate NPVs for the overall range of underlying values is shown in Table 6.4 for the nine investigated scenarios.

Table 6.4. Evaluation of the absolute difference graphs in test situation 5 for a constant risk-free rate, the implied forward rates, and the Ho-Lee modelled risk-free interest rates (see Table 5.93).

Real option	Initial yield curve	Best	Medium	Worst
case 1	inverted	forward	Ho-Lee	constant
case 1	flat	Ho-Lee	forward	constant*
case 1	normal	constant	forward	Ho-Lee
case 2	inverted	Ho-Lee	forward	constant
case 2	flat	Ho-Lee	forward	constant**
case 2	normal	constant	forward	Ho-Lee
case 3	inverted	forward	Ho-Lee	constant
case 3	flat	forward	constant***	Ho-Lee
case 3	normal	forward	Ho-Lee	constant

* very close to "forward" ** very close to "forward" *** only slightly worse than "forward"

Based on Table 6.4, the main results of test situation 5 are summarized in Table 6.5. These results were solely derived from the historical backtesting of the three real options cases for three chosen backtesting periods. Table 6.5 can be seen as the answer to the thesis' goal which was presented in Section 1.1. The goal was to add value by providing clear relationships that are both interesting from an academic and a practical point of view. Or citing Section 1.1:

> Who is the intended audience of this thesis? The thesis is written for two purposes. The first one is to enrich the field of academic research by providing reference literature that gives insight into a subject that has not yet been analyzed in detail, especially from the viewpoint of numerical analysis, simulation, and historical backtesting. To this end, the dependencies of the parameters in the stochastic term structure models and the various real options valuation tools will be described in detail. Moreover, the relationship of the real options value/project value in the various real options cases and the underlying value will be investigated thoroughly. The second purpose is to establish rules for practical application based on the relationships found. Expert knowledge of these relationships and rules is important for Corporate Fi-

nance practice in a world where interest rates are changing rapidly (as seen especially between 1999 and 2002).

This major gap in the literature on real options has been successfully closed by the results of this thesis.

Table 6.5. Summary of the main results for test situation 5: real options valuation in historical backtesting.

Area of analysis	Main results
Historical backtesting for flat initial yield curve	Independent of the analyzed real options case and of the underlying value the application of the implied forward rates yields (almost) the same NPV results as the application of the constant risk-free rate; however, for cases 1 and 2 the application of the Ho-Lee model yields better results than the application of the forward rates but worse results for case 3.
General rules depending on the shape of the initial yield curve	It is not possible to conclude from the backtesting results summarized in Table 6.4 that for a certain shape of the initial yield curve one method used to derive the risk-free interest rate is better than the other ones.
General rules depending on real options case	In case 3 it is always optimal to use the implied forward rates. This cannot be said for cases 1 and 2.
General evaluation of Ho-Lee simulated interest rates and implied forward rates approach	Table 6.5 shows that by and large the Ho-Lee model yields the best results in three scenarios, the implied forward rates approach in four scenarios and the constant risk-free rate only in two scenarios. In seven out of nine scenarios the application of either the implied forward rates or the Ho-Lee model is better suited to calculate the NPV of an investment project than using a constant risk-free interest rate. From the stand point of implementational convenience, Hull's implied forward rates approach is better suited in Corporate Finance practice than a stochastic term structure model like the Ho-Lee model. In all nine test scenarios analyzed the implied forward rates approach used for real options valuation never yields the worst NPV results whereas the currently prevailing practice to use a constant risk-free rate leads to the worst NPV results in two-thirds of the test scenarios.
Overall evaluation	Hull's approach of using the forward rates that are implied in the current yield curve has to be preferred in general over using a constant risk-free interest rate for real options valuation.

Source: Section 5.8.4 and Table 5.93

6.3 Economic Implications

This thesis analyzed real options valuation using numerical simulations and historical backtesting. The implications which have been pointed out so far were based on these numerical results without considering the economic background. Therefore, it now needs to be asked, what are the **economic** implications of interest rate modelling in real options valuation? In detail, two questions have to be addressed:

- What industries will benefit most from the results obtained in the thesis?
- How can interest rate modelling best be implemented in the day-to-day practice of Corporate Finance?

It could be clearly shown that the assumption of constant interest rates in real options valuation is not justifiable since interest rates fluctuate considerably in the long run. Although from a theoretical perspective the application of stochastic interest rates in real options valuation should be preferred over using the implied forward rates, an additional benefit could not be found in historical backtesting for the three cases analyzed. However, the real options analyzed in these three cases are typical for capital budgeting projects. They are most attractive in capital intensive projects, like natural-resource investments, land development projects, and flexible manufacturing projects. For instance, to build a manufacturing complex for semiconductor production is not only a long-term project but also very expensive with lots of high tech equipment needed.

However, since speed is key in the semiconductor business, it might be advisable not to continue building the manufacturing complex but to exercise the option to abandon, if a competitor offers a similar product earlier and/or cheaper. The building could be sold for a salvage value or, as a modification of real options case 1, parts of the high tech equipment could be sold and the remainder could be used for a new business or product line.

The option to expand the business (the only option in case 1) is useful e.g. for land development projects where the extent of the development is not determined yet. If a community wants to develop a new industrial complex, it may first start with a small land development plan and exercise the expand option if the project promises high returns. Since such an undertaking is long-term by nature, the interest rate will be a key evaluation parameter.

On the other hand, an option to defer is particularly valuable in natural resource investments, e.g. in mining and oil field exploitation, which are extremely long-term and cost intensive. The latter is especially true for off-shore oil drilling in the Gulf of Mexico or the North Sea. The option to defer, a learning option, allows learning to occur over a certain period of time, e.g. during the exploration work undertaken on a site prior to exploiting it. Having detailed knowledge about the size, shape location and depth of the oil field is crucial to develop the appropriate drilling strategy. Additionally, planning and building the off-shore platform is extremely capital extensive and long-term, resulting in huge investment cost years prior to receiving the first revenue. Clearly, incorporating a non-constant interest rate is beneficial to derive a reliable net present value. However, the other real options analyzed in this thesis are also applicable in the long-term oil drilling business. Therefore, applying the findings of the thesis should be of direct benefit to these industries.

What does this practically mean, i.e., how can interest rate modelling best be implemented in the day-to-day practice of Corporate Finance? Although it was shown that a constant risk-free rate should not be used in Corporate Finance, existing real options valuation need not be discarded. The Cox-Ross-Rubinstein binomial tree method as the ideal method to price even complex real options can easily be modified to accomodate for a non-constant risk-free rate. This was done by using the computer simulation tool developed for this thesis and can easily be achieved when corporations value an investment project that contains considerable flexibility. Therefore, the main result of the thesis, the application of the forward rate implied in the current yield curve, can quickly be adopted in Corporate Finance practice.

6.4 Outlook and Future Areas of Research

As summarized in the previous section, using a stochastically simulated risk-free rate, applying the Ho-Lee no-arbitrage model, or using Hull's approach of the implied forward rates for real options valuation turned out to be better suited for pricing real options than just simply applying a constant risk-free rate. As pointed out in Chapter 1, awareness of the interest rate as a value driver for real options is currently still limited, and interest rate development is only a minor aspect of sensitivity and scenario analysis. As seen in Chapter 1 as well, the interest rate level as a value driver in the real options approach is only a small concern for companies. Consequently, the results of this thesis will lead directly to practical applications in Corporate Finance and to improved capital budgeting through incorporating interest rate modelling in the real options valuation process.

In addition, this thesis examined various unresolved issues in the world of real options. Firstly, as was mentioned in Section 2.2, it was an important goal of this thesis to support future research, which covers the development of *generic options-based user-friendly software packages with simulation capabilities that can handle multiple real options as a practical aid to corporate planners*[1]. Therefore, this thesis is accompanied by a complex computer simulation program, used for all of the results displayed. This program can easily be modified to value other real options in addition to the three real options cases analyzed in this thesis. It allows more actual cases and real-life investment situations to be analyzed in detail. Secondly, by concentrating primarily on the analysis of complex real options that interact and are interdependent over time, this thesis also supports future research into an area emphasized by Trigeorgis as well[2]: Future research should be carried out into *modelling more rigorously the various growth and strategy options - e.g., synergies between projects taken together or interactions across time.*

However, there are still many aspects to investigate even within the context of the thesis topic. For example, only three cases have been analyzed in this thesis. More and different types of real options should be further analyzed as well. Additionally, the parameters for these real options cases and within the stochastic term structure models could be varied to an even greater extent than was possible in this thesis. This would provide an even deeper insight into the behaviour and interdependence of these parameters. It was, e.g., ascertained that the influence of the interest rate on an investment project is less than that of parameters such as the salvage factor and the expand factor. Moreover, the company has to have an exact idea of what the underlying value of the project should be, a factor which has a significant influence on the project, as could be seen. These parameters must therefore be estimated as accurately as possible, another important area of research into capital budgeting.

Another major task is to convince practitioners to apply not only the real options approach but also the findings of this thesis in capital budgeting. This process, however, is facilitated by today's complexity in decision-making as already pointed out in Chapter 1: More than half of the investment decisions made at both the headquarter level and within an operational unit have a degree of flexibility. This flexibility cannot be captured by the traditional DCF approach but only by the real options approach. Consequently, the real options approach will eventually become the method of choice for pricing investment projects and will replace the DCF method. This thesis enhances this process by conducting for the first time a detailed analysis of stochastic risk-free interest rates in the real options approach for various complex real options using simulations and historical backtesting. This clearly could encourage even greater use of the real options approach.

In summary, the future efforts in the field of real options should be twofold. Firstly, the superiority of the real options approach in capital budgeting must

[1] See Trigeorgis [132], page 375.
[2] See Trigeorgis [132], page 376.

be stressed by academics so that the real options approach becomes the tool of choice in capital budgeting. An even closer co-operation between academia and Corporate Finance practice will clearly foster this development. Secondly, the awareness of the interest rate as an important value driver within the real options approach must be stressed. This thesis could boost this development since it has shown that modelling the risk-free interest rate within the real options approach (either through implied forward rates or stochastically simulated rates using no-arbitrage models) captures flexibility inherent in the project much better than using a constant risk-free rate. Since the focus of this thesis was on practical applications, the rules derived can be used directly in practice. This successfully closes a major gap in literature and shows once again the superiority of the real options approach over the DCF method in capital budgeting.

List of Abbreviations and Symbols

$\mathbb{N}$	set of natural numbers
$\mathbb{N}_0$	set of natural numbers including zero
$\mathbb{R}$	set of real numbers
$\mathbb{R}^+$	set of positive real numbers
$\mathbb{R}_0^+$	set of positive real numbers including zero
$\mathbb{R}^-$	set of negative real numbers
$\mathbb{R}^d$	d-dimensional Euclidian space, $d \in \mathbb{N}$
$\mathbb{R}^{d,r}$	set of $(d \times r)$-matrices with real numbers
M^T	transpose of matrix M
v^T	transpose of vector v
$a \wedge b$	$\min(a, b)$
C^2	set of all twice smoothly differentiable functions in one variable
Π_3	set of all cubic functions with values in $\mathbb{R}$
1_A	indicator function for set A, i.e.:
$1_A(x)$	$= \begin{cases} 1 \text{ if } x \in A \\ 0 \text{ if } x \notin A \end{cases}$
$L_2(\Omega, P)$	$= \left\{ f : \Omega \longrightarrow \mathbb{R}^d \,\middle|\, \int_\Omega \lvert f(\omega) \rvert^2 \, dP(\omega) < \infty \right\}$
$\sigma(X_s \mid 0 \leq s \leq t)$	σ-field generated by the random variables $X_s, 0 \leq s \leq t$
$\mathcal{B}$	one-dimensional Borel σ-field
$P(\cdot)$	probability
$P(\cdot \mid \cdot)$	conditional probability
$E(\cdot)$	expected value
$E(\cdot \mid \cdot)$	conditional expected value
$Var(\cdot)$	variance

$Var(\cdot\|\cdot)$	conditional variance
$Cov(\cdot,\cdot)$	covariance of two random variables
$N(\mu,\sigma^2)$	normal distribution with mean μ and variance σ^2
$N(\cdot)$	cumulative normal distribution function
$\longrightarrow$	*converges to*
$\stackrel{d}{=}$	*is distributed as*

References

1. Alverz, L.H.R. & E. Koskela: *Irreversible Investment and Interest Rate Variability*, Working Paper, Turku School of Economics and Business Administration, Turku, Finland, May 6, 2002.

2. Amram, M. & N. Kulatilaka: *Real Options - Managing Strategic Investment in an Uncertain World*, Harvard Business School Press, 1999.

3. Babbel, D.F. & C.B. Merrill: *Valuation of Interest-Sensitive Financial Instruments*, Society of Actuaries, 1996.

4. Barone-Adesi, G. & R.E. Whaley: *Efficient Analytic Approximation of American Option Values*, The Journal of Finance, Vol. 42 No. 2, June 1987, pp. 301-320.

5. Baxter, M: *General Interest Rate Models and the Universality of HJM*, published in *Mathematics of Derivatives Securities* by M. Dempster & S. Pliska (editors), Cambridge University Press, 1997, pp. 315-335.

6. Björk, T.: *Arbitrage Theory in Continuous Time*, Oxford University Press, 1998.

7. Black, F.: *How we came up with the Option Formula*, Risk Magazine, December 1997, pp. 143-148.

8. Black, F., E. Derman & W. Toy: *A One-Factor Model of Interest Rates and its Application to Treasury Bond Options*, Financial Analysts Journal, January-February 1990, pp. 33-39.

9. Black, F. & M.C. Scholes: *The Pricing of Options and Corporate Liabilities*, Journal of Political Economy, No. 81, 1973, pp. 637-659.

10. Blomeyer, E.C.: *An Analytic Approximation for the American Put Price for Options on Stocks with Dividends*, Journal of Financial and Quantitative Analysis, Vol. 21 No. 2, June 1986, pp. 229-233.

11. Börsch-Supan, W.: *Skript zur Vorlesung Numerische Mathematik I*, School of Mathematics, Johannes Gutenberg-University in Mainz, Germany, 1989.

12. Borkovec, M. & C. Klüppelberg: *Extremal Behaviour of Diffusion Models in Finance*, Berichte zur Stochastik und verwandten Gebieten, Johannes Gutenberg-University in Mainz, Germany, January 1997.

13. Boyle, P.P.: *Options: A Monte Carlo Approach*, Journal of Financial Economics, Vol. 4, 1977, pp. 323-338.

14. Boyle, P.P.: *A Lattice Framework for Option Pricing with Two State Variables*, Journal of Financial and Quantitative Analysis, Vol. 23, 1988, pp. 1-26.

15. Brennan, M.J.: *The Pricing of Contingent Claims in Discrete Time Models*, The Journal of Finance, Vol. 34 No. 1, 1979, pp. 53-68.

16. Brennan, M.J. & E.S. Schwartz: *The Valuation of American Put Options*, The Journal of Finance, Vol. 32 No. 2, 1977, pp. 449-462.

17. Brennan, M.J. & E.S. Schwartz: *Finite Difference Methods and Jump Processes Arising in the Pricing of Contingent Claims: A Synthesis*, Journal of Financial and Quantitative Analysis, September 1978, pp. 461-475.

18. Brennan, M.J. & E.S. Schwartz: *A Continuous-Time Approach to the Pricing of Bonds*, Journal of Banking and Finance, Vol. 3, 1979, pp. 135-155.

19. Brennan, M.J., & E.S. Schwartz: *An Equilibrium Model of Bond Pricing and a Test of Market Efficiency*, Journal of Financial and Quantitative Analysis, Vol. 17 No. 3, September 1982, pp. 301-329.

20. Brennan, M.J. & E.S. Schwartz: *Evaluating Natural Resource Investments*, Journal of Business, Vol. 58 No. 2, 1985, pp. 135-157.

21. Brennan, M.J. & L. Trigeorgis: *Project Flexibility, Agency, and Competition*, Oxford University Press, 2000.

22. Bronstein, I.N. & K.A. Semendjajew: *Taschenbuch der Mathematik*, 24th Edition, Harri Deutsch Thun, Frankfurt, Germany, 1989.

23. Brown, R.H. & S.M. Schaefer: *The Term Structure of Real Interest Rates and the Cox-Ingersoll-Ross Model*, Journal of Financial Economics, Vol. 35, 1994, pp. 3-42.

24. Bubsy, J.S. & C.G.C. Pitts: *Real Options and Capital Investment Decisions*, Management Accounting, November 1997, pp. 38-39.

25. Bühler, A.: *Einfaktormodelle der Fristenstruktur: Theoretische und Empirische Betrachtungen*, Bank- und finanzwirtschaftliche Forschungen, No. 211, Paul Haupt, 1995.

26. Bühler, A. & H. Zimmermann: *A Statistical Analysis of the Term Structure of Interest Rates in Switzerland and Germany*, Journal of Fixed Income, Vol. 6, December 1996, pp. 55-67.

27. Carr, P.: *The Valuation of Sequential Exchange Opportunities*, The Journal of Finance, Vol. XLIII No. 5, December 1988, pp. 1235-1256.

28. Carr, P.: *The Valuation of American Exchange Options with Application to Real Options*, published in: *Real Options and Capital Investment*, pp. 109-120, Ed. L. Trigeorgis, Praeger, 1995.

29. Carverhill, A.P. & K. Pang: *Efficient and Flexible Bond Option Valuation in the Heath, Jarrow, and Morton Framework*, Journal of Fixed Income, Vol. 5, 1995, pp. 70-77.

30. Chan, K.C., G.A. Karolyi, F.A. Longstaff & A.B. Sanders: *The Volatility of Short-Term Interest Rates: An Empirical Comparison of Alternative Models of the Term Structure of Interest Rates*, The Journal of Finance, Vol. 47, 1992, pp. 1209-1227.

31. Clewlow, L. & C. Strickland: *Implementing Derivatives Models*, John Wiley & Sons, 1998.

32. Cochrane, J.H.: *Asset Pricing*, Princton University Press, 2001.

33. Copeland, T. & V. Antikarov: *Real Options - A Practitioner's Guide*, Texere, 2001.

34. Cox, J.C., J.E. Ingersoll Jr. & S.A. Ross: *A Theory of the Term Structure of Interest Rates*, Econometrica, Vol. 53 No. 2, Spring 1985, pp. 385-407.

35. Cox, J.C. & S.A. Ross: *The Valuation of Options for Alternative Stochastic Processes*, Journal of Financial Economics, Vol. 4, 1976, pp. 145-166.

36. Cox, J.C., S.A. Ross & M. Rubinstein: *Option Pricing: A Simplified Approach*, Journal of Financial Economics, Vol. 7, 1979, pp. 229-263.

37. Cox, J.C. & M. Rubinstein: *Options Markets*, Prentice Hall, 1985.

38. Dean, J.: *Capital Budgeting*, Columbia University Press, 1951.

39. Deutsch, H.-P.: *Derivate und interne Modelle*, Schäffer-Poeschel, Stuttgart, Germany, 2001.

40. Dixit, A.K. & R.S. Pindyck: *Investment Under Uncertainty*, Princeton University Press, 1993.

41. Duffie, D. & R. Kan: *Multi-Factor Term Structure Models*, Phil. TranR. Soc. London, Vol. 347, 1994, pp. 577-586.

42. Fabozzi, F.J.: *Fixed Income Analysis for the CFA Program*, Editor Frank J. Fabozzi Associates, 2000.

43. Fong, H.G. & O.A. Vasicek: *Interest Rate Volatility as a Stochastic Factor*, Working Paper, Gifford Fong Associates, 1992.

44. Geske, R.: *The Valuation of Compound Options*, Journal of Financial Economics, Vol. 7 No. 1., 1979, pp. 63-81.

45. Geske, R. & H.E. Johnson: *The American Put Option Valued Analytically*, The Journal of Finance, Vol. 39 No. 5, December 1984, pp. 1511-1524.

46. Geyer, A.L.J. & S. Pichler: *A State-Space Approach to Estimate and Test Multi-Factor Cox-Ingersoll-Ross Models of the Term Structure*, Working Paper of the Wirtschaftsuniversität und Technische Universität Wien, September 1996.

47. Gitman, L.J. & J.R. Forrester: *A Survey of Capital Budgeting Techniques Used by Major U.S. Firms*, Financial Management, Vol. 6 No. 3, Fall 1977, pp. 66-71.

48. Göing, A.: *Estimation in Financial Models*, Master's Thesis, ETH Zürich, School of Mathematics (Risklab), 1996.

49. Hayes, R. & W. Abernathy: *Managing our way to Economic Decline*, Harvard Business Review, Vol. 58 No. 4, 1980, pp. 381-408.

50. Hayes, R. & D. Garvin: *Managing as if Tomorrow Mattered*, Harvard Business Review, Vol. 60 No. 3, 1982, pp. 71-79.

51. Heath, D., R. Jarrow & A. Morton: *Bond Pricing and the Term Structure of Interest Rates: A Discrete Time Approximation*, Journal of Financial and Quantitative Analysis, Vol. 25 No. 4, December 1990, pp. 419-440.

52. Heath, D., R. Jarrow & A. Morton: *Contingent Claim Valuation with Random Evolution of Interest Rates*, Review of Futures Markets, Vol. 9, 1991, pp. 55-76.

53. Heath, D., R. Jarrow & A. Morton: *Bond Pricing and the Term Structure of Interest Rates: A new Methodology for Contingent Claims Valuation*, Econometrica, Vol. 60 No. 1, January 1992, pp. 77-105.

54. Heath, D., R. Jarrow & A. Morton: *Easier Done Than Said*, Risk Magazine, Vol. 5 No. 9, 1992, pp. 77-80.

55. Hertz, D.: *Risk Analysis in Capital Investment*, Harvard Business Review, Vol. 42, January-February 1964, pp. 95-106.

56. Ho, T.S.Y.: *Evolution of Interest Rate Models: A Comparison*, Journal of Derivatives, Summer 1995, pp. 9-20.

57. Ho, T.S.Y. & S.-B. Lee: *Term Structure Movements and Pricing Interest Rate Contingent Claims*, The Journal of Finance, Vol. 41, 1986, pp. 1011-1029.

58. Ho, T.S.Y., R. Stapleton & M. Subrahmanyam: *The Valuation of American Options With Stochastic Interest Rates: A Generalization of the Geske-Johnson Technique*, The Journal of Finance, Vol. 52 No. 2, 1997, pp. 827-840.

59. Hommel, U. & H. Lehmann: *Die Bewertung von Investitionsprojekten mit dem Realoptionsansatz - Ein Methodenüberblick*, published in: *Realoptionen in der*

Unternehmenspraxis, pp. 113-129, Ed. U. Hommel, M. Scholich & R. Vollrath, Springer, 2001.

60. Hommel, U., M. Scholich & P.N. Baecker: *Reale Optionen*, Springer, 2003.

61. Hommel, U., M. Scholich & R. Vollrath: *Realoptionen in der Unternehmenspraxis*, Springer, 2001.

62. Hull, J.C.: *Options, Futures, and Other Derivatives*, Prentice Hall, 3rd Edition, 1997.

63. Hull, J. & A. White: *The use of the Control Variate Technique in Option Pricing*, Journal of Financial and Quantitative Analysis, Vol. 23 No. 3, September 1988, pp. 237-251.

64. Hull, J. & A.White: *Valuing Derivative Securities Using the Explicit Finite Difference Method*, Journal of Financial and Quantitative Analysis, Vol. 25 No. 1, 1990, pp. 87-99.

65. Hull, J. & A. White: *Pricing Interest-Rate-Derivative Securities*, The Review of Financial Studies, Vol. 3 No. 3, 1990, pp. 573-592.

66. Hull, J. & A.White: *One-Factor Interest Rate Models and the Valuation of Interest-Rate Derivative Securities*, Journal of Financial and Quantitative Analysis, Vol. 28 No. 2, 1993, pp. 235-254.

67. Hull, J. & A. White: *Numerical Procedures for Implementing Term Structure Models I: Single-Factor Models*, Journal of Derivatives, Vol. 2, Fall 1994, pp. 7-16.

68. Hull, J. & A. White: *Numerical Procedures for Implementing Term Structure Models II: Two-Factor Models*, Journal of Derivatives, Vol. 2, Fall 1994, pp. 37-48.

69. Ingersoll Jr., J.E. & S.A. Ross: *Waiting to Invest: Investment and Uncertainty*, Journal of Business, Vol. 65 No. 1, 1992, pp. 1-29.

70. Jarrow, R.A.: *Modelling Fixed-Income Securities and Interest Rate Options*, McGraw-Hill, 1996.

71. Johnson, H.E.: *An Analytic Approximation for the American Put Price*, Journal of Financial and Quantitative Analysis, Vol. 18 No. 1, 1983, pp. 141-148.

72. Karatzas, I. & S.E. Shreve: *Brownian Motion and Stochastic Calculus*, Springer, 1994.

73. Karlin, S. & H.M. Taylor: *A Second Course in Stochastic Processes*, Academic Press, 1981.

74. Kester, W.C.: *Turning Growth Options into Real Assets*, published in: *Capital Budgeting under Uncertainty*, Ed. R. Aggarwal, Prentice-Hall, 1993.

75. Klammer, T.P.: *Empirical Evidence of the Adoption of Sophisticated Budgeting Techniques*, Journal of Business, Vol. 45 No. 3, 1972, pp. 66-71.

76. Klammer, T.P. & M.C. Walker: *The Continuing Increase of Sophisticated Budgeting Techniques*, California Management Review, Vol. 27 No. 1, Fall 1984, pp. 137-148.

77. Kloeden, P.E. & E. Platen: *Numerical Solution of Stochastic Differential Equations*, Springer, 1995.

78. Kolb, R.W.: *Futures, Options, & Swaps*, Blackwell Publishers Inc., 2000.

79. Leippold, M. & Th. Heinzl: *Zinsstrukturmodelle*, published in *Value at Risk im Vermögensverwaltungsgeschäft, Otto Bruderer und Konrad Hummler*, Stämpfli Bern, Switzerland, 1997, pp. 137-174.

80. Leithner, S.: *Valuation and Risk Management of Interest Rate Derivative Securities*, Bank- und finanzwirtschaftliche Forschungen, No. 163, Paul Haupt, 1992.

81. Lieskovsky, J., R. Onkey, M. Schulmerich, C.C. Teng & J. Wee: *Pricing and Hedging Asian Options on Copper*, Working Paper, MIT Sloan School of Management, Laboratory of Financial Engineering, December 2000.

82. Litterman, R. & J. Scheikman: *Common Factors Affecting Bond Returns*, Journal of Fixed Income, Vol. 1, 1991, pp. 54-62.

83. Longstaff, F.A. & E.S. Schwartz: *Interest Rate Volatility and the Term Structure: A Two-Factor General Equilibrium Model*, The Journal of Finance, Vol. 47, 1992, pp. 1259-1282.

84. MacMillan, L.W.: *Analytic Approximation for the American Put Option*, Advances in Futures and Options Research, Vol. 1, 1986, pp. 119-139.

85. Magee, J.: *How to use Decision Trees in Capital Investment*, Harvard Business Review, Vol. 42, September-October 1964, pp. 79-96.

86. Majd, S. & R.S. Pindyck: *Time to Build, Option Value, and Investment Decisions*, Journal of Financial Economics, Vol. 18, 1987, pp. 7-27.

87. Margrabe, W.: *The Value of an Option to Exchange one Asset for Another*, The Journal of Finance, Vol. 33 No. 1, March 1978, pp. 177-186.

88. Mason, S.P. & R.C. Merton: *The Role of Contingent Claims Analysis in Corporate Finance*, published in: *Recent Advances in Corporate Finance*, Ed. E. Altman and M. Subrahmanyam, Irwin, 1985.

89. Mason, S.P. & L. Trigeorgis: *Valuing Managerial Flexibility*, Midland Corporate Finance Journal, Vol. 5 No. 1, 1987, pp. 14-21.

90. McCracken, D. & W. Dorn: *Numerical Methods and Fortran Programming*, John Wiley & Sons, 1969.

91. McDonald, R.L. & D.R. Siegel: *Investment and the Valuation of Firms when there is an Option to Shut Down*, International Economic Review, Vol. 26 No. 2, June 1985, pp. 331-349.

92. McDonald, R.L. & D.R. Siegel: *The Value of Waiting to Invest*, The Quarterly Journal of Economics, November 1986, pp. 707-727.

93. Merton, R.C.: *The Theory of Rational Option Pricing*, Bell Journal of Economics and Management Science, Vol. 4, Spring 1973, pp. 141-183.

94. Merton, R.C.: *The Pricing of Corporate Debt: The Risk Structure of Interest Rates*, The Journal of Finance, Vol. 29, 1974, pp. 449-470.

95. Merton, R.C.: *Option Pricing when Underlying Stock Returns are Discontinuous*, Journal of Financial Economics, Vol. 3, 1976, pp. 125-144.

96. Miltersen, K.R.: *Valuation of Natural Resource Investments with Stochastic Convenience Yields and Interest Rates*, published in: *Project Flexibility, Agency, and Competition*, pp. 183-204, Ed. M.J. Brennan & L. Trigeorgis, Oxford University Press, 2000.

97. Miltersen, K.R. & E.S. Schwartz: *Pricing of Options on Commodity Futures with Stochastic Term Structures of Convenience Yields and Interest Rates*, Finance Working Paper No. 5-97, The John E. Anderson Graduate School of Management at UCLA, Journal of Financial and Quantitative Analysis, 1998.

98. Musiela, M. & M. Rutkowski: *Martingale Methods in Financial Modelling*, Springer, 1997.

99. Myers, S.C.: *Determinantes of Corporate Borrowing*, Journal of Financial Economics, Vol. 5 No. 2, 1977, pp. 147-176.

100. Myers, S.C.: *Finance Theory and Financial Strategy*, Interfaces, Vol. 14, January-February 1984, pp. 126-137.

101. Myers, S.C. & S. Majd: *Abandonment Value and Project Life*, Advances in Futures and Options Research, Vol. 4, 1990, pp. 1-21.

102. Nelson, D.B. & K. Ramaswamy: *Simple Binomial Processes as Diffusion Approximation in Financial Models*, The Review of Financial Studies, Vol. 3 No. 3, 1990, pp. 393-430.

103. Øksendal, B.: *Stochastic Differential Equations*, Springer, 1995.

104. Paddock, J.L., D.R. Siegel & J.L. Smith: *Option Valuation of Claims on Real Assets: The Case of Offshore Petroleum Leases*, The Quarterly Journal of Economics, August 1988, pp. 479-508.

105. Parkinson, M.: *Option Pricing: The American Put*, Journal of Business, Vol. 50 No. 1, 1977, pp. 21-36.

106. Pindyck, R.S.: *Investments of Uncertain Cost*, Journal of Financial Economics, Volume 34, 1993, pp. 53-76.

107. Pojezny, N.: *Multidimensional Models of American Compound Options in Research and Development*, Master's Thesis in Business Administration, ebs-European Business School, Oestrich-Winkel, Germany, 2002.

108. Pritsch, G. & J. Weber: *Die Bedeutung des Realoptionsansatzes aus Controlling Sicht*, in: *Realoptionen in der Unternehmenspraxis*, pp. 13-43, Ed. U. Hommel, M. Scholich & R. Vollrath, Springer, 2001.

109. Rebonato, R.: *Interest-Rate Option Models*, 2nd Edition, John Wiley & Sons, 2000.

110. Reichert, A.K., J.S. Moore & E. Byler: *Financial Analysis Among Large U.S. Corporations: Recent Trends and the Impact of the Personal Computer Industry*, Journal of Business, Vol. 15 No. 4, 1993, pp. 469-485.

111. Richard, S.: *An Arbitrage Model of the Term Structure of Interest Rates*, Journal of Financial Economics, Vol. 6, 1978, pp. 33-57.

112. Roberts, K. & M. Weitzman: *Funding Criteria for Research, Development, and Exploration Projects*, Econometrica, Vol. 49 No. 5, 1981, pp. 1261-1288.

113. Rogers, L.C.G. & D. Williams: *Diffusions, Markov Processes and Martingales Volume 2*, John Wiley & Sons, 1994.

114. Rudolf, M.: *Zinsstrukturmodelle*, Physica, 2000.

115. Sandmann, K.: *The Pricing of Options with an Uncertain Interest Rate: A Discrete Time Approach*, Mathematical Finance, Vol. 3, 1993, pp. 201-216.

116. Sandmann, K. & D. Sondermann: *A Term Structure Model and the Pricing of Interest Rate Derivatives*, Discussion Paper No. B-180, University of Bonn, Germany, 1991.

117. Schulmerich, M.: *Statistische Verfahren für Diffusionsprozesse mit Anwendung auf stochastische Zinsmodelle der Finanzmathematik*, Master's Thesis in Mathematics, School of Mathematics, Johannes Gutenberg-University Mainz, Germany, 1997.

118. Schulmerich, M.: *Einsatz und Pricing von Realoptionen: Einführung in grundlegende Bewertungsansätze*, published in: *Reale Optionen*, pp. 63-96, Ed. U. Hommel, M. Scholich & P.N. Baecker, Springer, 2003.

119. Selby, M.J.P. & C.R. Strickland: *Computing the Fong and Vasicek Pure Discount Bond Price Formula*, Journal of Fixed Income, September 1995, pp. 78-84.

120. Schaefer, S.M. & E.S. Schwartz: *A Two-Factor Model of the Term Structure: An Approximate Analytical Solution*, Journal of Financial and Quantitative Analysis, Vol. 19, 1984, pp. 413-424.

121. Schwartz, E.S. & M. Moon: *Evaluating Research and Development Investments*, published in: *Project Flexibility, Agency, and Competition*, pp. 85-106, Ed. L. Trigeorgis & M.J. Brennan, Oxford University Press, 2000.

122. Schweser, P.: *Schweser Study Program for the CFA Level 2 exam*, Book 4, 2001.

123. Schweser, P.: *Schweser Study Program for the FRM exam*, Book 2, 2002.

124. Stoer, J.: *Numerische Mathematik II*, 3rd Edition, Springer, 1990.

125. Stoer, J.: *Numerische Mathematik I*, 6th Edition, Springer, 1993.

126. Stulz, R.M.: *Options on the Minimum or the Maximum of Two Risky Assets: Analysis and Application*, Journal of Financial Economics, Vol. 10 No. 2, 1982, pp. 161-185.

127. Tobler, J.: *Schätzung von Zinsstrukturen für den Schweizer Franken-Kapitalmarkt unter Berücksichtigung von Friktionen*, Paul Haupt, 1996.

128. Trigeorgis, L.: *A Conceptual Options Framework for Capital Budgeting*, Advances in Futures and Options Research, Vol. 3, 1988, pp. 145-167.

129. Trigeorgis, L: *A Real-Options Application in Natural-Resource Investments*, Advances in Futures and Options Research, Vol. 4, 1990, pp. 153-164.

130. Trigeorgis, L.: *A Log-Transformed Binomial Analysis Method for Valuing Complex Multi-Option Investments*, Journal of Financial and Quantitative Analysis, Vol. 26 No. 3, September 1991, pp. 309-326.

131. Trigeorgis, L.: *Real Options in Capital Investment*, Praeger, 1995.

132. Trigeorgis, L.: *Real Options*, The MIT Press, 1996.

133. Vasicek, O.: *An Equilibrium Characterization of the Term Structure*, Journal of Financial Economics, Vol. 5, 1977, pp. 177-188.

134. Vollrath, R.: *Die Berücksichtigung von Handlungsflexibilität bei Investitionsentscheidungen - Eine empirische Untersuchung*, published in: *Realoptionen in der Unternehmenspraxis*, pp. 45-77, Ed. U. Hommel, M. Scholich & R. Vollrath, Springer, 2001.

135. Wilmott, P.: *Derivatives*, John Wiley & Sons, 2000.

Index

Printing and Binding: Strauss GmbH, Mörlenbach

Lecture Notes in Economics and Mathematical Systems

For information about Vols. 1–464
please contact your bookseller or Springer-Verlag

Vol. 465: T. J. Stewart, R. C. van den Honert (Eds.), Trends in Multicriteria Decision Making. X, 448 pages. 1998.

Vol. 466: A. Sadrieh, The Alternating Double Auction Market. VII, 350 pages. 1998.

Vol. 467: H. Hennig-Schmidt, Bargaining in a Video Experiment. Determinants of Boundedly Rational Behavior. XII, 221 pages. 1999.

Vol. 468: A. Ziegler, A Game Theory Analysis of Options. XIV, 145 pages. 1999.

Vol. 469: M. P. Vogel, Environmental Kuznets Curves. XIII, 197 pages. 1999.

Vol. 470: M. Ammann, Pricing Derivative Credit Risk. XII, 228 pages. 1999.

Vol. 471: N. H. M. Wilson (Ed.), Computer-Aided Transit Scheduling. XI, 444 pages. 1999.

Vol. 472: J.-R. Tyran, Money Illusion and Strategic Complementarity as Causes of Monetary Non-Neutrality. X, 228 pages. 1999.

Vol. 473: S. Helber, Performance Analysis of Flow Lines with Non-Linear Flow of Material. IX, 280 pages. 1999.

Vol. 474: U. Schwalbe, The Core of Economies with Asymmetric Information. IX, 141 pages. 1999.

Vol. 475: L. Kaas, Dynamic Macroeconomics with Imperfect Competition. XI, 155 pages. 1999.

Vol. 476: R. Demel, Fiscal Policy, Public Debt and the Term Structure of Interest Rates. X, 279 pages. 1999.

Vol. 477: M. Théra, R. Tichatschke (Eds.), Ill-posed Variational Problems and Regularization Techniques. VIII, 274 pages. 1999.

Vol. 478: S. Hartmann, Project Scheduling under Limited Resources. XII, 221 pages. 1999.

Vol. 479: L. v. Thadden, Money, Inflation, and Capital Formation. IX, 192 pages. 1999.

Vol. 480: M. Grazia Speranza, P. Stähly (Eds.), New Trends in Distribution Logistics. X, 336 pages. 1999.

Vol. 481: V. H. Nguyen, J. J. Strodiot, P. Tossings (Eds.). Optimation. IX, 498 pages. 2000.

Vol. 482: W. B. Zhang, A Theory of International Trade. XI, 192 pages. 2000.

Vol. 483: M. Königstein, Equity, Efficiency and Evolutionary Stability in Bargaining Games with Joint Production. XII, 197 pages. 2000.

Vol. 484: D. D. Gatti, M. Gallegati, A. Kirman, Interaction and Market Structure. VI, 298 pages. 2000.

Vol. 485: A. Garnaev, Search Games and Other Applications of Game Theory. VIII, 145 pages. 2000.

Vol. 486: M. Neugart, Nonlinear Labor Market Dynamics. X, 175 pages. 2000.

Vol. 487: Y. Y. Haimes, R. E. Steuer (Eds.), Research and Practice in Multiple Criteria Decision Making. XVII, 553 pages. 2000.

Vol. 488: B. Schmolck, Ommitted Variable Tests and Dynamic Specification. X, 144 pages. 2000.

Vol. 489: T. Steger, Transitional Dynamics and Economic Growth in Developing Countries. VIII, 151 pages. 2000.

Vol. 490: S. Minner, Strategic Safety Stocks in Supply Chains. XI, 214 pages. 2000.

Vol. 491: M. Ehrgott, Multicriteria Optimization. VIII, 242 pages. 2000.

Vol. 492: T. Phan Huy, Constraint Propagation in Flexible Manufacturing. IX, 258 pages. 2000.

Vol. 493: J. Zhu, Modular Pricing of Options. X, 170 pages. 2000.

Vol. 494: D. Franzen, Design of Master Agreements for OTC Derivatives. VIII, 175 pages. 2001.

Vol. 495: I. Konnov, Combined Relaxation Methods for Variational Inequalities. XI, 181 pages. 2001.

Vol. 496: P. Weiß, Unemployment in Open Economies. XII, 226 pages. 2001.

Vol. 497: J. Inkmann, Conditional Moment Estimation of Nonlinear Equation Systems. VIII, 214 pages. 2001.

Vol. 498: M. Reutter, A Macroeconomic Model of West German Unemployment. X, 125 pages. 2001.

Vol. 499: A. Casajus, Focal Points in Framed Games. XI, 131 pages. 2001.

Vol. 500: F. Nardini, Technical Progress and Economic Growth. XVII, 191 pages. 2001.

Vol. 501: M. Fleischmann, Quantitative Models for Reverse Logistics. XI, 181 pages. 2001.

Vol. 502: N. Hadjisavvas, J. E. Martínez-Legaz, J.-P. Penot (Eds.), Generalized Convexity and Generalized Monotonicity. IX, 410 pages. 2001.

Vol. 503: A. Kirman, J.-B. Zimmermann (Eds.), Economics with Heterogenous Interacting Agents. VII, 343 pages. 2001.

Vol. 504: P.-Y. Moix (Ed.),The Measurement of Market Risk. XI, 272 pages. 2001.

Vol. 505: S. Voß, J. R. Daduna (Eds.), Computer-Aided Scheduling of Public Transport. XI, 466 pages. 2001.

Vol. 506: B. P. Kellerhals, Financial Pricing Models in Con-tinuous Time and Kalman Filtering. XIV, 247 pages. 2001.

Vol. 507: M. Koksalan, S. Zionts, Multiple Criteria Decision Making in the New Millenium. XII, 481 pages. 2001.

Vol. 508: K. Neumann, C. Schwindt, J. Zimmermann, Project Scheduling with Time Windows and Scarce Resources. XI, 335 pages. 2002.

Vol. 509: D. Hornung, Investment, R&D, and Long-Run Growth. XVI, 194 pages. 2002.

Vol. 510: A. S. Tangian, Constructing and Applying Objective Functions. XII, 582 pages. 2002.

Vol. 511: M. Külpmann, Stock Market Overreaction and Fundamental Valuation. IX, 198 pages. 2002.

Vol. 512: W.-B. Zhang, An Economic Theory of Cities.XI, 220 pages. 2002.

Vol. 513: K. Marti, Stochastic Optimization Techniques. VIII, 364 pages. 2002.

Vol. 514: S. Wang, Y. Xia, Portfolio and Asset Pricing. XII, 200 pages. 2002.

Vol. 515: G. Heisig, Planning Stability in Material Requirements Planning System. XII, 264 pages. 2002.

Vol. 516: B. Schmid, Pricing Credit Linked Financial Instruments. X, 246 pages. 2002.

Vol. 517: H. I. Meinhardt, Cooperative Decision Making in Common Pool Situations. VIII, 205 pages. 2002.

Vol. 518: S. Napel, Bilateral Bargaining. VIII, 188 pages. 2002.

Vol. 519: A. Klose, G. Speranza, L. N. Van Wassenhove (Eds.), Quantitative Approaches to Distribution Logistics and Supply Chain Management. XIII, 421 pages. 2002.

Vol. 520: B. Glaser, Efficiency versus Sustainability in Dynamic Decision Making. IX, 252 pages. 2002.

Vol. 521: R. Cowan, N. Jonard (Eds.), Heterogenous Agents, Interactions and Economic Performance. XIV, 339 pages. 2003.

Vol. 522: C. Neff, Corporate Finance, Innovation, and Strategic Competition. IX, 218 pages. 2003.

Vol. 523: W.-B. Zhang, A Theory of Interregional Dynamics. XI, 231 pages. 2003.

Vol. 524: M. Frölich, Programme Evaluation and Treatment Choise. VIII, 191 pages. 2003.

Vol. 525: S. Spinler, Capacity Reservation for Capital-Intensive Technologies. XVI, 139 pages. 2003.

Vol. 526: C. F. Daganzo, A Theory of Supply Chains. VIII, 123 pages. 2003.

Vol. 527: C. E. Metz, Information Dissemination in Currency Crises. XI, 231 pages. 2003.

Vol. 528: R. Stolletz, Performance Analysis and Optimization of Inbound Call Centers. X, 219 pages. 2003.

Vol. 529: W. Krabs, S. W. Pickl, Analysis, Controllability and Optimization of Time-Discrete Systems and Dynamical Games. XII, 187 pages. 2003.

Vol. 530: R. Wapler, Unemployment, Market Structure and Growth. XXVII, 207 pages. 2003.

Vol. 531: M. Gallegati, A. Kirman, M. Marsili (Eds.), The Complex Dynamics of Economic Interaction. XV, 402 pages, 2004.

Vol. 532: K. Marti, Y. Ermoliev, G. Pflug (Eds.), Dynamic Stochastic Optimization. VIII, 336 pages. 2004.

Vol. 533: G. Dudek, Collaborative Planning in Supply Chains. X, 234 pages. 2004.

Vol. 534: M. Runkel, Environmental and Resource Policy for Consumer Durables. X, 197 pages. 2004.

Vol. 535: X. Gandibleux, M. Sevaux, K. Sörensen, V. T'kindt (Eds.), Metaheuristics for Multiobjective Optimisation. IX, 249 pages. 2004.

Vol. 536: R. Brüggemann, Model Reduction Methods for Vector Autoregressive Processes. X, 218 pages. 2004.

Vol. 537: A. Esser, Pricing in (In)Complete Markets. XI, 122 pages, 2004.

Vol. 538: S. Kokot, The Econometrics of Sequential Trade Models. XI, 193 pages. 2004.

Vol. 539: N. Hautsch, Modelling Irregularly Spaced Financial Data. XII, 291 pages. 2004.

Vol. 540: H. Kraft, Optimal Portfolios with Stochastic Interest Rates and Defaultable Assets. X, 173 pages. 2004.

Vol. 541: G.-y. Chen, X. Huang, X. Yang, Vector Optimization. X, 306 pages. 2005.

Vol. 542: J. Lingens, Union Wage Bargaining and Economic Growth. XIII, 199 pages. 2004.

Vol. 543: C. Benkert, Default Risk in Bond and Credit Derivatives Markets. IX, 135 pages. 2004.

Vol. 544: B. Fleischmann, A. Klose, Distribution Logistics. X, 284 pages. 2004.

Vol. 545: R. Hafner, Stochastic Implied Volatility. XI, 229 pages. 2004.

Vol. 546: D. Quadt, Lot-Sizing and Scheduling for Flexible Flow Lines. XVIII, 227 pages. 2004.

Vol. 547: M. Wildi, Signal Extraction. XI, 279 pages. 2005.

Vol. 548: D. Kuhn, Generalized Bounds for Convex Multistage Stochastic Programs. XI, 190 pages. 2005.

Vol. 549: G. N. Krieg, Kanban-Controlled Manufacturing Systems. IX, 236 pages. 2005.

Vol. 550: T. Lux, S. Reitz, E. Samanidou, Nonlinear Dynamics and Heterogeneous Interacting Agents. XIII, 327 pages. 2005.

Vol. 551: J. Leskow, M. Puchet Anyul, L. F. Punzo, New Tools of Economic Dynamics. XIX, 392 pages. 2005.

Vol. 552: C. Suerie, Time Continuity in Discrete Time Models. XVIII, 229 pages. 2005.

Vol. 553: B. Mönch, Strategic Trading in Illiquid Markets. XIII, 116 pages. 2005.

Vol. 554: R. Foellmi, Consumption Structure and Macroeconomics. IX, 152 pages. 2005.

Vol. 555: J. Wenzelburger, Learning in Economic Systems with Expectations Feedback (planned) 2005.

Vol. 556: R. Branzei, D. Dimitrov, S. Tijs, Models in Cooperative Game Theory. VIII, 135 pages. 2005.

Vol. 557: S. Barbaro, Equity and Efficiency Considerations of Public Higer Education. XII, 128 pages. 2005.

Vol. 558: M. Faliva, Topics in Dynamic Model Analysis. X, 144 pages. 2005.

Vol. 559: M. Schulmerich, Real Options Valuation. XVI, 357 pages. 2005.